T0236341

The Site Calculations Pocket Reference

The Site Calculations Pocket Reference

2nd Edition

Ed Hannan

John Wiley & Sons, Inc.

Library of Congress Cataloging-in-Publication Data:

Hannan, Ed.
    The site calculations pocket reference / Ed Hannan. —2nd ed.
        p. cm.
    Includes bibliographical references and index.
    ISBN-13: 978-0-471-73002-6 (pbk.)
    ISBN-10: 0-471-73002-5 (pbk.)
        1. Building sites—Design–Mathematics—Handbooks, manuals, etc. 2. Civil engineering—Mathematics—Handbooks, manuals, etc. 3. Engineering mathematics—Formulae—Handbooks, manuals, etc. I. Title.

TH375.H365 2006
624.02'12—dc22   2005044610

SECOND EDITION

10 9 8 7 6 5 4 3 2 1

This book is dedicated to:

My father, Ed Hannan, Sr., who taught me, among many other things, that proper planning is the key to proper execution.

Bob Martin, my friend and mentor in the golf business, who would appreciate this book more than anyone.

Contents

Section 3 Irrigation, Water Supply, and Materials **27**

Acknowledgments

I would like to express my sincere appreciation to those who contributed to this effort:

Ms. Diane Dayawan Dayrit, who kept me on the right path and who did all those things I hate to do.

Civil Engineering:

Mr. Mario Lopez of the Turf Company

Mr. Richard Davis and Mr. Ron Bacon of FHT

Consulting Engineers and Surveyors

Agronomy:

Mr. Don Roberts of Woerner Southern Turf Nurseries

Mr. Brad Dozler of International Seeds

Drainage:

Mr. John Adams of Advanced Drainage Systems

Irrigation:

Mr. Ken Kline of the Toro Company, who is always there with information and support.

I would also like to thank the NCDC and the USDA for their help in gathering much-needed information. Many thanks to you all!

Preface

How many times have you been in a meeting planning strategies for actual or hypothetical situations and needed that one piece of information, or formula, or chart, to make a determination, only to find out that it's unavailable or back at the office? Quite a few, I bet!

My background is civil engineering and golf course construction, and I cannot tell you how many times I have been in that situation.

This book was put together to be a good source of information for site construction of all kinds and for civil engineering in general. From my own experience I have compiled the most frequently used charts, formulas, and conversions used in general site construction, along with practical examples. The book is intended to combine, in one handy pocket-size publication, all aspects of site work, thus eliminating the time and effort spent to tap the many different sources of information required.

For those of us who work outside the United States, I have included the metric conversions for the English data included. Whether you are moving earth, irrigating, landscaping, or calculating water features, pump capacities, or drainage, there is information in this guide to help you in initial design, renovation, or additions in site work.

I hope you find it as convenient and helpful as I have.

Ed Hannan

The Site Calculations Pocket Reference

Section One

EARTHWORKS AND MATERIALS

1.1   SOILS—ANGLES OF REPOSE

Degrees from Horizontal

Soil	Dry	Moist	Wet
Sand	20–35	30–45	20–40
Loam	20–40	25–45	35–55
Gravel	30–50		

"Angle of repose" is the slope in which a given material will maintain without sliding.

1.2   VOLUME OF SAND OR GRAVEL IN A STOCKPILE

Assuming Angle of Repose is 37 Degrees

Height		Volume	
Feet	Meters	Cubic Yards	Cubic Meters
10	3.048	68	52.02
15	4.572	230	175.95
20	6.096	550	420.75
25	7.62	1,060	810.90
30	9.144	1,840	1,407.60
35	10.668	2,900	2,218.50
40	12.192	4,350	3,327.75
45	13.716	6,200	1,889.76
50	15.24	8,500	6,502.50
60	18.288	14,800	11,322.00
65	19.812	18,750	14,343.75
70	21.336	23,400	17,901.00
75	22.86	28,800	22,032.00
80	24.384	34,600	26,469.00
85	25.988	41,800	31,977.00
90	27.432	49,700	38,020.50
95	28.956	58,300	44,599.50
100	30.480	68,350	52,287.75

$$\text{Volume} = \frac{\text{Area of base} \times \text{Height}}{3}$$

1.3   APPROXIMATE SWELLING FACTORS FOR VARIOUS SOIL TYPES

Calculated from 100% Volume in Place to Excavated and Loose Material for Transport

Soil Type	90% Relative Compaction (Factor)	Loose Material (Factor)
Sand	1	1.15
Sandy loam	1	1.20
Clay loam	1	1.30
Clay	1	1.35

It will require haulage of 1,350 cu. yds. of loose clay material (1,000 × 1.35 = 1,350 cu. yds.) to achieve 1,000 cu. yds. of clay fill compacted.

1.4   APPROXIMATE SHRINKAGE FACTORS FOR VARIOUS SOIL TYPES

Loose Backfill Subject to Normal Settling

Soil Type	Loose Material	Normal Shrinkage (Factor)
Sand	1	1.18
Sandy loam	1	1.23
Clay loam	1	1.43
Clay	1	1.48

1.5   SOIL PERMEABILITY

As a General Rule

Soil Type	Water Applied (in.)	Depth Covered (in.)
Sands	1	12
Loam	1	6
Clay	1	3

The results will differ with any combination of the above or between. We suggest you have a proper test done to determine the exact permeability of your soil.

Example

1 in. of water applied to loam will penetrate 6 in.

1.6 WATER RETENTION FOR VARIOUS SOILS

Soil Type	Moisture (%)	Wilting Point (%)	Inches	Water to Be Added per 12" of Soil (gal.)	1,000 SF
Sand	2.6	1.8	0.31	194	12" deep
Sandy loam	6.9	4.2	0.78	487	
Fine sandy loam	9.2	5.2	0.95	593	
Loam	12.7	6.3	1.01	631	
Clay loam	18.4	10.0	1.37	856	
Silty clay loam	24.4	14.3	1.45	906	
Clay	45.9	29.6	1.42	887	

1.7 WEIGHTS OF MATERIALS

Earth	Pounds/Cubic Foot	Pounds/Cubic Yard	Cubic Yards/Ton	Tons/Cubic Yard
Clay (wet)	105–120	2,835–3,240	0.71–0.62	1.42–1.62
Clay (dry)	90–110	2,430–2,970	0.82–0.67	1.22–1.49
Granite	170	4,590	0.44	2.30
Gravel	120–135	3,240–3,645	0.62–0.55	1.62–1.82
Loam (dry—compact)	90–100	2,430–2,700	0.82–0.74	1.22–1.35
Loam (dry—loose)	75–90	2,025–2,430	0.99–0.82	1.01–1.22
Peat moss (compact)	15–18	405–486	4.9–4.1	0.20–0.24
Peat moss (loose)	3–5	81–135	24.7–14.8	0.04–0.07
Sand (dry)	95–110	2,565–2,970	0.78–0.67	1.28–1.49
Sand (wet)	120–130	3,240–3,510	0.62–0.57	1.62–1.76
Silt (wet)	130–145	3,510–3,915	0.57–0.51	1.76–1.96

1.8 WATER-HOLDING ESTIMATES OF A VARIETY OF SOILS[a]

	Soil				
	Very Coarse, Gravelly Coarse	Coarse Sand, Fine Sand, Loamy Coarse Sand	Moderately Coarse Loamy Fine Sand, Sandy Loam, Fine Sandy Loam	Medium and Fine Fine Sandy Loam, Loam Silt, Loam Silt Sandy Clay, Silty Clay, Clay	Moderately Fine Sandy Clay Loam, Clay Loam, Silty Clay Loam
AW—in./ft.	0.5"	1.0"	1.5"	2.0"	2.2"
AW—gal./cub. ft.	½	¾	1	1⅓	1⅓

SOURCE: From the Soil Conservation Service Handbook

[a]AW = Available water.

5

1.9 TRIANGLE OF PHYSICAL CHARACTERISTICS OF VARIOUS SOILS

Soils are a mixture of particles of different sizes. The smaller the particle, the heavier the soil is and the greater the quantity of water it retains. This is determined by the USDA triangle of physical characteristics of soil:

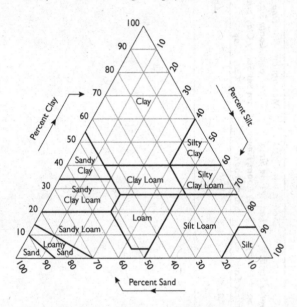

1.10 ESTIMATED RUNOFF VOLUME[a] FROM A 24-HOUR, 7-INCH RAIN EVENT

Land Use	Runoff Volume (in.)
Woodland	2.2
Golf course	2.7
Contour-terraced row crop	3.7
Straight-row crop	4.7

SOURCE: USDA Natural Resources Conservation Service
[a] Runoff varies depending on land use. Crop production tends to allow greater amounts of flow than other land uses.

1.11   AVERAGE SOIL INFILTRATION RATES FOR VARIOUS SOILS BY PERCENT OF SLOPE[a]

	Infiltration Rate (Ir) in Inches/Hour for Various Percents of Slope				
Soil Texture, Type	0–4%	5–8%	8–12%	12–16%	Over 16%
Coarse sand	1.25	1.00	0.75	0.50	0.31
Medium sand	1.06	0.85	0.64	0.42	0.27
Fine sand	0.94	0.75	0.56	0.38	0.24
Loamy sand	0.88	0.70	0.53	0.35	0.22
Sandy loam	0.75	0.60	0.45	0.30	0.19
Fine sandy loam	0.63	0.50	0.38	0.25	0.16
Very fine sandy loam	0.59	0.47	0.35	0.24	0.15
Loam	0.54	0.43	0.33	0.22	0.14
Silt loam	0.50	0.40	0.30	0.20	0.13
Silt	0.44	0.35	0.26	0.18	0.11
Sandy clay	0.31	0.25	0.19	0.12	0.08
Clay loam	0.25	0.20	0.15	0.10	0.06
Silt clay	0.19	0.15	0.11	0.08	0.05
Clay	0.13	0.10	0.08	0.05	0.03

SOURCE: Derived from USDA information
[a] Note: Rates based on full cover. These figures decrease with time and percent of cover.

1.12   COMPUTATION FOR VOLUME OF EXCAVATED MATERIAL (ENGLISH)

Depth	Cubic Yards per Square Foot	Depth	Cubic Yards per Square Foot
2"	0.006	4'	0.148
4"	0.012	5'	0.185
6"	0.018	6'	0.222
8"	0.025	7'	0.259
10"	0.031	8'	0.296
1'	0.037	9'	0.332
2'	0.074	10'	0.369
3'	0.111		

No swellage factor has been applied.

Example

(20' wide × 30' long) × 4' deep = 600 × 0.148 = 88.8 cu. yd.

Other useful conversions:

$$1" = 2.54 \text{ cm}$$
$$1 \text{ cu. yd.} = 0.765 \text{ m}^3 \text{ (or) m}^3 \times 1.31 = \text{cu. yd.}$$
$$1 \text{ m}^2 = 10.764 \text{ SF}$$
$$\text{SF} \times 0.0929 = \text{m}^2$$

1.13 COMPUTATION FOR VOLUME OF EXCAVATED MATERIAL (METRIC)

Depth, Centimeter/Meter	m^3 per m^2
10	0.10
15	0.15
20	0.20
25	0.25
30	0.30
35	0.35
40	0.40
45	0.45
50	0.50
55	0.55
60	0.60
65	0.65
70	0.70
75	0.75
80	0.80
85	0.85
90	0.90
1 meter	1.0

(Depth ÷ 100) × Width (meters) × Length (meters) = m^3

1.14 COMPUTATION OF MATERIALS VOLUME FROM TRENCH EXCAVATION (ENGLISH)

Cubic Yards per 100 Lineal Feet

Depth (in.)	Width (in.)						
	6"	18"	24"	30"	36"	42"	48"
6"	1.9	2.8	3.7	4.6	5.6	6.6	7.4
12"	3.7	5.6	7.4	9.3	11.1	13.0	14.8
18"	5.6	8.3	11.1	13.9	16.7	19.4	22.3
24"	7.4	11.1	14.8	18.5	22.2	26.0	29.6
30"	9.3	13.8	18.5	23.2	27.8	32.4	37.0
36"	11.1	16.6	22.2	27.8	33.3	38.9	44.5
42"	13.0	19.4	25.9	32.4	38.9	45.4	52.0
48"	14.8	22.2	29.6	37.0	44.5	52.0	59.2

Conversion Formula:

1.15 COMPUTATION OF MATERIALS VOLUME FROM TRENCH EXCAVATION (METRIC)

m³ per 100 L.M.

Depth	Width of Trench (cm)						
	15	30	45	60	75	90	10
15 cm	2.25	4.5	6.75	9.0	11.25	13.5	15.75
30 cm	4.5	9.0	13.5	18.0	22.5	27.0	31.5
45 cm	6.75	13.5	20.25	27.0	33.75	40.5	47.25
60 cm	9.0	18.0	27.0	36.0	45.0	54.0	63.0
75 cm	11.25	22.5	33.75	45.0	56.25	67.5	78.75
90 cm	13.5	27.0	40.5	54.0	67.5	81.0	94.5
1 m	15.75	31.5	47.75	61.0	74.75	88.0	101.5

(Depth ÷ 100) × (Width ÷ 100) × Length = Cubic meters

1.16 ANGLES OF SLOPES

Angles (in degrees and minutes)	Slope Equivalent Ratio
18-25	3 to 1
26-35	2 to 1
33-42	1.5 to 1
45-00	1 to 1
53-00	0.75 to 1
56-20	0.67 to 1
63-30	0.5 to 1

1.17 SLOPE MEASUREMENT PLAN (HORIZONTAL) TO TRUE MEASURE

Slope Ratio Horizontal to Vertical	Map Measure	Multiplied by Factor	True Measure
¾ to 1	×	1.6670	=
1 to 1	×	1.4142	=
1½ to 1	×.	1.2019	=
2 to 1	×	1.1180	=
1½ to 1	×	1.0770	=
3 to 1	×	1.0541	=
4 to 1	×	1.0198	=

Here is how to use the chart:

1. Measure the distance on the map.
2. Determine slope ratio:
   Distance (Length) ÷ Elevation (Height) = Slope ratio
3. Refer to chart under slope ratio.
4. Multiply by factor to find true measure.

Example

Map distance is 30 ft.
Elevation of mound is 10 ft.
Slope ratio 30 ft. ÷ 10 ft. = 3 (3 1)
True measure = 30 ft. × 1.0541 = 31.62 ft.

True measure is the actual surface measurement taking into account various slopes.

1.18 APPROXIMATE EQUIVALENCIES: SLOPE/GRADE/DEGREE

Slope Ratio	% Grade	Angle Degrees (approximate)
Flat	Flat	0
100 to 1	1%	1
20 to 1	5%	3
10 to 1	10%	6
5 to 1	20%	11
4 to 1	25%	13
	30%	16
3 to 1	33⅓%	17
	40%	22
2 to 1	50%	26
	60%	31
	80%	38
1 to 1	100%	45
Vertical	Vertical	90

1.19 SAFE LIMIT RESTRICTIONS (GENERAL GUIDE)

% Grade	Restriction Guide
100% (1 1)	Maximum cut slope
50% (2 1)	Maximum fill slope
33½% (3 1)	Maximum for simple construction
30%	Maximum for practical landscape care (use retaining walls where practical)
20%	Maximum lawn slope
15%	Consider ground cover or shrubs to replace lawn
12%	Maximum driveway slope
10%	Steep grade for vehicles, maximum for snow and ice
5%	Optimum drain slope from buildings
3%	Optimum slope for grassed surfaces such as golf course fairways
2%	Optimum for sports fields
1%	Minimum landscape slope

1.20 SOIL COVERAGE

Square Feet and Square Meters per Cubic Yard

Depth (in.)	Coverage in	
	Square Feet	*Square Meters*
⅛	2,592	240.79
¼	1,296	120.39
⅜	864	80.265
½	648	60.199
⅝	518	48.122
¾	432	40.132
1	324	30.099
2	162	15.049
3	108	10.033
4	81	7.524
5	67	6.224
6	54	5.016
8	40	3.716
10	33	3.065
12	27	2.508

Example
Area to be covered is 3,000 SF @ 4 in.; from the chart: 1 cu. yd. will cover 81 SF @ 4 in.

$$\frac{3,000\,\text{SF}}{81} = 37.03 \text{ cu. yds. required}$$

$$\text{SF} \times 0.0929 = \text{Square meter}$$

$$1 \text{ in.} = 2.54 \text{ cm}$$

$$1 \text{ cu. yd.} = 0.765 \text{ cu. m}^3$$

$$1 \text{ cu. m}^3 = 1.31 \text{ cu. yds.}$$

$$1 \text{ sq. yd.} = 9 \text{ SF}$$

1.21 TOP DRESSING (ENGLISH)

Coverages in Cubic Feet and Cubic Yards

Depth (in.)	For 1,000 SF		
	In Cubic Feet	*In Cubic Yards*	*For 1 Acre in Cubic Yards*
⅛	10.53	0.39	17
¼	21.00	0.78	34
⅜	30.50	1.17	51
½	42.00	1.56	68
⅝	52.50	1.95	85
¾	63.00	2.34	102
1	84.00	3.12	136
2	168.00	6.24	272

Example

To cover 2,000 SF @ ½ in., from the chart: ½ in. coverage will require 1.56 cu. yd. per 1,000 SF. $1.56 \times 2 = 3.12$ cu. yd.

$$\text{Acres} \times 0.4047 = \text{Hectares}$$
$$\text{Hectares} = 2.417 \text{ acres}$$
$$m^3 \times 1.31 = \text{Cubic yards}$$
$$\text{Cubic yards} \times 0.765 = \text{Cubic meters}$$
$$1 \text{ cu. yd.} = 27 \text{ cu. ft.}$$
$$\text{Square meter} = 10.764 \text{ SF}$$
$$1 \text{ hectare} = 10,000 \text{ m}^2$$

1.22 TOP DRESSING (METRIC)

Coverages in Cubic Meters

Depth (in.)	For 1,000 m²		
	In Centimeters	*In Cubic Meters*	*For 1 Hectare in Cubic Meters*
⅛	0.317	3.17	31.7
¼	0.635	6.35	63.5
⅜	0.952	9.52	95.2

Coverages in Cubic Meters (continued)

Depth (in.)	For 1,000 m²		For 1 Hectare in Cubic Meters
	In Centimeters	In Cubic Meters	
½	1.270	12.70	127.0
⅝	1.587	15.87	158.7
¾	1.905	19.05	190.5
1	2.540	25.40	254.0
2	5.080	50.80	508.0
3	7.620	76.20	762.0
4	10.160	101.60	1016.0
5	12.700	127.00	1270.0
6	15.240	152.40	1524.0
8	20.320	203.20	2032.0
10	25.400	254.00	2540.0
12	30.480	304.80	3048.0

Centimeters × 10 = mm (millimeters)

Depth (meters) × Width (meters) × Length (meters) = Cubic meters

Useful conversion; Hectare = 10,000 m²

1.23 PRACTICAL EXAMPLES OF SPECIALIZED SMALL AREAS OF FILL (SAND TRAP)

Bunker Sand Placement: Quantity of Material by Area

Example 1

Depth requirement: 6 in. SF ÷ 9 = Square yards

Be sure to refer to Section 1.4 for shrinkage factor.

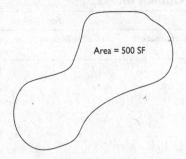

Area = 500 SF

Example 2

Sand has a shrinkage factor of 1.18; therefore, to obtain 6 in. after shrinkage, you would need 6 in. × 1.18 = 7.08 in. initially.

$$\frac{\text{Depth (decimal of feet)} \times \text{Area (square feet)}}{27} = \text{Cubic yards}$$

$$\frac{0.5 \times 500 \text{ SF}}{27} = 9.25 \text{ cu. yd.}$$

Shrinkage factor 9.25 cu. yd. × 1.18 = 10.91 cu. yd.

Cubic yards × 0.765 = Cubic meters

$$9.25 \times 0.765 = 7.07 \text{ m}^3$$

1.24 CALCULATION OF AREA OF AN ODD-SHAPED AREA

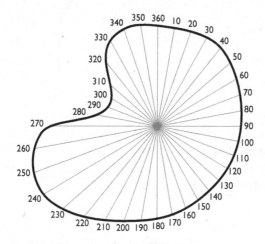

First:

   1. Set up in center of area.

   2. Measure radius every 10 degrees (36 measurements).

   3. Add total measurements and divide by 36.

   4. This will give you an average radius.

   5. Use the formula for area of a circle: πR^2 ($\pi = 3.14$)

Or:

   1. Measure perimeter.

   2. Use chart in Section 6.6 to determine the area.

Or:

1. Measure by cross section.

1.25 CALCULATION OF FILL MATERIALS FOR AN ODD-SHAPED AREA (GOLF GREEN)

Example

Materials placement: Quantity of material by area
Area of a golf green

6500 SF

Depth of greens material requirement: 12 in.
Sand shrinkage factor × 1.18

$$\frac{\text{Depth (decimal of foot)} \times \text{Area (square feet)}}{27} = \text{Cubic yards}$$

$$\frac{1.0 \times 6{,}500 \text{ SF}}{27} = 240.74 \text{ cu. yd.}$$

Shrinkage factor $240.74 \times 1.18 = 284.07$ cu. yd.

Cubic yards $\times 0.765 = $ Cubic meters

Determine area as illustrated in Section 1.24.

1.26   CHART OF VOLUMES BY AREA

Golf Green: Based on 14 In. Fill in Cavity for Shrinkage to 12 In. per (1,000 SF in cubic yards/cubic meters)

	3	3.5	4.0	4.5	5.0	5.5	6.0	6.5
Cubic yards	131	153	175	197	219	240	262	284
Cubic meters	100	117	134	151	168	184	200	217

	7.0	7.5	8.0	8.5	9.0	9.5	10.0
Cubic yards	306	328	350	371	393	415	437
Cubic meters	234	251	268	284	301	317	334

Formulas

$$\frac{\text{Depth (decimal of feet)} \times \text{Area}}{27} = \text{Cubic yards}$$

$$\text{Square meters} \times 10.764 = \text{Square feet}$$

$$\text{Cubic yards} \times 0.765 = \text{Cubic meters}$$

Section Two

DRAINAGE

2.1 MANNING'S "n" VALUE FOR DESIGN

(Storm and Sanitary Sewer and Culverts)

Pipe Type	"n"
A.D.S. corrugated polyethylene pipe	
3"–6" diameter	0.015
8" diameter	0.016
10" diameter	0.017
12"–15" diameter	0.018
18"–36" diameter	0.020
A.D.S.N-12	0.012
Concrete pipe	0.013
Corrugated metal pipe (2⅔" × ½" corrugation)	
Annular	
Plain	0.024
Paved invert	0.020
Fully paved (smooth lined)	0.013
Helical	
Plain 15" diameter	0.013
Plain 18" diameter	0.015
Plain 24" diameter	0.018
Plain 36" diameter	0.021
Spiral-rib	0.012
Plastic pipe (SDR, S&D, etc.)	0.011
Vitrified clay	0.013

2.2 CIRCULAR PIPE FLOW CAPACITY

Full-Flow (cubic feet per second) Double-Wall (smooth wall) Manning's "n" = 0.012

Diameter (in.)	Conveyance Factor[a]	Percent Slope (feet per 100 feet)															
		0.02	0.05	0.10	0.20	0.35	0.50	0.75	1.00	1.25	1.58	1.75	2.0	2.5	5.0	10.0	20.0
3	0.957	0.014	0.021	0.030	0.043	0.057	0.068	0.083	0.096	0.107	0.12	0.13	0.14	0.15	0.21	0.30	0.43
4	2.062	0.029	0.046	0.065	0.092	0.122	0.146	0.179	0.206	0.231	0.25	0.27	0.29	0.33	0.46	0.65	0.92
5	3.738	0.053	0.084	0.118	0.167	0.221	0.264	0.324	0.374	0.418	0.46	0.49	0.53	0.59	0.84	1.18	1.67
6	6.079	0.086	0.136	0.192	0.272	0.360	0.430	0.526	0.608	0.680	0.74	0.80	0.86	0.96	1.36	1.92	2.72
8	13.091	0.185	0.293	0.414	0.585	0.774	0.926	1.134	1.309	1.464	1.60	1.73	1.85	2.07	2.93	4.14	5.85
10	23.74	0.34	0.53	0.75	1.06	1.40	1.68	2.06	2.37	2.65	2.91	3.14	3.36	3.75	5.31	7.51	10.61
12	38.60	0.55	0.86	1.22	1.73	2.28	2.73	3.34	3.86	4.32	4.73	5.11	5.46	6.10	8.63	12.21	17.26
15	69.98	0.99	1.56	2.21	3.13	4.14	4.95	6.06	7.00	7.82	8.57	9.26	9.90	11.06	15.65	22.13	31.30
18	113.80	1.61	2.54	3.60	5.09	6.73	8.05	9.86	11.38	12.72	13.94	15.15	16.09	17.99	25.45	35.99	50.89
21	171.65	2.43	3.84	5.43	7.68	10.16	12.14	14.87	17.17	19.19	21.02	22.71	24.28	27.14	38.38	54.28	76.77
24	245.08	3.47	5.48	7.75	10.96	14.50	17.33	21.22	24.51	27.40	30.02	32.42	34.66	38.75	54.80	77.50	109.6
27	335.51	4.74	7.50	10.61	15.00	19.85	23.72	29.06	33.55	37.51	41.09	44.38	47.45	53.06	75.0	106.1	150.0
30	444.35	6.28	9.94	14.05	19.87	26.29	31.42	38.48	44.44	49.68	54.42	58.78	62.84	70.25	99.4	140.5	198.7
36	722.57	10.22	16.16	22.85	32.31	42.75	51.09	62.58	72.26	80.79	88.50	95.59	102.19	114.25	161.6	228.5	323.1
42	1089.90	15.41	24.37	34.47	48.74	64.5	77.1	94.4	109.0	121.9	133.5	144.2	154.1	172.3	243.7	334.7	487.4
48	1556.10	22.01	34.80	49.21	69.59	92.1	110.0	134.8	155.6	174.0	190.6	205.9	220.1	246.0	348.0	492.1	695.9

[a] Conveyance factor = $(1.486 \times R^{1/2} \times A)/n$

Example

See Section 2.5.

2.3 CIRCULAR PIPE FLOW CAPACITY

Full-Flow (cubic feet per second) Single Wall (spiral) Manning's "n" = 0.015

Dia-meter (in.)	Con-ver-gence Factor[a]	Percent Slope (feet per 100 feet)															
		0.02	0.05	0.10	0.20	0.35	0.50	0.75	1.00	1.25	1.58	1.75	2.0	2.5	5.0	10.0	20.0
3	0.766	0.017	0.017	0.024	0.034	0.045	0.054	0.066	0.077	0.086	0.09	0.10	0.11	0.12	0.17	0.24	0.34
4	1.649	0.023	0.037	0.052	0.074	0.098	0.117	0.143	0.165	0.184	0.20	0.22	0.23	0.26	0.37	0.52	0.74
5	2.991	0.042	0.067	0.095	0.134	0.177	0.211	0.259	0.299	0.334	0.37	0.40	0.42	0.47	0.67	0.95	1.34
6	4.863	0.069	0.109	0.154	0.217	0.288	0.344	0.421	0.486	0.544	0.60	0.64	0.69	0.77	1.09	1.54	2.17
8	10.473	0.148	0.234	0.331	0.468	0.620	0.741	0.907	1.047	1.171	1.28	1.39	1.48	1.66	2.34	3.31	4.68
10	18.99	0.27	0.42	0.60	0.85	1.12	1.34	1.64	1.90	2.12	2.33	2.51	2.69	3.00	4.25	6.00	8.49
12	30.88	0.44	0.69	0.98	1.38	1.83	2.18	2.67	3.09	3.45	3.78	4.08	4.37	4.88	6.90	9.76	13.81
15	55.98	0.79	1.25	1.77	2.50	3.31	3.96	4.85	5.60	6.26	6.86	7.41	7.92	8.85	12.52	17.70	25.04
18	91.04	1.29	2.04	2.88	4.07	5.39	6.44	7.88	9.10	10.18	11.15	12.04	12.87	14.39	20.36	28.79	40.71
21	137.32	1.94	3.07	4.34	6.14	8.12	9.71	11.89	13.73	15.35	16.82	18.17	19.42	21.71	30.71	43.43	61.41
24	196.06	2.77	4.38	6.20	8.77	11.60	13.86	16.98	19.61	21.92	24.01	25.94	27.73	31.00	43.84	62.00	87.68
27	268.41	3.80	6.00	8.49	12.00	15.88	18.98	23.24	26.84	30.01	32.87	35.51	37.96	42.44	60.0	84.90	120.0
30	355.48	5.03	7.95	11.24	15.90	21.03	25.14	30.79	35.55	39.74	43.54	47.03	50.27	56.21	79.5	112.4	159.0
36	578.05	8.17	12.93	18.28	25.85	34.20	40.87	50.06	57.81	64.63	70.80	76.47	81.75	91.40	129.3	182.8	258.5
42	872.0	12.33	19.50	27.57	38.99	51.6	61.7	75.5	87.2	97.5	106.8	115.3	123.3	137.9	195.0	275.7	389.9
48	1,244.9	17.61	27.84	39.37	55.67	73.6	88.0	107.8	124.5	139.2	152.5	164.7	176.1	196.8	278.4	393.7	556.7

[a] Conveyance factor = $(1.486 \times R^{1/2} \times A)/n$

Example

See Section 2.5.

2.4 PIPE SIZE CONVERSION (DRAINAGE) ENGLISH TO METRIC

English	Metric	English	Metric
4"	100 mm	28"	700 mm
6"	150 mm	30"	750 mm
8"	200 mm	32"	800 mm
10"	250 mm	36"	900 mm
12"	300 mm	40"	1,000 mm
14"	350 mm	44"	1,100 mm
16"	400 mm	48"	1,200 mm
18"	450 mm	52"	1,300 mm
20"	500 mm	56"	1,400 mm
24"	600 mm	60"	1,500 mm

2.5 EXAMPLE OF CIRCULAR PIPE FLOW CAPACITY CALCULATION

Once the amount of water expected (i.e., 100-year storm for design purposes) is determined, refer to Section 2.2 or 2.3 to determine what size pipe you need for a particular area by determining the slope of the pipe and then going down the chart until you find the volume you want to drain.

Example

The 100-year storm is 8 in. The catchment area is 1,000 SF. (Rainfall) 8 in. × Area (based on 1,000 SF) = 666 cu. ft. using the single wall chart in Section 2.3. The slope of the pipe will be 0.5% (feet per 100 ft.) using single wall pipe. Go to the chart (single wall) and at the top of the chart under "Percent Slope" find the column under 0.50. Determine your time parameter, for example, 15 min. × 60 sec. = 900 sec. The chart is in cubic feet per second. We have 666 cu. ft. to move: (666 cu. ft./900 sec.) = 0.74. According to the chart, an 8 in. pipe at 0.50% slope would move 0.741 cubic feet per second (cfs), which would remove the water in under 15 minutes.

When draining a large area broken into several catchment areas, you must add all lateral lines and the duty of the main catchment area (junction box) to determine the size of the main outfall pipe.

Example

The diagram for pipe flow capacity calculation on the following page shows three catchment areas, A, B, and C, all having an area of 1,000 SF for the purpose of this demonstration. The 100-year storm is 8 in. to be drained in 15 minutes. Catchment areas A, B, and C will each have a capacity of 666 cu. ft.: 666 × 3 = 1,998 cu. ft.

15 min. = 900 sec.

The lateral lines from areas A and B have already been determined from the previous example at 8 in.

$$\frac{1,998 \text{ cu. ft.}}{900 \text{ sec.}} = 2.22$$

The pipe will be laid at a gradient of 0.5%. The single wall chart shows that it will take a 12 in. pipe slightly longer than 15 minutes to handle the duty. Therefore, the main outfall line from catchment basin C will be 12 in.

2.6 DIAGRAM FOR PIPE FLOW CAPACITY CALCULATION

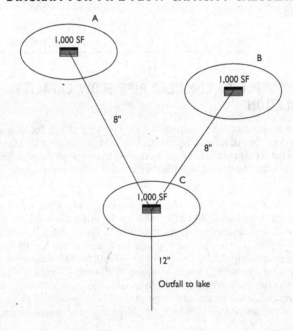

2.7   DRAINAGE CALCULATION: WATER QUANTITY BY AREA Per 1,000 SF Area

Rainfall (fraction of in./ft.)	Decimal Equivalent	Cubic Feet of Water
⅛	0.01037	10.37
¼	0.02075	20.75
½	0.04150	41.50
¾	0.06225	62.25
1	0.083	83
2	0.166	166
3	0.250	250
4	0.333	333
5	0.416	416
6	0.500	500
7	0.583	583
8	0.666	666
9	0.750	750
10	0.833	833
11	0.916	916
12	1.000	1,000

Formula

Depth (decimal of foot)

$$\frac{\times \text{Area}\left(\text{length} \times \text{width}\right)}{27} = \text{Cubic feet}$$

Determine area by: Length × Width Area of Circle

2.8 PIPE CONVERSION FACTORS (NATURAL GAS, IRRIGATION, DRAINAGE, ETC.)

English (NPS) to Metric (DN)

NPS	DN	NPS	DN	NPS	DN
⅛"	6 mm	3"	80mm	20"	500mm
³⁄₁₆"	7mm	3½"	90mm	24"	600mm
¼"	8mm	4"	100mm	28"	700mm
⅜	10mm	4½"	115mm	30"	750mm
½"	15mm	5"	125mm	32"	800mm
⅝"	18mm	6"	150mm	36"	900mm
¾"	20mm	8"	200mm	40"	1000mm
1"	25mm	10"	250mm	44"	1100mm
1¼"	32mm	12"	300mm	48"	1200mm
1½"	40mm	14"	350mm	52"	1300mm
2"	50mm	16"	400mm	56"	1400mm

Section Three

IRRIGATION, WATER SUPPLY, AND MATERIALS

3.1 PIPE DATA INTRODUCTION

The basic essential for proper hydraulic design of any sprinkler irrigation system is the proper calculation of pressure loss from friction through the piping. Proper pipe sizing is of the utmost importance in order to achieve a good combination of economy, and fitting life and the correct operating pressure at each and every sprinkler.

It is important to note here that all pressure loss charts in this manual have been calculated on the conservative side by using a lower C factor than even many pipe manufacturers use. It is our belief that using a C factor of 150 for plastic pipe reflects an absolute maximum flow under ideal circumstances and leaves no safety factor whatsoever.

The use of different C factors for pipe is as follows:

PVC pipe. Given a 140 value for C as it is extremely smooth and it is chemically inert so it does not deteriorate internally.

Asbestos-cement, copper, and polyethylene pipes. Given a 140 value for C as they also have smooth inside walls. AC and PE do not deteriorate in most conditions.

Cast-iron and galvanized steel pipes. Given a 100 value for C as assumes an average 10-year-old pipe. As both pipes are highly susceptible to deterioration caused by minerals in the water, the roughness will vary with the locale and the irrigation use of the pipe. In some areas, the pipe will reach the 100 value in 2 years, while in some other areas, the pipe may not reach the 100 value for 15 to 20 years. Generally, if water is continually flowing through the pipe, deterioration is reduced compared to having water at rest in the pipe.

To adjust the friction losses to a different C value, multiply the loss shown in the friction loss tables by the factor selected from the following table:

C Factor Given in Table	To Change to C Constant of:							
	80	90	100	110	120	130	140	150
100	1.50	1.22	1.00	0.84	0.72	0.63	0.54	0.47
120			1.41	1.18	1.00	0.86	0.76	0.66
140				1.60	1.34	1.15	1.00	0.87
150					1.54	1.32	1.15	1.00

The velocity of flow is of almost equal importance to friction loss and is a factor all too often overlooked by many designers. To fully realize the importance of velocity, please study the information on surges and water hammer detailed in Section 3.7. These factors play an important part in the hydraulic design of every sprinkler system.

3.2   LAKE LINING CALCULATION

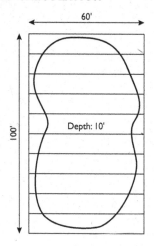

Calculating Lake Liner Size

Liner length = Lake length + Twice the lake depth + 2' (0.6 m)

Liner width = Lake width + Twice the lake depth + 2' (0.6 m)

Result of this example: 122' × 82' = 10,004 st SF

Volume of Water by Depth (per 1000 sf):

Depth	Gallons
¼"	155.75
½"	311.5
¾"	467.25
1"	623
2"	1246
6"	3738
1'	7476

Gallons per Acre/inch = 27,157
This chart can also be used to determine irrigation calculations.
For example: to produce a ½" rain over 1,000 sf, you would require 311.5 gal.
Per acre at ½" would require 13,578.5 gal.

3.3 PRESSURE LOSS FROM FRICTION PER 100 FT. OF PIPE (POUNDS PER SQUARE INCH)
SDR 26/Class 160 PVC 1120, PVC 1220, PVC 2120, C = 140

Flow (gpm)	½	¾	1	1¼	1½	2	2½	3	4
1	0.26	0.07							
2	0.89	0.26							
3	1.86	0.52							
4	3.12	0.90	0.27						
5	4.76	1.37	0.40						
6	6.62	1.90	0.57						
7	8.82	2.52	0.76						
8	11.26	3.21	0.98	0.31					
9	14.10	4.05	1.19	0.41					
10	17.11	4.90	1.48	0.45	0.23				
11	20.37	5.82	1.77	0.55	0.29				
12	23.87	6.83	2.08	0.65	0.33				
13		7.94	2.42	0.75	0.39				
14		9.11	2.76	0.84	0.44				
15		10.36	3.17	0.95	0.50				
20		17.67	5.41	1.64	0.84	0.29			
25		26.70	8.28	2.45	1.28	0.42			
30			11.56	3.34	1.78	0.60	0.23		
35			15.47	4.45	2.37	0.81	0.32		

40	19.76	5.65	3.02	1.02	0.40		
50	30.55	8.74	4.67	1.59	0.63		
60		11.95	6.40	2.18	0.88		
70			8.52	2.91	1.16	0.46	
80			10.82	3.68	1.48	0.57	
90			13.47	4.57	1.84	0.72	
100			16.57	5.62	2.24	0.88	0.25
110				6.59	2.63	1.02	0.30
120				7.81	3.16	1.20	0.36
130				9.16	3.68	1.42	0.42
140				10.38	4.18	1.62	0.49
150				11.81	4.76	1.84	0.55
160				13.34	5.38	2.09	0.64
170				14.81	5.97	2.31	0.70
180				16.48	6.63	2.56	0.77
190				18.18	7.31	2.81	0.85
200				20.08	8.07	3.10	0.95
210					8.96	3.45	1.04
220					9.80	3.80	1.15
230					10.48	4.06	1.23
240					11.30	4.36	1.32
250					12.11	4.67	1.41

(continued)

3.3 PRESSURE LOSS FROM FRICTION PER 100 FT. OF PIPE (POUNDS PER SQUARE INCH) SDR 26/Class 160 PVC 1120, PVC 1220, PVC 2120, C = 140 (continued)

Flow (gpm)	½	¾	1	1¼	1½	2	2½	3	4
260							12.90	4.99	1.50
270							14.17	5.47	1.65
280							15.23	5.89	1.78
290							15.91	6.15	1.85
300							16.86	6.50	1.95
320								7.57	2.27
340								8.30	2.50
360								9.18	2.77
380								10.20	3.09
400								11.41	3.45

* ½ and ¾ Class 160 not included in SR 26. Above figures show losses for ½" Class 315 and ¾" Class 200 wall thickness. Flows below heavy lines may develop surge pressures, which could cause damage. Use with caution.

32

3.4   PRESSURE CONVERSION CHART PSI-Kg/Cm-BAR

	1–40			41–80			81–200	
PSI	Kg/Cm	BAR	PSI	Kg/Cm	BAR	PSI	Kg/Cm	BAR
1	0.0703	0.0069	41	2.8833	2.8275	81	5.6962	5.5861
2	0.1406	0.1379	42	2.9536	2.8965	82	5.7665	5.655
3	0.211	0.2069	43	3.0239	2.9654	83	5.8368	5.724
4	0.2813	0.2759	44	3.06942	3.0344	84	5.9072	5.793
5	0.3516	0.3448	45	3.1646	3.1034	85	5.9775	5.8619
6	0.4219	0.4138	46	3.2349	3.1723	86	6.0478	5.9309
7	0.4923	0.4827	47	3.3052	3.2413	87	6.1181	5.9998
8	0.5626	0.5517	48	3.3755	3.3103	88	6.1885	6.0688
9	0.6329	0.6207	49	3.4459	3.3792	89	6.2588	6.1378
10	0.7032	0.6896	50	3.5162	3.4482	90	6.3291	6.2067
11	0.7736	0.7586	51	3.5865	3.5172	91	6.3994	6.2757
12	0.8439	0.8276	52	3.6568	3.5861	92	6.4698	6.3447
13	0.9142	0.8965	53	3.7271	3.6551	93	6.5401	6.4136
14	0.9845	0.9655	54	3.7975	3.724	94	6.6104	6.4826
15	1.0549	1.0345	55	3.8678	3.793	95	6.6807	6.5516
16	1.1252	1.1034	56	3.9381	3.862	96	6.7511	6.6205
17	1.1955	1.1724	57	4.0084	3.9309	97	6.8214	6.6895
18	1.2658	1.2413	58	4.0788	3.9999	98	6.8917	6.7585
19	1.3361	1.3103	59	4.1491	4.0689	99	6.962	6.8274
20	1.4065	1.3793	60	4.2194	4.1378	100	7.0323	6.8964
21	1.4768	1.4482	61	4.2897	4.2068	105	7.384	7.2412
22	1.5471	1.5172	62	4.3601	4.2758	110	7.7356	7.586
23	1.6174	1.5862	63	4.4304	4.3447	115	8.0872	7.9308
24	1.6878	1.6551	64	4.5007	4.4137	120	8.4388	8.2757
25	1.7581	1.7241	65	4.571	4.4826	125	8.7904	8.6205
26	1.8284	1.7931	66	4.6414	4.5516	130	9.1421	8.9653
27	1.8987	1.862	67	4.7117	4.6206	135	9.4937	9.3101
28	1.9691	1.931	68	4.782	4.6895	140	9.8453	9.6549
29	2.0394	1.9999	69	4.8523	4.7585	145	10.197	9.9997
30	2.1097	2.0689	70	4.9226	4.8275	150	10.549	10.345
31	2.18	2.1379	71	4.993	4.8964	155	10.9	10.689
32	2.2504	2.2068	72	5.0633	4.9654	160	11.252	11.034
33	2.3207	2.2758	73	5.1336	5.0344	165	11.603	11.379
34	2.391	2.3448	74	5.2039	5.1033	170	11.955	11.724

(continued)

3.4 PRESSURE CONVERSION CHART PSI-Kg/Cm-BAR (continued)

	1–40			41–80			81–200	
PSI	Kg/Cm	BAR	PSI	Kg/Cm	BAR	PSI	Kg/Cm	BAR
35	2.4613	2.4137	75	5.2743	5.1723	175	12.307	12.069
36	2.5316	2.4827	76	5.3446	5.2412	180	12.658	12.413
37	2.602	2.5517	77	5.4149	5.3102	185	13.01	12.758
38	2.6723	2.6206	78	5.4852	5.3792	190	13.361	13.103
39	2.7426	2.6896	79	5.5556	5.4481	195	13.713	13.448
40	2.8129	2.7586	80	5.6259	5.5171	200	14.065	13.793
205	14.416	14.138	510	35.865	35.172	910	63.994	62.757
210	14.768	14.482	520	36.568	35.861	920	64.698	63.447
215	15.12	14.827	530	37.271	36.551	930	65.401	64.136
220	15.471	15.172	540	37.975	37.24	940	66.104	64.826
225	15.823	15.517	550	38.678	37.93	950	66.867	65.516
230	16.174	15.862	560	39.381	38.62	960	67.511	66.205
235	16.526	16.206	570	40.084	39.309	970	68.214	66.895
240	16.878	16.551	580	40.788	39.999	980	68.917	67.585
245	17.229	16.896	590	41.491	40.689	990	69.62	68.274
250	17.581	17.241	600	42.194	41.378	1000	70.323	68.964
255	17.932	17.586	610	42.897	42.068	1010	71.027	69.653
260	18.284	17.931	620	43.601	42.758	1020	71.73	70.343
265	18.636	18.275	630	44.304	43.447	1030	72.433	71.033
270	18.987	18.62	640	45.007	44.137	1040	73.136	71.722
275	19.339	18.965	650	45.71	44.826	1050	73.84	72.412
280	19.691	19.31	660	46.414	45.516	1060	74.543	73.102
285	20.042	19.655	670	47.117	46.206	1070	75.246	73.791
290	20.394	19.999	680	47.82	46.895	1080	75.949	74.481
295	20.745	20.344	690	48.525	47.585	1090	76.653	75.171
300	21.097	20.689	700	49.226	48.275	1100	77.356	75.86
310	21.8	21.379	710	49.93	48.964	1120	78.762	77.239
320	22.504	22.068	720	50.663	49.654	1140	80.169	78.69
330	23.207	22.758	730	51.336	50.344	1160	81.575	79.998
340	23.91	23.448	740	52.039	51.033	1180	82.982	81.377
350	24.613	24.137	750	52.743	51.723	1200	84.338	82.757
360	25.316	24.827	760	53.446	52.412	1220	85.795	84.136

(continued)

1–40			41–80			81–200		
PSI	Kg/Cm	BAR	PSI	Kg/Cm	BAR	PSI	Kg/Cm	BAR
370	26.02	25.517	770	54.149	53.102	1240	87.201	85.515
380	26.723	26.206	780	54.852	53.792	1260	88.608	86.894
390	27.426	26.896	790	55.556	54.481	1280	90.014	88.274
400	28.129	27.586	800	56.259	55.171	1300	91.421	89.653
410	28.833	28.275	810	56.962	55.861	1320	92.827	91.032
420	29.536	28.965	820	57.665	56.55	1340	94.223	92.411
430	30.239	29.654	830	58.368	57.24	1360	95.64	93.791
440	30.942	30.344	840	59.072	57.93	1380	97.046	95.17
450	31.646	31.034	850	59.775	58.619	1400	98.453	96.549
460	32.349	31.723	860	60.478	59.309	1420	99.859	97.929
470	33.052	32.413	870	61.181	59.998	1440	101.27	99.308
480	33.755	33.103	880	61.885	60.688	1460	102.67	100.69
490	34.459	33.792	890	62.588	61.378	1480	104.08	102.07
500	35.162	34.482	900	63.291	62.067	1500	105.49	103.45

3.5 PIPE CONVERSION FACTORS (Natural Gas, Irrigation, Drainage, Etc.). English (NPS) to Metric (DN)

NPS	DN	NPS	DN	NPS	DN
⅛"	6 mm	3"	80 mm	20"	500 mm
³⁄₁₆"	7 mm	3½"	90 mm	24"	600 mm
¼"	8 mm	4"	100 mm	28"	700 mm
⅜"	10 mm	4½"	115 mm	30"	750 mm
½"	15 mm	5"	125 mm	32"	800 mm
⅝"	18 mm	6"	150 mm	36"	900 mm
¾"	20 mm	8"	200 mm	40"	1,000 mm
1"	25 mm	10"	250 mm	44"	1,100 mm
1¼"	32 mm	12"	300 mm	48"	1,200 mm
1½"	40 mm	14"	350 mm	52"	1,300 mm
2"	50 mm	16"	400 mm	56"	1,400 mm
2½"	65 mm	18"	450 mm	60"	1,500 mm

3.6 VELOCITY OF FLOW (FEET PER SECOND)

Bell and Ring PVC Pipe for 160 PSI Service

Flow (gpm)	1	1¼	1¼	2½	2	3	4	6
16	4.57	2.78	2.12					
18	5.14	3.13	2.39					
20	5.72	3.48	2.66	1.70				
25	7.15	4.35	3.32	2.12				
30	8.58	5.22	3.98	2.55	1.74			
35	10.01	6.09	4.65	2.97	2.03			
40	11.44	6.96	5.31	3.40	2.32			
45	12.84	7.82	5.96	3.82	2.61			
50	14.30	8.70	6.64	4.25	2.90			
55		9.55	7.29	4.67	3.19			
60		10.44	7.97	5.10	3.48			
65			8.61	5.52	3.77			
70			9.29	5.95	4.06	2.74		
75			9.94	6.37	4.35	2.94		
80			10.62	6.80	4.64	3.13		
85			11.26	7.22	4.93	3.33		
90			11.95	7.64	5.22	3.52		
95			12.59	8.07	5.51	3.72		
100			13.28	8.49	5.80	3.92	2.37	
110				9.34	6.37	4.31	2.60	
120				10.19	6.95	4.70	2.84	
130				11.04	7.53	5.09	3.08	
140				11.89	8.11	5.48	3.31	
150				12.74	8.69	5.87	3.55	
160				13.59	9.27	6.26	3.79	
170				14.44	9.85	6.66	4.02	
180				15.29	10.43	7.05	4.26	
190				16.14	11.01	7.44	4.50	
200				16.99	11.59	7.83	4.73	
220					12.75	8.61	5.21	
240					13.91	9.40	5.68	
260					15.07	10.18	6.15	2.83
280					16.23	10.96	6.63	3.05
300					17.39	11.75	7.10	3.27
320						12.53	7.58	3.49

(continued)

Flow (gpm)	I	1¼	1¾	2½	2	3	4	6
340						13.31	8.05	3.71
360						14.10	8.52	3.92
380						14.88	9.00	4.14
400						15.66	9.47	4.36
450						17.62	10.65	4.91
500						19.58	11.84	5.45
550							13.02	6.00
600								6.54
650								7.09
700								7.63

Useful Conversions:

1 gallon = 231 cu. in.

Gallons per minute (gpm) = 226.8 Liters per second

Cubic yard = 202 gallons

Cubic feet per second × 448.8 = GPM

Cubic feet per minute × 7.48 = gpm

Cubic feet per second = GPM/449

Cubic feet × 0.0283 = Cubic meters

Cubic meters × 35.37 = Cubic feet

Cubic yards × 0.765 = Cubic meters

Feet per second × 3.28 = Meters per second

Cubic meters × 0.06 = Cubic inches

Cubic feet × 28.316 = Liters

Cubic foot = 1,728 cu. in.

Cubic yard = 46,656 cu. in.

3.7 SURGE PRESSURES AND WATER HAMMER

Surge pressures and water hammer refer to various degrees of the same phenomenon. When water moves through pipes, kinetic energy is created in relation to the mass of water and the velocity at which it is moving. When the flow is stopped, this kinetic energy exerts itself in the form of a momentary increase in pressure or pressure rise. The intensity of this surge depends on the speed of closure of the valve that stops the flow in addition to the factors mentioned above.

The most commonly used formula for calculating pressure rise is

$$P = 0.070 \frac{VL}{T}$$

where P = Pressure rise (psi) above static pressure
V = Velocity of flow (ft./sec.)
L = Length of pipe (ft.) on pressure side of valve
T = Closing time of valve (sec.)

Surge pressure at a given velocity is therefore dependent on both the length of pipe and the closing time of the valve. For an example of these effects, the following calculations are shown for:

1-in. Class 315 PVC pipe:
Velocity of flow at 8.13 ft./sec. (25 gpm)
Static pressure at 160 psi (max. recommended for pressure lines)

With 100 ft. of pipe to valve, and valve closure time of 10 seconds.

$$P = 0.70 \frac{(8.13 \times 100)}{10} = 0.070 \frac{(813)}{10}$$
$$= 0.070 \times 81.3 = 5.69 \text{ psi pressure rise}$$

+ 160.00   psi static pressure
165.69   psi total surge pressure

Using the same velocity and pressure conditions, 10 ft. of pipe, and a valve closure time of 1 second:

$$P = 0.070 \frac{(8.13 \times 100)}{1} = 0.070 \frac{(813)}{1}$$
$$= 0.070 \times 813 = 56.9 \text{ psi pressure rise}$$

+ 160.00 psi static pressure
216.9   psi total surge pressure

Surge of this magnitude causes creep in PVC pipe and contributes to eventual fatigue of the pipe.

Again assuming the same conditions along with 1,000 ft. of pipe, and a valve closure time of 10 seconds:

$$P = 0.070 \frac{(8.13 \times 1,000)}{10} = 0.70 \frac{(8,130)}{10}$$
$$= 0.070 \times 813 = 56.9 \text{ psi pressure rise}$$

+ 160.00 psi static pressure
216.9   psi total surge pressure

With the same conditions, 1,000 ft. of pipe, and a valve closure time of 1 second:

$$P = 0.070 \frac{(8.13 \times 1,000)}{1} = 0.070 \frac{(8,130)}{1}$$

$$= 0.070 \times 8,130 = 569.1 \text{ psi pressure rise}$$

$$\underline{+ \quad 160.00} \qquad \text{psi static pressure}$$

$$729.9 \qquad \text{psi total surge pressure}$$

$$\downarrow$$

Severe, damaging water hammer

If length of line to valve cannot be changed, water hammer must be reduced to or below the safe working pressure of 315 psi either by increasing pipe size for the required flow to reduce velocity or by slowing the closure time for the valve, or both.

Use extreme caution where long main lines and a fast closing valve are used.

3.8 COMPARATIVE FLOW CAPACITIES FOR PIPE

Pipe Size (in.)	½	¾	1	1¼	1½	2	2½	3	4	6	8
½	1	2½	5	8	13	25					
¾		1	2	3	5	10	17	27			
1			1	2	3	5	9	14	28		
1¼				1	1½	3	5	8	17		
1½					1	2	3	5	11	28	
2						1	1½	3	5½	15	29
2½							1	1½	3	9½	19
3								1	2	5½	11
4									1	3	5½
6										1	2

Figures in this chart represent approximate comparison to flow at the same pressure loss. As actual inside diameters of pipes vary in comparison in various schedules or ratings, this chart can only be used as a general guide.

The following is a good chart for calculating fill-up time for tanks, lakes, and the like by pipe size. Maximum flow is usually 5 feet per sec (5 ft/sec.); however, you can go marginally past that with an open discharge.

GPM	Velocity in ft./sec. 1"	Pressure Drop 1"	Velocity in ft./sec. 1 1/4"	Pressure Drop 1 1/4"	Velocity in ft./sec. 1 1/2"	Pressure Drop 1 1/2"	Velocity in ft./sec. 2"	Pressure Drop 2"	Velocity in ft./sec. 2 1/2"	Pressure Drop 2 1/2"	Velocity in ft./sec. 3"	Pressure Drop 3"
2	0.57	0.06										
3	0.86	0.14										
4	1.14	0.23	0.54	0.04	0.55	0.04						
5	1.43	0.35	0.73	0.08	0.69	0.06						
6	1.72	0.49	0.91	0.12	0.83	0.08						
8	2.29	0.84	1.09	0.16	1.10	0.14	0.71	0.05	0.60	0.03		
10	2.86	1.27	1.45	0.28	1.38	0.21	0.89	0.07	0.90	0.06	3"	3"
15	4.29	2.68	1.81	0.42	2.07	0.45	1.33	0.15	1.20	0.10		
20	5.72	4.57	2.72	0.89	2.76	0.77	1.77	0.26	1.50	0.15	0.84	0.04
25	4"	4"	3.63	1.51	3.45	1.16	2.21	0.39	1.81	0.22	1.05	0.06
30	0.99	0.04	4.54	2.28	4.15	1.62	2.65	0.54	2.10	0.29	1.26	0.09
35	1.10	0.05	5.45	3.2	4.83	2.16	3.10	0.72	2.41	0.37	1.49	0.11
40	1.23	0.06	6.35	4.25	5.52	2.76	3.54	0.92	2.71	0.45	1.68	0.15
45	1.48	0.09	7.26	5.45	6.20	3.43	3.98	1.14	3.01	0.56	1.90	0.18
50	1.72	0.11	8.17	6.77	6.90	4.17	4.42	1.39	3.61	0.88	2.11	0.22
60	1.97	0.15	9.08	8.23	8.29	5.84	5.3	1.95	4.21	1.04	2.57	0.31
70	2.22	0.19	6"	6"	8"	8"	6.19	2.59	4.82	1.32	2.96	0.41
80	2.46	0.23					6.77	3.32	5.42	1.64	3.38	0.53
90	3.08	0.35					7.10	4.12	6.02	2.88	3.80	0.66
100	3.70	0.49	1.14	0.03			7.95	5.01	7.50	3.00	4.21	0.81
125	4.31	0.65	1.42	0.05					9.03	3.24	5.27	1.21
150	4.93	0.84	1.71	0.07					10.05	3.64	6.33	1.7
175	5.54	1.04	1.99	0.10	1.17	0.02	10"	10"			7.37	2.25
200	6.16	1.27	2.27	0.13	1.34	0.03					8.42	2.9
225	6.78	1.51	2.56	0.16	1.45	0.04					9.48	3.58
250	7.39	1.77	2.84	0.19	1.68	0.05	1.08	0.02			10.6	4.36
275	8.62	2.36	3.13	0.23	1.84	0.06	1.19	0.02	12"	12"	11.6	5.21
300	9.85	3.03	3.41	0.27	2.01	0.07	1.29	0.02				
350			3.98	0.36	2.35	0.10	1.51	0.03	1.07	0.01		
400			4.55	0.46	2.68	0.12	1.73	0.04	1.23	0.02		

	11.09	3.77	5.12	0.57	3.02	0.16	1.94	0.06	1.38	0.02	14"	14"
450												
500	12.32	4.58	5.69	0.70	3.35	0.19	2.16	0.06	1.53	0.03	1.32	0.02
550	13.55	5.46	6.26	0.83	3.69	0.23	2.37	0.08	1.69	0.03	1.43	0.02
600	14.78	6.41	6.82	0.98	4.02	0.27	2.59	0.09	1.84	0.04	1.54	0.02
650			7.39	1.13	4.36	0.31	2.81	0.11	1.99	0.04	1.65	0.02
700			7.96	1.30	4.69	0.36	3.02	0.12	2.15	0.05	1.76	0.03
750			8.53	1.48	5.03	0.41	3.24	0.14	2.30	0.06	1.87	0.03
800			9.1	1.67	5.36	0.46	3.45	0.15	2.45	0.06	1.98	0.04
850			9.67	1.87	5.70	0.51	3.67	0.18	2.61	0.07	2.09	0.04
900			10.24	2.07	6.04	0.57	3.88	0.19	2.76	0.09	2.20	0.05
950			10.81	2.29	6.37	0.63	4.10	0.22	2.91	0.09	2.42	0.06
1000			11.37	2.52	6.71	0.70	4.32	0.24	3.07	0.10	2.64	0.06
1100			12.51	3.00	7.38	0.83	4.75	0.28	3.37	0.12	2.86	0.07
1200			13.65	3.53	8.05	0.96	5.18	0.33	3.68	0.15	3.08	0.09
1300			14.79	4.10	8.72	1.13	5.61	0.40	3.99	0.17	3.30	0.10
1400			15.92	4.70	9.39	1.30	6.04	0.44	4.29	0.19	3.52	0.11
1500					10.06	1.48	6.47	0.51	4.60	0.22	3.97	0.14
1600					10.73	1.66	6.91	0.57	4.91	0.25	4.41	0.16
1800					12.07	2.07	7.77	0.71	5.52	0.31	4.85	0.20
2000					13.41	2.51	8.63	0.86	6.13	0.38	5.29	0.23
2200					14.75	3.00	9.50	1.03	6.75	0.44	5.73	0.27
2400					16.09	3.52	10.36	1.21	7.36	0.52	6.17	0.31
2600							11.22	1.40	7.98	0.61	6.61	0.35
2800							12.09	1.61	8.59	0.70	7.05	0.40
3000							12.95	1.83	9.20	0.80	7.49	0.44
3200							13.81	2.06	9.82	0.90	7.94	0.49
3400							14.68	2.30	10.43	1.00	8.38	0.54
3600							15.54	2.56	11.04	1.11	8.82	0.60
3800							16.40	2.83	11.6	1.23		
4000									12.27	1.35		

3.9 UNIFORMITY AND EFFICIENCY

Uniformity relates to how evenly water is spread over the irrigated area. *Uniformity coefficient* and *distribution uniformity* are measurements of this uniformity. Efficiency deals not only with how uniformly the water is spread, but also with losses such as percolation beyond the root zone, wind drift, evaporation, and so on.

3.10 UNIFORMITY COEFFICIENT

Uniformity coefficient values are determined by catching discharge from sprinklers in evenly spaced cans and evaluating the catchment mathematically. One mathematical description widely used was developed by J. E. Christiansen, who called his value the *uniformity coefficient*. This is commonly referred to as the coefficient of uniformity or CU. The coefficient is expressed as a percentage and is dependent on rotation uniformity, speed of rotation (for rotary type sprinklers), geometric layout pattern, nozzle pressure, and spacing distances. Christiansen's uniformity is determined by the following:

$$CU = 100\left(1.0 - \frac{\sum x}{mn}\right)$$

where CU = Uniformity coefficient in percent

x = Difference between individual catchment and the mean catchment

m = Mean catchment

n = Number of catchments

Values of 80% are considered minimum for field crops and turfs, and 85% for closely spaced, high-value crops. Closer spacings generally result in higher CU values.

3.11 SYSTEM DISTRIBUTION UNIFORMITY

The CU value describes the unevenness of how water falls from a sprinkler or sprinklers in a geometric pattern. Distribution uniformity (DU) relates to the differences in discharge between sprinklers in a system due primarily to pressure differentials.

There are different formulas describing distribution uniformity with some advocating using the absolute minimum catch and others the lowest 25% of the catchment. Generally the distribution uniformity value is approximately equal to

$$\frac{(\text{Minimum})}{\text{Average}} \times 100$$

For a sprinkler system, the system distribution uniformity approximately equals

$$(\text{Sprinkler gpm uniformity}) \times (\text{Catch can uniformity}) \times 100$$

where the sprinkler and catch can uniformities are expressed as decimals. For example, if a CU of catch can uniformity is 80%, the minimum sprinkler discharge is 2.5 gpm, and the average sprinkler discharge is 3.0 gpm, then

System distribution uniformity =

$$\frac{(205 \text{ gpm})}{3.0 \text{ gpm}} \times 0.80 \times 100 = 67\%$$

3.12 EFFICIENCIES

There are a multitude of ways to measure water application efficiencies (WAE). A simple approach is to look at the water used by plants (satisfy transpiration and leaching requirements) to the total water applied times 100 to convert to a percent.

$$\text{WAE} = \frac{\text{Water used by plants}}{\text{Water applied}} \times 100$$

WAE values are usually less than system distribution uniformities because

1. Most systems have evaporation losses.
2. Many systems have wind losses.
3. Systems are rarely shut off just when the soil moisture deficiency has been satisfied, so deep percolation occurs.

3.13 DESIGNING IRRIGATION SYSTEM CAPABILITIES

The typical formula used to determine the necessary volume of water or flow rate to apply is as follows:

$$\text{Gross need} = \frac{\text{Net need for ET leaching}}{\left(\frac{\text{WAE}}{100}\right)}$$

ET stands for evapotranspiration and is discussed in Section 3.15. The value of WAE must account for the DU, evaporation losses, and a reasonable estimate of management (timing).

Example

Find the gross depth that must be applied and the necessary sprinkler gallons per minute. Sprinklers spaced at 40×40 ft. Estimated WAE is 75%. Sprinklers can run for 4 hours and must apply a net of 1.5 in.

Formula

Gross = Net/(WAE/100)

gpm = (Gross inches) × (Area in square feet)/(Hours × 96.3)

Gross = 1.5 in. (75%/100) = 2.0 in. needed

gpm = (2.0 × 40' × 40')/(4 hr. × 96.3) = 8.3 gpm

Example

Find the necessary flow rate. An irrigation system must be designed to supply a net of 0.3 in./day to meet the peak summer ET requirement. Estimated WAE is 70%. Area is 2.3 acres. The maximum operating period is 8 hours.

Formula

Gross = Net/(WAE/100)

gpm = (Gross inches) × (Area in square feet)/(Hours × 96.3)

Gross = 0.3 in./(70%/100) = 0.43% needed

gpm = (0.43 × 2.3 acres × 43,560 SF)/(96.3 × 8 hr.)

= 56 gpm

3.14 THE SOIL RESERVOIR

Soil is composed of sand, silt, and clay *particles*. The percentage of each of these particles, called "factions" or *separates*, determines the *texture* of a specific soil. See Section 1.9.

Combined with decayed organic matter, the separates arrange into larger, more complex particles called soil *structure*. Spaces between the separates, called *pore spaces*, can contain air and water. A soil most favorable to plant root growth will contain approximately 50% pore space.

When water first moves into soil, its molecules form a film around the soil particles by electrical attraction, called *adhesion*. As more water is applied, *cohesion* (the water molecules' attraction to each other) allows the film to enlarge and fill the pore spaces. Further additions of water allow *gravity* to drain it to deeper layers. Water is absorbed by soil at a speed measured as *infiltration rate* (Ir), which varies with soil texture and structure. See Section 1.11.

Depending upon soil texture, drainage is rapid at first but decreases over time until there is a balance between gravity and cohesion. At this point the soil is holding all the water possible and is at *field capacity* (FC).

3.15 NATURAL WATER LOSS

The main cause of water loss is solar *evaporation*. Water moves by capillary action to the soil surface and evaporates. Additionally, water absorbed by the roots of a plant is passed upward through the plant's tissue and evaporates from the stomata of the leaves. This *transpiration* process helps to cool the plant and moves nutrients within it. These related processes are called *evapotranspiration* (ET), which is the key to efficient sprinkler system scheduling. The *ET rate* is the measurement of the amount of water loss.

As plants transpire and evaporation draws water from soil, the film of water becomes thinner and its adhesion to the soil particles is stronger than the roots' ability to extract it, slowing the ET rate and causing water stress visible in the plant. This is called the *permanent wilting point* (PWP). As the reservoir is depleted, the ET rate does not slow until the approach of the PWP.

The difference in the amount of water in the soil at field capacity and the amount contained at the PWP—the water that is usable by the plant roots—is the *available water* (AW), as:

$$FC - PWP = AW$$

See Section 1.8.

Root depth will vary with soil texture, depth, substrata layers, and cultural regime. As an example, a turf-type tall fescue grown in a uniform soil will have a rooting depth of 15 to 18 in. if the cutting height is 4 in. Lowering the cut to 3 in. shows a proportional reduction in the root depth. The depth of the roots defines the *soil reservoir*, and when multiplied by the available water per root, it gives the *soil reservoir capacity* (SRC), as:

$$Root\ depth \times AW = SRC$$

Keeping the soil at field capacity by using a set watering schedule without considering the ET rate can damage roots and be extremely water-wasteful. Allowing plants to constantly approach PWP puts them in stress that can make them susceptible to disease and poor quality.

For continued healthy growth, an efficient schedule plans irrigation well before PWP. This is the *allowable depletion* (AD). Experimentation and judgment are needed to select a reasonable AD level of the soil reservoir capacity, which is based on the value of the planting, plant type, soil texture, and the season. A medium-value turf is typically managed at 50% AD, whereas a low-value turf could have an AD of 75% of the SRC.

3.16 GUIDES TO DETERMINING ET RATE

The ET rate is affected by the atmospheric factors of solar radiation, air temperature, humidity, and wind velocity. Thus, in a hot and dry area during the normal growing season, a series of cloudy, cool days with high humidity would dramatically lower the season's average ET rate. The percent of soil shaded by plants affects the ET rate as the leaf surfaces intercept the sunlight. Measurements indicate that when 70 to 80% of ground cover is present, maximum ET rates may be assumed.

There is an excellent correlation between plant ET and evaporation from a free water surface. The U.S. Weather Bureau Class A Pan is the standard reference for evaporation ET and pan evaporation is expressed as a *crop coefficient* (Kp).

Check with your local county agent, co-operative extension specialist, or soil conservation office for the Kp values of the plants you are irrigating. In many areas of the country, either Class A pan reading or an ET rate for turfgrass is shown in newspapers or broadcast daily on radio or TV.

Another guide is *historic data*, which, modified for local current conditions, is an excellent point of departure in your planning. The best source of such data is the book 490-1358: *Rainfall Evapotranspiration Data*, published by Toro, which gives the average rainfall over a period of 30 years for 342 climate zones throughout the United States and Canada.

Soil moisture sensors are placed in the plant's root zone and monitor the amount of available water. If the root zone has adequate water, the soil moisture sensor will not allow a scheduled irrigation cycle to occur. Using soil moisture sensors with automatic controllers has proven to be effective in reducing water use by as much as 40% compared to using ET as the basis of irrigation scheduling.

3.17 MINIMIZING SYSTEM LOSSES

Runoff is a serious irrigation loss that can be minimized by ensuring that the precipitation rate (Pr) does not exceed the infiltration rate (Ir) of the soil. See Section 1.11. If a summer thundershower drops 1 in. of rain in 20 minutes, as much as 75% may be runoff; the remainder is added to the soil reservoir as *effective rainfall*. An efficient schedule will allow for such adjustment and show that more water may be required.

Deep *percolation* below the soil reservoir is wasteful and is caused by excessive or nonuniform application to the planted area. When planning sprinkler spacing, consider their positioning in relation to manufacturer's recommendation and seasonal *wind speed*.

3.18 OTHER CONSIDERATIONS

A *hydraulic analysis* of the system to avoid pressure differentials and a periodic *pressure test* will greatly contribute to efficiency.

At the beginning of your growing season, estimate the amount of water *stored* in the soil and irrigate if necessary to bring it to field capacity.

Where off-season rainfall is low, the *leaching* of accumulated salts from the root zone may be necessary. While this need is normally satisfied naturally by winter rains or the spring snowmelt, additional leaching, if it is required, will affect the overall water requirement. Samples sent to the local soil laboratory for a *test* will reveal salt content and indicate any leaching requirement.

3.19 IRRIGATION WATER REQUIREMENTS

The speed at which water is applied to soil from sprinkler sources is measured as the *precipitation rate* (Pr) and can be calculated using the following formula, which gives total gallonage as inches/hour, where GPM = U.S. gallons per minute.

General Formula
$$\frac{\text{Total gpm of area irrigated} \times 96.3}{\text{Area irrigated in square feet}} = \text{Pr}$$

Formula for Square Sprinkler Pattern
$$\frac{\text{gpm of full circle sprinkler} \times 96.3}{\text{Head spacing squared}} = \text{Pr}$$

Formula for Triangular Sprinkler Pattern
$$\frac{\text{gpm of full circle sprinkler} \times 96.3}{\text{Head spacing squared}} = \text{Pr}$$

Irrigation water requirements
$$= \text{ET rate} + \text{System losses} - \text{Effective rainfall}$$
Irrigation efficiency
$$= \frac{\text{Water available to plants}}{\text{Total water applied}}$$

3.20 SCHEDULING PROCEDURES

Based on the Water Budget Method of scheduling, the step-by-step procedure shown below will result in maximum automatic sprinkler efficiency:

Step 1: Gather Site Information

Season	Summer (April–October)
Number of days	214
Environment	Sacramento Valley, CA
Topography	Average 6% slope
Crop	Medium value turf
Rooting depth	12" (one foot)
Soil type	Soil fine sandy loam
Precipitation rate	System design is 0.4"/hr.

Step 2: Determine the ET Rate

$$\text{Estimated ET rate} = \frac{\text{Seasonal ET}}{\text{Days in season}}$$

$$\frac{\text{Seasonal ET}}{\text{Number of days}} \quad \frac{40.7 \text{ in.}}{21} = 0.19 \text{ in./day}$$

Estimated ET rate = 0.19 in./day

Step 3: Determine Irrigation Frequency

$$\text{Irrigation frequency} = \frac{\text{AD}}{\text{Estimated ET rate}}$$

AW from Section 1.8 = 2.0 in./ft.

To obtain soil reservoir capacity (SRC), multiply by root depth:

1 ft. × 2.0 in. ft. = 2.0 in. ft.

AD for medium-value turf assumed at 50%:

SRC 2.0 in. × 50% = 1.0 AD

$$\frac{\text{AD 1.0 in.}}{\text{Estimated ET rate 0.9 in./day}} = 5.26 \text{ days between irrigations}$$

Adjustment to even number of days gives the following:

Adjusted irrigation frequency = 5 days
Adjusted AD = 5 days at 1.9 in./day = 0.95 in.

Step 4: Determine Water Requirement

$$\text{Water requirement} = \frac{AD}{\text{System efficiency}}$$

System efficiency from Section 3.21: Hot and dry = 70% (0.70).

$$\frac{AD\ 0.95\ \text{in.}}{\text{System efficiency}\ 0.70} = 1.36\ \text{in. water requirement}$$

Water requirement = 1.36 in.

Step 5: Determine Station Run Time

$$\text{Station run time} = \frac{\text{Water application}}{\text{System Pr}}$$

Pr should be less than or equal to Ir to minimize runoff.

Assign Ir as maximum water application.
Use Ir from Section 1.11 for soil type and slope, as: Ir = 0.47 in./hr.
Therefore, 0.47 in. = water application.

Summer growing season is April to October. The same sequence is used for winter or can be customized by factoring in your local conditions.

$$\frac{\text{Water application}\ 0.47\ \text{in.}}{\text{System Pr}\ 0.4\ \text{in./hr.}} = 1.175\ \text{hr. run time}$$

Station run time = 1.175 hr. or 70.5 min.

Step 6: Determine Irrigation Cycle

$$\text{Irrigation} = \frac{\text{Water requirement}}{\text{Water application}}$$

To obtain the number of times the station must operate per day to satisfy the water requirement:

$$\text{Water requirement}\ \frac{1.36\ \text{in.}}{0.47\ \text{in.}} = 2.89\ \text{cycles}$$

Adjust to even number = 3 cycles per day

The result so far gives the automatic sprinkler system schedule as:

Irrigation frequency—To replace AD every	5 days
Station run time—Each cycle	70.5 min.
Irrigation cycles—Number of starts to meet water requirement	3 cycles

Step 7: Adjust Schedule to Equipment

Example

If controllers' maximum station run time is 60 minutes, then adjust to 4 cycles and divide water requirement by cycles.

$$\text{Adjusted water requirement} = \frac{\text{Water requirement}}{\text{Adjusted cycles}}$$

$$\text{Water requirement} \frac{1.36 \text{ in.}}{4} = .34 \text{ in. (water application)}$$

$$\frac{\text{Adjusted water application } 0.34 \text{ in.}}{\text{System Pr. } 0.4 \text{ in./hr.}} = 0.85 \text{ hr.}$$

Adjusted station run time = 0.85 hr. or 51 min., resulting in an automatic sprinkler system schedule of:

Irrigation frequency—To replace AD every	5 days
Station run time—Each cycle	51 minutes
Irrigation cycle—Number of starts to meet water requirement	4 cycles

Note: If Pr is greater than Ir, then irrigation cycles should be separated by 1 hour to minimize runoff. The time interval between cycles is executed by the controller, which sequences through all other station run times before completing the remaining cycles. Elapsed time between cycles will depend upon controller setting and specification.
See Summary (Section 3.21).

System designers must consider prevailing wind speed when deciding sprinkler spacing. By determining correct precipitation pattern overlap and ensuring even water application, designers use the wind as a design tool. *Typical* controller schedules for growing and nongrowing seasons are widely spaced even in this California example.

	Sample Controller Schedule			
Season	*Station*	*Operating Time (minutes)*	*Cycles*	*Frequency (days)*
Summer—				
April to October	1	51	4	5
Winter—				
November to March	1	60	3	16

3.21 SUMMARY

The water budget method of irrigation considers known and predicted factors in the environment and balances them to the components of an automatic sprinkler system, producing an efficient schedule that satisfies crop water requirement and reduces water and man-hour waste. It is the understanding and application of these basic principles that leads to better water management and successful design. We know of one example where its application to an existing system by landscape superintendents has resulted in man-hour and dollar savings as high as 43% in the first year! Other examples show that because this relatively simple method was applied by the system designer, the result was a remarkably low annual operating cost in new installations.

While it is applied here for turfgrass, the method is equally valid without modification, where rooting depths and ET rates are available, for most landscape crops, geographic areas, and seasonal variations.

3.22 SPRINKLER SPACING

Square Spacing
No wind	55% of diameter
4 mph wind	50% of diameter
6, 4 kph wind	50% of diameter
8 mph wind	45% of diameter
12, 8 kph	45% of diameter

Triangular Spacing
No wind	60% of diameter
4 mph wind	55% of diameter
6, 4 kph wind	55% of diameter
8 mph wind	50% of diameter
12, 8 kph	50% of diameter

Single-Row Spacing
No wind	50% of diameter
4 mph wind	50% of diameter
6, 4 kph wind	50% of diameter
8 mph wind	45% of diameter
12, 8 kph	45% of diameter

Design in consideration of the worst wind conditions.

3.23 PRECIPITATION RATE FORMULAS

Square-spaced sprinklers in pattern:

$$\frac{\text{gpm of full circle} \times 96.3}{(\text{Spacing})^2}$$

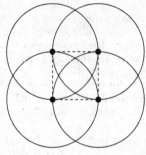

Triangular-spaced sprinklers in pattern:

$$\frac{\text{gpm of full circle} \times 96.3}{(\text{Spacing})^2 (0.866)}$$

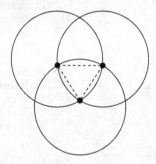

Area of flow:

$$\frac{\text{Total gpm of zone} \times 96.3}{\text{Total irrigated square feet of zone}}$$

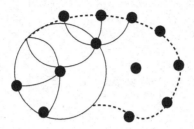

Single row:

$$\frac{\text{gpm of full circle} \times 96.3}{(\text{Spacing})(\text{Scallop})}$$

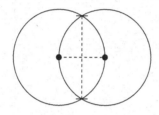

3.24 SPRINKLER EFFICIENCY AND SPACING UNDER CERTAIN CONDITIONS

Estimated Sprinkler Efficiencies by Climate Zone

Climate	Average Efficiencies
Low desert	60%
High desert	65%
Hot, dry	70%
Moderate	75%
Cool, humid	80%

Maximum Sprinkler Spacing for Wind

Wind Speed (mph)	Square Spacing	Triangular Spacing
0–7	50%	55%
8–10	45%	50%
Over 10	40%	45%

SOURCE: Based on Toro recommended sprinkler diameters.

3.25 APPROXIMATE NUMBER OF SPRINKLERS PER ACRE

Triangular Spacing		Square Spacing			
Spacing (ft.)	Heads	Spacing (ft.)	Heads	Spacing (ft.)	Heads
10	504	70	10.5	8.5	552	60.5	12
15	224	75	9	13	258	65	10
20	125	80	8	17	151	69	9
25	80	85	7	21.5	97	73.5	8
30	56	90	6.2	26	65	78	7
35	42	95	5.6	30	48	82	6.4
40	31	100	5.0	34.5	38	86.5	5.8
45	25	105	4.5	39	29	91	5.3
50	20	110	4.2	43	24	95	4.8
55	16.5	115	3.8	47.5	20	99.5	4.4
60	14	120	3.5	52	16	104	4.1
65	12	125	3.3	56	14	108	3.9

3.26 EXAMPLE—HOW TO DETERMINE PIPE SIZE PER REQUIREMENT

You need to add more sprinklers to an existing irrigation system. Each sprinkler provides 30 gpm at optimum operation.

1. Determine how many of the additional sprinklers you will operate simultaneously.

2. Refer to the friction loss chart of Section 3.3 and find the appropriate gpm required.

3. Scan across the page until you find the pipe size suitable "above" the heavy line. (The heavy line represents the maximum recommended friction loss figure 6 ft./ sec.)

4. This will be the pipe size necessary to service the new sprinklers. Remember to adjust the main feed (pipe size) to the new area if necessary.

You have decided to run 4 sprinklers at one time to fit your watering time window.

1. Number of sprinklers run at once: 4 × 30 gpm = 120 gpm.

2. Review friction loss chart; find 120 gpm.

3. Move across the page until you find a pipe size above the heavy line. In this case it will be 3 in.

The friction loss chart also indicates the maximum capacities of each size pipe.

For example, a 2 ½ in. pipe can handle 300 gpm. However, severe friction loss occurs past 100 gpm (as per chart).

Note: Doubling the pipe size increases capacity 4 times.

3.27 PUMP DEFINITIONS

Net Positive Suction Head (NPSH) required refers to the pressure required to force a given flow rate of water into the entrance of the pump. When the pump is located above the water surface, this force comes from the weight of the atmosphere. When the pump is located below the water surface, it comes from both the weight of the atmosphere and the weight of the water above the centerline of the pump entrance.

The amount of *NPSH available* must always be greater than the NPSH required. The amount available is calculated by subtracting the total dynamic suction head *(TDSH)* (elevation head, friction head, velocity head, and vapor pressure of water) from 33.9 ft. at sea level.

Since velocities in suction pipes are low and water temperatures are generally below 60°F in irrigation, they are usually ignored. The *NPSH available* at sea level is then determined by subtracting the sum of the elevation head between the water surface and the centerline of the pump entrance, and the friction head of all piping on the suction side of the pump from 33 ft. For every 1,000 ft. elevation above sea level, the *NPSH available* is reduced by about 1 ft.

3.28 PRACTICAL SUCTION LIFTS AT VARIOUS ELEVATIONS

Elevation	Barometer Reading (psi)	Theoretical Suction Lift (ft.)	Practical Suction Lift (ft.)
Sea level	14.7	34.0	22
¼ mile (1,320') above sea level	14.0	32.4	21
½ mile (2,640') above sea level	13.3	30.8	20
¾ mile (3,960') above sea level	12.7	29.2	18
1 mile (5,280') above sea level	12.0	27.8	17
1¼ miles (6,600') above sea level	11.4	26.4	16
1½ miles (7,920') above sea level	10.9	25.1	15
2 miles (10,560') above sea level	9.9	22.8	14

3.29 PUMP COST AND EFFICIENCY DATA

Cost per Hour Comparison between Single- and Three-Phase Pump Motors

Motor (H.P.)	Cost per Hour for Continuous Operation at a Rate of $.01 per Kilowatt-Hour[a]	
	Single Phase	Three Phase
¼	0.0031	—
½	0.0054	0.0052
¾	0.0076	0.0077
1	0.0099	0.0096
1½	0.0150	0.0142
2	0.0197	0.0183
3	0.0295	0.0270
5	0.0465	0.0451
7½	—	0.0675
10	—	0.0900

[a]If prevailing rate is more than $.01, multiply the rate figure by 100 and use the product to multiply by the above given cost figure.

3.30 COST OF PUMPING WATER PER 1,000 U.S. GALLONS PUMPED

$$\text{Cost} = \frac{\begin{array}{c}0.189 \times \text{Head in feet} \times \\ \text{Power cost per kilowatt-hour}\end{array}}{\text{Pump efficiency} \times \text{Motor efficiency} \times 60}$$

3.31 COST PER HOUR OF PUMPING UNDER CONTINUOUS CONDITIONS

$$\text{Cost} = \frac{\begin{array}{c}0.000189 \times \text{Head in feet} \times \\ \text{Power cost per kilowatt-hour}\end{array}}{\text{Pump efficiency} \times \text{Motor efficiency}}$$

3.32 FORMULA FOR FIGURING EFFICIENCY OF PUMP

$$\text{Pump efficiency} = \frac{\begin{array}{c}\text{U.S. gallons per minute rating} \times \\ \text{head rating in feet}\end{array}}{3,960 \times \text{Brake horsepower}}$$

3.33 FORMULA FOR DETERMINING PUMP HORSEPOWER

$$\text{Horsepower} = \frac{\text{gpm} \times \text{Total head required}}{3,960 \times \text{Pump efficiency}}$$

3.34 PUMP CALCULATION: SIZE FOR DUTY

Formula

$$\text{Horsepower} = \frac{Q\,(\text{gpm})\,\text{TDH}\,(\text{Head rating in feet})}{3,960 \times \text{Pump efficiency}}$$

Example
You need to determine what size (H.P.) pump is required to recharge a lake in a given period of time.

Given: An irrigation system is using 500,000 U.S. gallons/day.

Therefore, you need to determine what size pump will provide a minimum 500,000 gallons/day (open discharge) to maintain the same capacity.

1,000,000 gallons/day = 694 gpm

500,000 gallons/day = 694 ÷ 2 = 347 gpm

We will require 40 psi for open discharge.

TDH = 40 × 2.31 = 92.4

We will assume a pump efficiency of 60%. Therefore:

$$\frac{347 \times 92.4}{3,960 \times 0.60} = \frac{32,062.80}{2,376} = 13.49 \text{ H.P.}$$

A 13.49 H.P. motor will provide 500,000 U.S. gallons per 24 hours.

3.35 MISCELLANEOUS IRRIGATION DATA

Areas of circle are to each as the squares of their diameter.

Doubling the diameter of a pipe or cylinder increases its capacity four times.

Friction of liquids in pipes increases as the squares of the velocity.

Velocity in feet per minute necessary to discharge a given volume of water in a given time equals

$$\frac{\text{Cubic feet of water} \times 144}{\text{Area of pipe in square inches}}$$

Precipitation rate of heads in a set pattern in inches per hour:

For square spacing: $\dfrac{\text{gpm of full circle heads} \times 96.3}{\text{Head spacing squared}}$

For triangular spacing: $\dfrac{\text{gpm of full circle heads} \times 96.3}{\text{Head spacing squared} \times 0.866}$

3.36 EXAMPLE—IRRIGATION SYSTEM DESIGN FOR SPORTS FIELDS

Irrigation Design Criteria

- Determine static pressure and water source(s) for site. To determine maximum available water, select a flow that does not allow a pressure loss greater than 10% of static pressure.
- Determine field sizes and placement on-site.
- Select sprinkler that can be buried up to ½ in. below grade and have a stainless steel sleeving on the riser and nozzle assembly. In addition, the head should accept a rubber cover kit for the surface. The sprinkler head shall be of gear-driven rotary type, with a surface not to exceed 2.5 in.

- As a minimum, irrigate the playing field surfaces only. Use square spacing, not to exceed 50% of diameter. In cases where winds consistently exceed 5 mph, derate spacing to 45% of diameter. Quarter-circle sprinklers, half-circle sprinklers, and full-circle sprinklers should be zoned separately.

- Locate all control valves off the playing field, at a minimum of 5 ft. from the edge. All valves should be installed in a valve box with locking cover. All valves should have a bed of pea gravel 6 in. deep below the valve. Select only a control valve with flow control and manual bleed capabilities. In all climates, PVC pipe should be used for the mainline or pressurized line. Selection of the pipe will be based on economy, but as a guideline, static pressure should not exceed two-thirds of the pressure rating. In northern climates, polyethylene pipe with a pressure rating of 120 psi can be used for all lateral lines sized 2 in. and smaller. Only stainless steel clamps should be used with poly pipe.

- All sprinkler heads should be installed on a swing joint of the same size as the sprinkler head inlet. The swing joint should be constructed so as to tighten when it moves down.

- Controllers should be pedestal mounted, unless a permanent structure exists on the site. Manufacturers' guidelines for lightning protection and proper grounding should be followed. Each controller should be equipped with freeze and rain shutoff devices.

- All valve wiring should be done with UF direct burial wire, with sizing depending on length of the wire from controller to valve. Color coding should indicate common wire—white—and power (hot) wires—red or black. All wire connections should be waterproof. Slack should be provided every 100 ft. and at the valve.

- The irrigation designer must provide watering requirements based on evapotranspiration needs and develop a schedule of controller adjustments for what frequency changes are required and for the actual changes.

- Local backflow prevention codes should be followed by the irrigation designer.

- The designer must provide an installation detail sheet with the complete design. If the projects will be bid to irrigation contractors, then the irrigation designer is required to review installation specifications and modify to reflect local conditions.

3.37 MATERIALS FOR SPRINKLER SYSTEMS

The materials for the design of sprinkler systems discussed in this section are those an irrigation designer would select so as to establish the level of quality and performance required by an owner. A contractor who wishes to substitute equipment made by alternate manufacturers should be required to submit, in duplicate, specification material that proves equal performance, life, and maintenance, two weeks prior to bid opening date. The owner should reserve the right to disallow any equal substitutions.

- **Copper piping.** Piping should be Type K, hard copper, unless code dictates otherwise.
- **Polyethylene pipe.** All polyethylene pipe specifications on the plan should be flexible nontoxic polyethylene made from 100% virgin polyethylene material. All polyethylene pipe specified should not exceed a nominal size of 2 in. and should not be smaller than a nominal size of ¾ in., and all sizes should have a minimum rating of 120 psi. All polyethylene pipe should be continuously and permanently marked with the manufacturer's name, material, size, and schedule.
- **PVC pipe.** All PVC pipe specified on plan should be virgin, high-impact, polyvinyl chloride (PVC) pipe, having a minimum working pressure rating of Class 200. Static pressure is not to exceed two-thirds of operating pressure of pipe. All PVC pipe should be continuously and permanently marked with the manufacturer's name, material, size, and schedule or type. The pipe should conform to U.S. Department of Commerce standard CS 207-60, or latest revision. Material should conform to all requirements of Commercial standard CS 256-63, or latest revision.
- **Copper pipe fittings.** All copper pipe fittings should be solder-type cast solder-joint fittings.
- **Polyethylene pipe fittings.** Plastic-type or insert fittings and/or brass saddle tees or compression-type fittings should only be used as specified by the designer on the plans.
- **PVC pipe fittings.** Use only Solvent Weld, Sch 40 fittings. PVC saddles or compression-type fittings should only be used as specified by designer on the plans.
- **Swing-joint risers.** All sprinkler heads should be mounted on a swing joint consisting of three 1 in. 90 Ell FPT × FPT PVC Sch 40 fittings, three 1 × 3 in. Sch 80 threaded nipples, and one 1 in. × 8 in. Sch nipple threaded. Teflon tape should be used on all threaded connections, with no more than two wraps. Elbows and nipple should be assembled such that downward pressure will tighten the swing joint. Contractor may also choose to use premanufactured swing joints, such as Lasco or Spears products.
- **Electric remote-control valves.** These valves should be of globe/angle configuration with a stainless steel banded female pipe thread inlet and outlet. The diaphragm should be of fabric-reinforced rubber construction to retain flexibility and provide maximum strength throughout its area. The valve should have a manual flow control.

The valve should be normally closed by internal water pressure. A 22 in. solenoid lead wire should be attached to a 24 VAC, 50/60 Hz solenoid with waterproof molded coil capable of being removed. Valve should have a self-cleaning metering pin to protect bleed ports and to purge contaminants.

- **Valve boxes.** Valve boxes should be provided for each remote-control valve or group of remote-control valves. Valve boxes should be plastic with locking covers, such as a Carson or Ametek valve box.

- **Field wiring.** All field wiring from the controller to the remote-control valves should be UF# 14 direct burial wires, unless otherwise noted by designer on the plan. A distinctive color, such as white, should be selected for the common wire and a second color, such as red, for the power wire. All wires must be furnished in minimum 2,500 ft. reels and spliced only at valve or tee locations. All wire splices should be placed in a valve box as described in *Valve boxes*, above. Expansion coils should be provided at each remote-control valve and at least every 300 ft. of run. Expansion coils are easily formed by wrapping at least five turns of wiring around a 1 in. pipe, then withdrawing the pipe.

- **Quick coupling valves and keys.** The contractor should provide the owner two quick coupling keys and two swivel hose ells.

- **Backflow prevention devices.** Backflow prevention requirements must meet local codes. Devices and proper installation per code should be noted on plans.

- **Sprinkler heads.** All heads on the playing fields, both full- and part-circle heads, should be gear drive rotary and designed with an integral check valve for control on line drainage. The sprinkler should be mounted up to ½ in. below the final finish grade. Water distribution should be via two nozzles mounted in a 2 in. diameter nozzle turret, encased in stainless steel. The dual nozzles should elevate 3½ in. when in operation. Retraction should be achieved by a heavy-duty stainless steel retraction spring. The sprinklers should have a riser seal and a wiper. Rotation should be accomplished by a sealed, oil-packed gear assembly isolated from the water. The sprinkler housing should be of high-impact molded plastic with a 1 in. NPT connection. The sprinkler should be constructed such that it is serviceable from the top; in other words, the drive assembly, screen, and all internal components should be accessible through the top of the sprinkler without distributing case installation. Radius reduction should be adjustable up to 25% by means of a radius adjustment screw accessible from the top of the cap when the sprinkler is properly installed. As an option, rubber cover kits can be used. Since matched or balanced precipitation rates are not possible with these sprinklers, all full-circle heads and part-circle heads need to be zoned separately. Spacing of heads should not exceed the manufacturer's maximum recommendation. It is the contractor's responsibility to verify the spacing is within these limits. If there is a spacing problem, the contractor is to cease work and report such problem in writing to the owner. The report should include description of problem and a possible solution, and a satisfactory solution must be negotiated with the contractor.

- **Control equipment.** The sprinkler controller should be an 8-, 12-, 16-, or 24-station model, which will be indicated on the drawing, providing for complete automatic operation of the system. The controller should have an 8-, 12-, 16-, or 24-station capacity with a maximum run time of 60 minutes per station. Independent drip stations should have an operating range of 2 to 18 hours. Changes of the station timing and program start time should be easily made without interfering with set program start times through the use of faceplate knobs and pins. The controller operation should provide for rapid advance between stations and manual operation for each station. The controller should provide for schedules up to one week, with six days scheduling available on an optional day wheel. The controller should be equipped with a manual wheel for activation of a manual watering cycle. The controller should be programmable for up to 11 start times per day on turf and shrub stations. The controller should be capable of operating 24 VAC electric remote-control valves via a 40 VA transformer. The controller will result in lost time, but it should not affect program retention in any way. No battery backup should be required. Clear step-by-step instructions, as well as an owner's manual, should be included with each controller. All programming information should be visible on the faceplate. The controller housing should be constructed of metal with a locking door. The contractor should be required to provide surge protection and grounding as per instructions with the controller. The pedestal should be securely fastened to it with the appropriate mounting nuts and bolts, as covered by the installation instruction of the manufacturer. Templates supplied by the manufacturer will often allow construction of the pad before installation of the pedestal.

- **Drains.** Where a system must be winterized to prevent freeze damage, an air hose connection of approved design will be required at the location or locations noted on plans. The entire system can be drained by blowing it out with compressed air, not to exceed 70 psi. In addition, automatic drain valves may be utilized to drain lateral lines. Their locations will be determined by the elevation change and noted on the drawings. These automatic drain valves are to be installed with a rock sump of appropriate size and depth for soil types and amount of water to be drained. The automatic drain valve might be a King Brothers brand.

3.38   MAINTENANCE PROCEDURE FOR IRRIGATION SYSTEM

Prior to the first game of the day on a sports field, a team of two individuals will locate sprinkler row markers at the end of the field. They will walk the row of sprinklers to search for any sprinkler heads stuck in the up position. If a head is spotted, generally tapping the head with the foot will cause it to retract. Row and distance from sideline should be noted. During the next regular maintenance scheduled, the head should be disassembled to look for dirt or debris in the head, or excessive seal wear. Appropriate actions to repair or replace the head should be taken.

The individual in charge of coordinating the regular mowing and fertilization schedules should be provided the schedules developed by the irrigation designer. This individual will be responsible for implementing the prescribed scheduling changes. With these scheduling changes, the field should use a minimum 30% less water across a growing season.

Once a month, each zone should be visually checked during operation for proper performance. Any problems should be corrected as quickly as possible,

3.39   WINTERIZATION IN SOUTHERN CLIMATES

Controllers should be placed in the off position. The fields should be checked every two to four weeks, depending on rainfall. If the fields appear to be dry, then semiautomatic operation of a cycle should occur. In periods of extreme cold weather, the water to the system should be turned off and the backflow assembly drained.

3.40   WINTERIZATION IN NORTHERN CLIMATES

Once freezing days and nights occur regularly, and the need or ability to apply water diminishes, the irrigation should be turned off. The controllers should be placed in the off position. The water to the system should be shut off and the system blown out. To properly blow out a system, an air compressor is hooked up immediately downstream of the backflow prevention device. Compressed air is forced into the mainline at the same operating pressure as the sprinkler heads, not to exceed 70 psi. The valve should be manually bled, which will allow air to open the valve and force water out of both the mainline and the lateral lines. If any manual drains are installed on piping, they should be opened.

3.41   SPRING START-UP

If the system water has been turned off and the piping system has been drained, then caution is required in the initial system recharging. Water should be turned on very slowly, allowing time for compressed air to escape through an opening such as quick coupler valve with a key in it, and open. The valves closest to the water source should be manually bled and operate for a few minutes. Once these steps are completed, the entire system is ready for operation. Now perform a visual check for any zone problem that may have developed during the winter. Take steps to correct problems. Set up the controller schedule.

3.42 ELECTRICAL TABLE FOR SINGLE-PHASE IRRIGATION WIRING

This table is based on the equation $I = V/R$. I is given in the last column. V, the voltage drop, is assumed to be IV. R is the resistance of 100 m of cable. The "Maximum Current" column lists the maximum permissible current for each cable size.

How to use this table: If the voltage drop allowable were doubled (i.e., $2V$), then the current would double. If the distance (i.e., resistance) were doubled, then the current would be halved.

Cable Size (mm²)	m V/A (m)	Ohms/100 m	Maximum Current	Current Based on Above
0.5	85	8.500	10	0.118
1.0	46	4.600	14	0.217
1.5	31	3.100	16	0.323
2.5	18	1.800	23	0.556
4	12	1.200	30	0.833
6	7.9	0.790	39	1.266
10	4.6	0.460	53	2.174
16	3.0	0.300	70	3.333
25	1.8	0.180	91	5.556
35	1.4	0.140	115	7.143
50	0.99	0.099	140	10.101
70	0.70	0.070	175	14.286
95	0.54	0.054	210	18.519
120	0.44	0.044	250	22.727
150	0.38	0.038	280	26.316
185	0.32	0.032	325	31.250
240	0.28	0.028	385	35.714

Note: The maximum current for cables varies with the application. Underground cable must be derated, since it cannot dissipate the heat as well as a nonenclosed cable. The maximum current and deratings vary between the standards.

3.43 ELECTRICAL TABLE FOR BASIC THREE-PHASE IRRIGATION WIRING

This table is based on the equation $I = V/R$. I is given in the last column. V, the voltage drop, is assumed to be IV. R is the resistance of 100 m of cable. The "Maximum Current" column lists the maximum permissible current for each cable size.

How to use this table: If the voltage drop allowable were doubled (i.e., $2V$), then the current would double. If the distance (i.e., resistance) were doubled, then the current would be halved.

Cable Size (mm²)	m V/A (m)	Maximum Current	Current Based on Above
1.0	40	11	0.25
1.5	27	14	0.37
2.5	16	20	0.625
4	10	28	1.000
6	6.8	35	1.471
10	4.0	49	2.500
16	2.6	63	3.846
25	1.6	83	6.250
35	1.2	100	8.333
50	0.86	125	11.628
70	0.61	155	16.393
95	0.47	185	21.277
120	0.38	220	26.316
150	0.33	250	30.303
185	0.28	285	35.714
240	0.24	340	41.667

3.44 VOLTAGE LOSS FOR VARIOUS WIRE SIZES PER 100 FT. OF COPPER WIRE

Amps				Two A.C. Wire Gauge			
	14	12	10	8	6	4	2
0.1	0.0506	0.0319	0.0200	0.0121	0.0076	0.0048	0.0030
0.2	0.1013	0.0637	0.0401	0.0243	0.0153	0.0096	0.0060
0.3	0.1519	0.0956	0.0601	0.0364	0.0229	0.0144	0.0091
0.4	0.2026	0.1274	0.0801	0.0485	0.0305	0.0192	0.0121
0.5	0.2532	0.1593	0.1002	0.0607	0.0382	0.0241	0.0151
0.6	0.3038	0.1911	0.1202	0.0728	0.0458	0.0289	0.0181
0.7	0.3545	0.2229	0.1402	0.0849	0.0534	0.0337	0.0211
0.8	0.4051	0.2548	0.1602	0.0970	0.0610	0.0385	0.0242
0.9	0.4558	0.2867	0.1803	0.1092	0.0687	0.0433	0.0272
1.0	0.5064	0.3185	0.2003	0.1213	0.0763	0.0481	0.0302
1.5	0.7596	0.4778	0.3005	0.1820	0.1145	0.0722	0.0453

2.0	1.0128	0.6370	0.4006	0.2426	0.1526	0.0962	0.0604
2.5	1.2660	0.7963	0.5008	0.3033	0.1908	0.1203	0.0755
3.0	1.5192	0.9555	0.6009	0.3639	0.2289	0.1443	0.0906
3.5	1.7724	1.1148	0.7011	0.4246	0.2671	0.1684	0.1057
4.0	2.0256	1.2740	0.8012	0.4852	0.3052	0.1924	0.1208
4.5	2.2788	1.4333	0.9014	0.5459	0.3434	0.2165	0.1359
5.0	2.5320	1.5925	1.0015	0.6065	0.3815	0.2405	0.1510
5.5	2.7852	1.7518	1.1017	0.6672	0.4197	0.2646	0.1661
6.0	3.0384	1.9110	1.2018	0.7278	0.4578	0.2886	0.1812
6.5	3.2916	2.0703	1.3020	0.7885	0.4960	0.3127	0.1963
7.0	3.5448	2.2295	1.4021	0.8491	0.5341	0.3367	0.2114

This table is based on the equation

$$E = \frac{KLI}{CM}$$

For one wire,

CM = Circular mills of wire
E = Voltage loss
K = Ohms (10.4 for copper)
L = Length
I = Amps

Wire Gauge	CM
14	4,107
12	6,530
10	10,380
8	16,510
6	26,250

Wire Conversion Chart

American Wire Gauge (AWG)	Metric (mm³)
14	2.5
12	4.0
10	6.0
8	10.0
6	15.0
4	25.0
2	35.0
0	50.0
00	70.0

3.45 WIRE SIZING CALCULATION FORM[a]

American Gauge		Metric	
100 ft.	100 ft.	100 m	100 m
2–0.0313	10–0.2159	35–0.12	6–0.68
4–0.0498	12–0.3185	25–0.16	4–1.0
6–0.0792	14–0.5064	16–0.26	2.5–1.6
8–0.1260	16–0.8230	10–0.40	1.5–2.7

[a]Voltage loss in each size wire per 1 amp for distance shown. Both wires in the circuit, power common and neutral, taken into consideration.

3.46 AMERICAN TO METRIC CABLE SIZE CONVERSION

Conductor Size (AWG)	Conductor Diameter (mm)	Conductor Cross-Sectional Area (mm²)	Conductor Resistance (π/km)	Conductor Size (AWG)	Conductor Diameter (mm)	Conductor Cross-Sectional Area (mm²)	Conductor Resistance (π/km)	Conductor Size (AWG)	Conductor Diameter (mm)	Conductor Cross-Sectional Area (mm²)	Conductor Resistance (π/km)
6/0	14.73	170.3	0.11	10	2.59	5.27	3.64	25	0.455	0.163	111
5/0	13.12	135.1	0.14	11	2.30	4.15	4.44	26	0.405	0.128	146
4/0	11.68	107.2	0.18	12	2.05	3.31	5.41	27	0.361	0.102	176
3/0	10.40	85.0	0.23	13	1.83	2.63	7.02	28	0.321	0.0804	232
2/0	9.27	67.5	0.29	14	1.63	2.08	8.79	29	0.286	0.0646	282
0	8.25	53.4	0.37	15	1.45	1.65	11.2	30	0.255	0.0503	350
1	7.35	42.4	0.47	16	1.29	1.31	14.7	31	0.227	0.0400	446
2	6.54	33.6	0.57	17	1.15	1.04	17.8	32	0.202	0.320	578
3	5.83	26.7	0.71	18	1.024	0.823	23.0	33	0.180	0.0252	710
4	5.19	21.2	0.91	19	0.912	0.653	28.3	34	0.160	0.0200	899
5	4.62	16.8	1.12	20	0.812	0.519	34.5	35	0.143	0.0161	1125
6	4.11	13.3	1.44	21	0.723	0.412	44.0	36	0.127	0.0123	1426
7	3.67	10.6	1.78	22	0.644	0.325	54.8	37	0.113	0.0100	1800
8	3.26	8.35	2.36	23	0.573	0.259	70.1	38	0.101	0.0078	2255
9	2.91	6.62	2.77	24	0.511	0.205	89.2	39	0.0897	0.00632	2860

Section Four

CONCRETE, RETAINING WALLS, STREETS, AND WEIR STRUCTURES

4.1 APPROXIMATE WEIGHTS OF MATERIALS REQUIRED PER CUBIC YARD OF CONCRETE

Weights Given Assume Surface-Dry Aggregate

Consistency	Proportions With Gravel $c' + s' + g' + w'$	Proportions With Stone $c' + s' + g' + w'$	Cement (sack)	Water (gal.)	With Gravel Sand (tons)	With Gravel Gravel (tons)	With Stone Sand (tons)	With Stone Gravel (tons)
1-In. Aggregate								
Water = 5 Gal. per Sack of Cement, 28-Day Strength = 5,000 psi								
Stiff	1 + 2.0 + 3.3 + 0.44	1 + 2.4 + 2.9 + 0.44	6.6	33	0.63	1.01	0.74	0.90
Medium	1 + 1.8 + 2.9 + 0.44	1 + 2.1 + 2.6 + 0.44	7.2	36	0.60	0.98	0.72	0.87
Wet	1 + 1.6 + 2.6 + 0.44	1 + 1.9 + 2.3 + 0.44	7.8	39	0.58	0.94	0.69	0.84
Water = 6 Gal. per Sack of Cement, 28-Day Strength = 4,000 psi								
Stiff	1 + 2.7 + 3.8 + 0.53	1 + 3.2 + 3.4 + 0.53	5.5	33	0.70	0.90	0.82	0.88
Medium	1 + 2.4 + 3.4 + 0.53	1 + 2.8 + 3.0 + 0.53	6.0	36	0.68	0.96	0.79	0.85
Wet	1 + 2.1 + 3.1 + 0.53	1 + 2.5 + 2.7 + 0.53	6.5	39	0.66	0.93	0.77	0.82
Water = 7.5 Gal. per Sack of Cement, 28-Day Strength = 3,000 psi								
Stiff	1 + 3.7 + 4.7 + 0.67	1 + 4.3 + 4.1 + 0.67	4.4	33	0.77	0.97	0.90	0.85
Medium	1 + 3.3 + 4.2 + 0.67	1 + 3.9 + 3.7 + 0.67	4.8	36	0.75	0.94	0.87	0.83
Wet	1 + 3.0 + 3.7 + 0.67	1 + 3.4 + 3.3 + 0.67	5.2	39	0.73	0.91	0.84	0.80

SOURCE: Values in this table are taken from publications of the Portland Cement Association.
Conversions: cu. yd. × 1.31 = cubic meter
kg/cm² × 14.223 = psi

4.2 APPROXIMATE WEIGHTS OF MATERIALS REQUIRED PER CUBIC YARD OF CONCRETE

Weights Given Assume Surface-Dry Aggregate

| | Proportions | | | Materials for 1 Cu.Yd. | | With Gravel | | With Stone | |
| | With Gravel $c'+s'+g'+w'$ | With Stone $c'+s'+g'+w'$ | Cement (sack) | Water (gal.) | | Sand (tons) | Gravel (tons) | Sand (tons) | Gravel (tons) |
Consistency									
2-In. Aggregate									
Water = 5 Gal. per Sack of Cement, 28-Day Strength = 5,000 psi									
Stiff	1 + 2.1 + 3.9 + 0.44	1 + 2.5 + 3.5 + 0.44	6.0	30		0.60	1.10	0.70	1.00
Medium	1 + 1.9 + 3.4 + 0.44	1 + 2.2 + 3.1 + 0.44	6.6	33		0.58	1.06	0.68	0.96
Wet	1 + 1.7 + 3.0 + 0.44	1 + 1.9 + 2.8 + 0.44	7.2	36		0.56	1.02	0.65	0.93
Water = 6 Gal. per Sack of Cement, 28-Day Strength = 4,000 psi									
Stiff	1 + 2.8 + 4.6 + 0.53	1 + 3.3 + 4.2 + 0.53	5.0	30		0.67	1.07	0.77	0.97
Medium	1 + 2.5 + 4.0 + 0.53	1 + 2.9 + 3.6 + 0.53	5.5	33		0.65	1.04	0.75	0.94
Wet	1 + 2.2 + 3.6 + 0.53	1 + 2.6 + 3.2 + 0.53	6.0	36		0.63	1.01	0.73	0.91
Water = 7.5 Gal. per Sack of Cement, 28-Day Strength = 3,000 psi									
Stiff	1 + 3.9 + 5.6 + 0.67	1 + 4.5 + 5.0 + 0.67	4.0	30		0.74	1.05	0.85	0.95
Medium	1 + 3.5 + 4.9 + 0.67	1 + 4.0 + 4.5 + 0.67	4.4	33		0.72	1.02	0.83	0.92
Wet	1 + 3.1 + 4.4 + 0.67	1 + 3.5 + 4.0 + 0.67	4.8	36		0.70	0.99	0.80	0.89

SOURCE: Values in this table are taken from publications of the Portland Cement Association.
Conversions: cu. yd. × 1.31 = cubic meter
kg/cm² × 14.223 = psi

4.3 APPROXIMATE WEIGHTS OF MATERIALS REQUIRED PER CUBIC YARD OF CONCRETE

Proportions by Volume Assume Surface-Dry Aggregate

| | Proportions | | | Materials for 1 Cu. Yd. | | | | | |
| | With Gravel | With Stone | | | With Gravel | | With Stone | |
Consistency	c' + s' + g' + w'	c' + s' + g' + w'	Cement (sack)	Water (gal.)	Sand (tons)	Gravel (tons)	Sand (tons)	Gravel (tons)
1-In. Aggregate								
Water = 5 Gal. per Sack of Cement, 28-Day Strength = 5,000 psi								
Stiff	1 + 2.1 + 3.1 + 0.67	1 + 2.5 + 2.8 + 0.67	6.6	33	0.52	0.76	0.62	0.67
Medium	1 + 1.9 + 2.7 + 0.67	1 + 2.2 + 2.4 + 0.67	7.2	36	0.50	0.73	0.59	0.65
Wet	1 + 1.7 + 2.4 + 0.67	1 + 2.0 + 2.2 + 0.67	7.8	39	0.49	0.70	0.57	0.63
Water = 6 Gal. per Sack of Cement, 28-Day Strength = 4,000 psi								
Stiff	1 + 2.8 + 3.6 + 0.8	1 + 3.3 + 3.2 + 0.8	5.5	33	0.58	0.74	0.68	0.66
Medium	1 + 2.5 + 3.2 + 0.8	1 + 3.0 + 2.9 + 0.8	6.0	36	0.56	0.72	0.66	0.64
Wet	1 + 2.3 + 2.9 + 0.8	1 + 2.6 + 2.6 + 0.8	6.5	39	0.54	0.70	0.64	0.62
Water = 7.5 Gal. per Sack of Cement, 28-Day Strength = 3,000 psi								
Stiff	1 + 3.9 + 4.5 + 1	1 + 4.6 + 3.9 + 1	4.4	33	0.64	0.73	0.74	0.64
Medium	1 + 3.5 + 4.0 + 1	1 + 4.1 + 3.5 + 1	4.8	36	0.62	0.70	0.72	0.62
Wet	1 + 3.1 + 3.5 + 1	1 + 3.6 + 3.1 + 1	5.2	39	0.60	0.68	0.70	0.60

SOURCE: Values in this table are taken from publications of the Portland Cement Association.
Conversion: cu. yd. × 1.31 = cubic meter

4.4 APPROXIMATE WEIGHTS OF MATERIALS REQUIRED PER CUBIC YARD OF CONCRETE

Proportions by Volume Assume Surface-Dry Aggregate

| | Proportions | | Materials for 1 Cu. Yd. | | | | | |
| | With Gravel $c' + s' + g' + w'$ | With Stone $c' + s' + g' + w'$ | Cement (sack) | Water (gal.) | With Gravel | | With Stone | |
Consistency					Sand (tons)	Gravel (tons)	Sand (tons)	Gravel (tons)
2-In. Aggregate								
Water = 5 Gal. per Sack of Cement, 28–Day Strength = 5,000 psi								
Stiff	1 + 2.2 + 3.7 + 0.67	1 + 2.6 + 3.4 + 0.67	6.0	30	0.49	0.82	0.58	0.75
Medium	1 + 2.0 + 3.2 + 0.67	1 + 2.3 + 3.0 + 0.67	6.6	33	0.48	0.79	0.56	0.72
Wet	1 + 1.7 + 2.9 + 0.67	1 + 2.0 + 2.6 + 0.67	7.2	36	0.46	0.76	0.54	0.70
Water = 6 Gal. per Sack of Cement, 28–Day Strength = 4,000 psi								
Stiff	1 + 3.0 + 4.3 + 0.8	1 + 3.4 + 4.0 + 0.8	5.0	30	0.55	0.80	0.64	0.73
Medium	1 + 2.6 + 3.8 + 0.8	1 + 3.0 + 3.5 + 0.8	5.5	33	0.53	0.78	0.62	0.70
Wet	1 + 2.3 + 3.4 + 0.8	1 + 2.7 + 3.1 + 0.8	6.0	36	0.52	0.76	0.60	0.68
Water = 7.5 Gal. per Sack of Cement, 28–Day Strength = 3,000 psi								
Stiff	1 + 4.1 + 5.3 + 1	1 + 4.7 + 4.8 + 1	4.0	30	0.61	0.79	0.70	0.71
Medium	1 + 3.6 + 4.6 + 1	1 + 4.2 + 4.2 + 1	4.4	33	0.59	0.76	0.68	0.69
Wet	1 + 3.2 + 4.2 + 1	1 + 3.7 + 3.8 + 1	4.8	36	0.58	0.74	0.66	0.67

SOURCE: Values in this table are taken from publications of the Portland Cement Association.
Conversions: cu. yd. × 1.31 = cubic meter
kg/cm² × 14.223 = psi

4.5 QUANTITIES OF PORTLAND CEMENT FOR CONCRETE ASTM: C-150

Concrete Classification	Pounds/Cubic Foot	Pounds/Cubic Yard	Tons/Cubic Yard
Class A (6 sk.)	94	564	0.282
Class B (5 sk.)	94	470	0.235
Class C (4 sk.)	94	376	0.188
Class D (7 sk.)	94	658	0.329

4.6 MATERIALS FOR CONCRETE PER CUBIC YARD

Mix	Cubic Feet of Concrete per Lineal Feet	Cubic Feet of Concrete per 100 Lineal Feet	Cubic Yards of Concrete per 100 Lineal Feet
1:2:3	6.18	0.63	0.89
1:2.5:3.5	5.56	0.69	0.91
1:3:4	4.94	0.74	0.97
1:3:5	4.43	0.68	1.02
1:3:6	4.02	0.61	1.13

4.7 QUANTITIES: CONCRETE FOR FOOTINGS

Footings Size (in.)	Cubic Feet of Concrete per Lineal Feet	Cubic Feet of Concrete per 100 Lineal Feet	Cubic Yards of Concrete per 100 Lineal Feet
6 × 12	0.50	50.0	1.9
8 × 12	0.67	66.67	2.5
8 × 16	0.89	88.89	3.3
10 × 12	0.83	83.33	3.1
10 × 16	1.11	111.11	4.1
10 × 18	1.25	125.00	4.6
12 × 12	1.00	100.00	3.7
12 × 16	1.33	133.33	4.9
12 × 20	1.67	166.67	6.1
12 × 24	2.00	200.00	7.4

4.8   CONCRETE SURFACE COVERAGE PER CUBIC YARD

In Various Thicknesses

Thickness (in.)	Square Feet	Thickness (in.)	Square Feet
1	324	5	65
1.5	216	5.5	59
2	162	6	54
2.5	130	6.5	50
3.0	108	7	46
3.5	93	7.5	43
4	81	8	40

4.9   REINFORCING STEEL REQUIREMENTS FOR CONCRETE ASTM STANDARDS

Bar Size Designation Number	Weight (lb./ft.)	Nominal Dimension, Round Section		
		Diameter (in.)	Cross-Sectional Area (sq. in.)	Perimeters (in.)
ASTM Standard Reinforcing Bars				
2	0.167	0.250	0.05	0.786
3	0.376	0.375	0.11	1.178
4	0.668	0.500	0.20	1.571
5	1.043	0.625	0.31	1.963
6	1.502	0.750	0.44	2.356
7	2.044	0.875	0.60	2.749
8	2.670	1.000	0.79	3.142
9	3.400	1.128	1.00	3.544
10	4.303	1.270	1.27	3.990
11	5.313	1.410	1.56	4.430

4.10 CONCRETE MIXER—AVERAGE OUTPUT IN CUBIC YARDS PER HOUR[a]

Mixer's Capacity (cu. ft.)	Batches/Hour	Average Output (cu. yd./hr.)
3.5	30	4.0
6.0	30	6.5
11.0	28	11.0
14.0	24	12.0
16.0	24	14.0
28.0	22	23.0

[a]Assuming one minute mixing time.

4.11 PROPERTIES OF PLYWOOD FORMING

Thickness (in.)	Number of Plies	Weight (lb./1,000 SF)	Weight (lb./4 × 8 ft. sheet)
¼	3	790	25
½	5	1,525	49
⅜	5	1,825	58
¾	7	2,225	71
1 ⅛	7	3,350	107

4.12 ALLOWABLE STRESSES FOR FORMWORK LUMBER

Species (construction grade)	Allowable Unit Stress (psi)[a]			
	Extreme Fiber Bending	Compression Parallel to Grain	Compression Perpendicular to Grain	Horizontal Shear
Douglas fir	1,500	1,200	390	120
Southern pine	1,500	1,350	390	120
Eastern spruce	1,200	900	300	95
Western hemlock	1,500	1,100	365	100

[a]Unit stresses may be increased up to one-third when short-term loading is involved.

4.13 APPROXIMATE BOARD FOOT CONTENT OF SAWED RAILROAD TIES

Grade	End Dimension	Length (ft.) Narrow Gauge:	Standard Gauges:	
	(in.)	6.5"	8.0'	8.5'
1	6 × 6	20	24	26
2	6 × 7	23	28	30
3	6 × 8	26	32	34
3	7 × 7	27	33	35
4	7 × 8	30	37	40
5	7 × 8	—	42	45

*a*Weights per tie are extremely variable, due to differences in age and water content. Range: Approximately 75 to 160 lb.

4.14 CALCULATIONS FOR RETAINING WALLS—CUBIC YARDS PER AREA BY DEPTH

	Depth (in.)					
Area	6	12	18	24	30	36
1 sq. ft.	0.018	0.037	0.055	0.074	0.092	0.111
1 sq. yd.	0.166	0.333	0.500	0.666	0.833	1.00

1 sq. yd.	= 9 SF
1 sq. yd. × .8361	= Square meter
1 m^2 × 1.196	= Square yards
1 cu. yd. × 0.765	= Cubic meter
1 m^3 × 1.31	= Cubic yards
1 SF × 0.0929	= Square meter
1 m^2 × 10.764	= Square feet
1 m^3 × 35.3	= Cubic feet

4.15 RECOMMENDED GUIDELINES FOR SUBDIVISION OR CITY STREET CONSTRUCTION

1. Compact top 12 in. of subgrade (soil) to 95% of standard proctor density.
2. Apply 8 in. minimum of crushed stone aggregate (or comparable) and compact to 95% of standard proctor density.

3. Apply 1½ in. (minimum) of bituminous binder or base. (Base is coarser than binder.)

4. Apply 1 in. (minimum) bituminous surface.

Be sure to reference the local roadway standard in the county, state, or locality where the work is to be performed. There is a wide variance of materials used worldwide, especially for subbase and base courses.

4.16 TYPICAL ROAD SECTION SKETCH

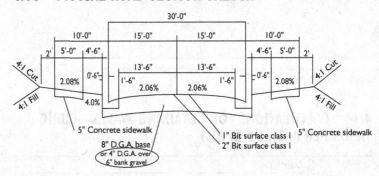

4.17 DISCHARGE FROM RECTANGULAR NOTCH WEIRS WITH END CONTRACTIONS

Figures in Table are in Gallons per Minute

	Length, L, of Weir (ft.)					Length, L, of Weir (ft.)		
Head, H (in.)	2	3	5	Additional gpm for Each Foot over 5 ft.	Head, H (in.)	3	5	Additional gpm for Each Foot over 5 ft.
¼	4.5	13.4	22.4	4.5	7¾	2,238	3,785	774
¼	12.8	38.2	63.8	12.8	8	2,338	3,956	814
¾	23.4	70.2	117	23.4	8¼	2,442	4,140	850
1	35.4	108	180	36.1	8½	2,540	4,312	890
1¼	49.5	150	250	50.4	8¾	2,656	4,511	929
1½	64.9	197	330	66.2	9	2,765	4,699	970
1¾	81	248	415	83.5	9½	2,876	4,899	1,011
2	98.5	302	506	102	9½	2,985	5,098	1,051
2¼	117	361	605	122	9¾	3,101	5,288	1,091
2¼	136	422	706	143	10	3,216	5,490	1,136
2¾	157	485	815	165	10½	3,480	5,940	1,230
3	178	552	926	187	11	3,716	6,355	1,320
3¼	200	624	1,047	211	11½	3,960	6,780	1,410
3½	222	695	1,167	236	12	4,185	7,165	1,495
3½	245	769	1,292	261	12¼	4,430	7,595	1,575
4	269	846	1,424	288	13	4,660	8,010	1,660
4¼	294	925	1,559	316	13½	4,950	8,510	1,780

(continued)

4.17 DISCHARGE FROM RECTANGULAR NOTCH WEIRS WITH END CONTRACTIONS (continued)

Head, H (in.)	Length, L, of Weir (ft.)			Additional gpm for Each Foot over 5 ft.
	2	3	5	
4½	318	1,006	1,696	345
4¾	344	1,091	1,835	374
5	370	1,175	1,985	405
5¼	396	1,262	2,130	434
5½	422	1,352	2,282	465
5¾	449	1,442	2,440	495
6	447	1,535	2,600	538
6¼		1,632	2,760	560
6½		1,742	2,920	596
6¾		1,826	3,094	630
7		1,923	3,260	668
7¼		2,029	3,436	702
7½		2,130	3,609	736

Head, H (in.)	Length, L, of Weir (ft.)		Additional gpm for Each Foot over 5 ft.
	3	5	
14	5,215	8,980	1,885
14½	5,475	9,440	1,985
15	5,740	9,920	2,090
15½	6,015	10,400	2,165
16	6,290	10,900	2,300
16½	6,565	11,380	2,410
17	6,925	11,970	2,520
17½	7,140	12,410	2,640
18	7,410	12,900	2,745
18½	7,695	13,410	2,855
19	7,980	13,940	2,970
19½	8,280	14,450	3,090

$Q = 3.33 \, (L - 0.2H) \, H^{1.5}$

where Q = cubic feet of water flowing per second

L = length of weir opening in feet (should be 4 to 8 times H)

H = head on weir in feet (should be measured at least 6' back from weir structure)

4.18 DISCHARGE FROM TRIANGULAR NOTCH WEIRS WITH END CONTRACTIONS

	Flow (gpm)			Flow (gpm)			Flow (gpm)	
Head, H (in.)	90° Notch	60° Notch	Head, H (in.)	90° Notch	60° Notch	Head, H (in.)	90° Notch	60° Notch
¼	0.07	0.04	6¼	214	124	14½	1,756	1,014
½	0.42	0.24	6½	236	136	15	1,912	1,104
¾	1.1	0.62	6¼	260	150	15½	2,073	1,197
1	2.19	1.27	7	284	164	16	2,246	1,297
1¼	3.83	2.21	7¼	310	179	16½	2,426	1,401
1½	6.05	3.49	7½	338	195	17	2,614	1,509
1¾	3.89	5.13	7¼	367	212	17½	2,810	1,623
2	12.4	7.16	8	397	229	18	3,016	1,741
2½	16.7	9.62	8¼	429	248	18½	3,229	1,864
2½	21.7	12.5	8½	462	267	19	3,452	1,993
2¼	27.5	15.9	8¼	498	287	19½	3,684	2,127
3	34.2	19.7	9	533	308	20	3,924	2,266
3¼	41.8	24.1	9¼	571	330	20½	4,174	2,410
3½	50.3	29	9½	610	352	21	4,433	2,560
3¾	59.7	34.6	9¾	651	376	21½	4,702	2,715
4	70.2	40.5	10	694	401	22	4,980	2,875
4¼	81.7	47.2	10½	784	452	22½	5,268	3,041
4½	94.2	54.4	11	880	508	23	4,565	3,213
4 ³⁄₂	103	62.3	11½	984	568	23½	5,873	3,391

(continued)

4.18 DISCHARGE FROM TRIANGULAR NOTCH WEIRS WITH END CONTRACTIONS (continued)

Head, H (in.)	Flow (gpm) 90° Notch	60° Notch	Head, H (in.)	Flow (gpm) 90° Notch	60° Notch	Head, H (in.)	Flow (gpm) 90° Notch	60° Notch
5	123	70.8	12	1,094	632	24	6,190	3,574
5¼	139	80	12½	1,212	700	24½	6,518	3,763
5½	156	89.9	13	1,337	772	25	6,855	3,958
5¾	174	100	13½	1,469	848			
6	193	112	14	1,609	929			

Based on the formula:

$$Q = (C)\, (\tfrac{8}{15})\, (L)\, (H)$$

in which Q = flow of water in cubic feet per second

  L = width of notch in feet at H distance above apex

  H = head of water above apex of notch in feet

  C = constant varying with conditions, 57 being used for this table

  a = reduction of width of the stream on either

For 90° notch the formula becomes

  $Q = 2.586\, H^{\frac{1}{2}}$

For 60° notch the formula becomes

  $Q = 14,408\, H^{\frac{1}{2}}$

Section Five

GRASSING, LANDSCAPING, FERTILIZATION, AND GROUND COVERS

5.1   TURFGRASS STOLONIZATION—WHAT IS A BUSHEL?

Warm-season turfgrasses are often harvested and transplanted in the form of "stolons," or "sprigs." By definition, these terms describe the parts of a creeping type of grass as a whole unit, such as Bermuda grass, seashore paspalum, and zoysia grass. A sprig or stolon consists of roots, rhizomes, stems, leaves, and nodes. Certain hybrid grasses do not bear viable and true-to-type seed. In order to maintain genetic integrity and known performance, these varieties may only be propagated by harvesting and transplanting the vegetative parts of the original plants.

There has always been a degree of confusion when dealing with the different units of measure for vegetatively propagated and transplanted turfgrass stolons. For lack of a better unit of measurement, stolons or sprigs are sold as "bushels." Recently, the Golf Course Builders Association of America and the American Society of Golf Course Architects have agreed to adopt the unit of measure for a bushel to be the same as other commodities. Now, one bushel is measured as 1.244 cubic feet. This is referred to as a U.S. Standard Bushel, and in accordance to the National Institute of Standards and Technology.

Certain vegetatively propagated turfgrass varieties are considered to be "Dwarf types" or "Super-Dwarf types" and have lower and slower growing habits that dictate these varieties be planted at higher rates than the Fairway types, which are faster- and taller-growing varieties. Higher rates of planting varieties in either category will aid faster grow-in.

Here are some recommended planting rates:

Area	Growth Types	Planting Rate in U.S. Bushels
Greens	Dwarf	3–5 bu./1,000 ft^2
	Super-Dwarf	4–6 bu./1,000 ft^2
Tees	Fairway	3–5 bu./1,000 ft^2
Fairways	Fairway	100–200 bu./acre
Roughs	Fairway	100–200 bu./acre

5.2   WHAT DOES "CERTIFIED" MEAN?

There are very few sources of clean, "certified" hybrid Bermuda grasses true to variety. The only way to be certain of trueness to variety in vegetatively propagated plants is by knowing the parents of the plant. In other words, if you are sure of the pedigree, you will know what to expect of the performance. "Certification" indicates that the heritage of the grass plant is known and documented—beginning with the mother block (foundation plants), which represents the grass breeder's selection, and culminating in the producer's field, which represents the registered plants—to ensure heritage. In a Certification Program, the production fields are inspected three times annually by government officials to guarantee purity. The end-user product is classified as "Certified" stolons after successfully passing inspections.

For international shipment of the "Certified" pedigree stolons, they must first be washed free of soil then inspected by the U.S. Department of Agriculture. The USDA then provides a "Phytosanitary Certificate" stating that the stolons meet the requirements of the importing country. In addition, the stolons are accompanied by a "Blue Tag Certification Card" from the state of Georgia.

5.3 ORNAMENTAL GRASSES

Out of bounds? No problem. One of the main concerns of golf course design and construction is how to use native and ornamental grasses in the surrounding landscape. Here is a list of some of the "grassy" plants that are commonly used in the out-of-bounds area of a golf course to add beauty and reduce maintenance.

Transplanted Ornamental Grasses for Dunes and Wetlands

Common Name	Botanical Name
Beach grass	*Ammophila* spp.
Salt grass	*Distichlis* spp.
Panic grass	*Panicum* spp.
Cordgrass	*Spartina* spp.
Sea oats	*Uniola* spp.
Saw grass	*Cladium* spp.
Sedges	*Cyperus* spp.
Rushes	*Juncus* spp.
Bullrush	*Scirpus* spp.
Seashore paspalum	*Paspalum Vaginatum*

Seeded Ornamental Grasses

Common Name	Botanical Name
Wheatgrass	*Agropyron* spp.
Bluestem	*Andropogon* spp.
Cow grass	*Axonopus* spp.
Buffalo grass	*Buchloë* spp.
Love grass	*Eragrostis* spp.
Indian ricegrass	*Oryzopsis* spp.
Switchgrass	*Panicum* spp.
Bahiagrass	*Paspalum* spp.
Little bluestem	*Schizachyrium* spp.
Indiangrass	*Sorgastrum* spp.

5.4 RECOMMENDED SEEDING RATES OF SOME GRASSES

Grass	Seed/lb. (millions)	Rate (lb./1,000 SF)	Rate (lb./acre)
Agrostis Alba (Red Top)	5.0	1	44
Agrostis Palustris (Creeping Bent grass)	8.0	1	44
Agrostis Tenuis (Colonial Bent grass)	8.5	1	44
Cynodon Dactylon (Bermuda grass)	1.8	2	88
Festuca Elatior (Meadow fescue)	0.23	6	264
Festuca Elatior Arundinacease (Tall fescue)	0.23	6	264
Festuca Rubra (Red fescue)	0.615	4	176
Lolium Domesticum (Domestic ryegrass)	0.23	8	352
Lolium Perenne (Perennial ryegrass)	0.23	8	352
Poa Pratensis (Kentucky bluegrass)	2.2	2	88
Poa Trivialis (Roughstalk blue)	2.5	2	88
Dichondra Repens[a] (Dichondra)	0.8	1	44
Trifolium Repens[a] (White duth clover)	0.7	2	88

[a] Not grass but often used as turf.

5.5 CHARACTERISTICS OF SOME TURFGRASSES

Grass, Botanical and Common Names	Purity (%)	Germination (%)	Germination Period (days)	Seed/Lb. (in millions)
Agrostis Alba (Red Top)	92	90	6–10	5.0
Agrostis Palustris (Creeping Bent grass)	95	90	7–14	8.0
Agrostis Tenuis (Colonial Bent grass)	95	90	7–14	8.5
Cynodon Dactylon (Bermuda grass)	97	85	14–21	1.8
Festuca Elatior (Meadow fescue)	95	90	6–10	0.23
Festuca Elatior Arundinacease (Tall fescue)	95	90	7–14	0.23
Festuca Rubra (Red fescue)	95	80	7–14	0.615
Lolium Domesticum (Domestic ryegrass)	98	90	5–10	0.23
Lolium Perenne (Perennial ryegrass)	98	90	5–10	0.23
Poa Pratensis (Kentucky bluegrass)	85	80	10–20	2.2
Poa Trivialis (Roughstalk blue)	85	80	10–20	2.5
Dichondra Repens[a] (Dichondra)	96	90	7–14	0.8
Trifolium Repens[a] (White duth clover)	96	95	5–10	0.7

[a] Not grass but often used as turf.

5.6   COVERAGE AREA PER BALE OF BEAN STRAW

One Bale of Bean Straw Equals 16.6 Cu. Ft.

Depth (in.)	Area Covered (SF)
1	200
2	100
4	50
8	25
12	16.6

5.7   GRASS STOLONS: DISTRIBUTION RATE BY MEANS OF SPRIGGING

| Spacing (sq. in.) | Bushels Required per Spacing | |
	Per 1,000 SF	Per Acre
6 O.C.	2.0	88
9 O.C.	1.25	55
12 O.C.	0.5	22

5.8   GRASS STOLONS: QUANTITIES REQUIRED FOR BROADCAST DISTRIBUTION[a]

Grass	Bushels Rate per 1,000 SF	Bushels Rate per Acre
St. Augustine grass	5.0	220
Hybrid Bermuda grass	2.0–4.0	88–176
Bent grass	4.0–6.0	176–264

[a] Soil conditions, season, quality of expected maintenance, and initial cost influence rate of stolonization desirable. Desired establishment is also a major factor.

5.9   COVERAGE AREAS FOR TURF AND OTHER GROUND COVERS VIA PLUGS

Plugs per Flat[a] and Square-Foot Coverage for Various Plug Sizes and Various Plug Spacings

Plug Size (sq. in.)	Plugs per Flat	Square-Feet Coverage per Spacings Indicated		
		6" O.C.	9" O.C.	12" O.C.
1.25	200	50	110	200
1.00	280	70	160	280
0.75	500	125	280	500
0.50	1,100	280	625	1,100

[a] Normal size flat. Example: Plan calls for square inch plugs 9" O.C. If area is 4,500 SF, from the chart: one flat will cover 160 SF 9" O.C.; therefore, 4500/160 = 28 flats needed.

5.10 FERTILIZATION (NITROGEN REQUIREMENTS)

Amount of Mixed Material Necessary to Apply 1 Lb. of Actual Nitrogen per Each 1,000 SF Surface

Percentage of N in Mix	Pounds of Mix for 1 lb. Actual N per 1,000 SF	Pounds of Mix for 1 lb. Actual N per 1,000 SF on Acre Basis	Equivalent Tons per Acre	Approximate Acreage Covered per Ton
1	100.00	4,356	2.178	0.46
2	50.00	2,178	1.096	0.92
3	33.30	1,452	0.726	1.38
4	25.00	1,089	0.544	1.84
5	20.00	871	0.435	2.30
6	16.60	726	0.363	2.75
7	14.30	622	0.311	3.21
8	12.50	545	0.273	3.66
9	11.10	484	0.242	4.13
10	10.00	436	0.218	4.58
11	9.00	400	0.200	5.04
12	8.30	363	0.182	5.50
14	7.10	311	0.155	6.45

(continued)

Amount of Mixed Material Necessary to Apply 1 Lb. of Actual Nitrogen per Each 1,000 SF Surface (continued)

Percentage of N in Mix	Pounds of Mix for 1 lb. Actual N per 1,000 SF	Pounds of Mix for 1 lb. Actual N per 1,000 SF on Acre Basis	Equivalent Tons per Acre	Approximate Acreage Covered per Ton
16	6.25	272	0.136	7.35
18	5.50	242	0.121	8.26
20	5.00	218	0.109	9.17
22	4.50	198	0.099	10.10
24	4.10	181	0.091	11.00
26	3.80	168	0.084	11.90
28	3.50	156	0.078	12.80
30	3.30	145	0.073	13.70
32	3.10	136	0.068	14.70
34	3.00	128	0.064	15.60
36	2.75	121	0.061	16.40
38	2.60	115	0.058	17.20
40	2.50	109	0.055	18.10
42	2.40	104	0.052	19.20
44	2.27	100	0.050	20.00
46	2.17	95	0.048	20.80
48	2.10	91	0.045	22.20
50	2.00	87	0.044	22.70

Example
If a 16-16-8 mix is to be used, and 3 lb. actual nitrogen is required, from the chart: 6.25 lb. of the mix is required for 1 lb. actual N; per 1,000 SF 6.25 × 3 = 18.75 lb. material required for N. On an acreage basis this would be 3 × 136 = 0.408 tons.

Helpful Metric Conversion Formulas

1 kg = 2.2 lb.

1 m^2 = 10.764 SF

U.S. ton = 0.90 metric ton

Metric ton = 1.1 U.S. tons

U.S. ton = 909 kg

1 hectare = 2.4 acres

1,000 kg = 1 metric ton

5.11 COVERAGE AREA PER FLAT OF GROUND COVERS SQUARE FEET PER FLAT

Spacing (in. O.C.)	Plants per Flat			
	50	64	81	100
4	5.56	7.11	9.00	11.11
6	12.50	16.00	20.25	25.00
8	22.22	28.44	36.00	44.44
9	28.12	36.00	45.56	56.25
10	34.72	44.44	56.25	69.44
12	50.00	64.00	81.00	100.00
16	88.89	113.77	144.00	177.77
18	112.50	144.00	182.25	225.00
24	200.00	256.00	324.00	400.00
30	312.50	400.00	506.25	625.00

5.12 LINERS AND HEDGE PLANTS: PLANTS REQUIRED PER 100 LIN. FT.

Spacing (ft.)	Number Required per 100 Lin. Ft.
0.33	300.0
0.50	200.0
0.67	150.0
1.00	100.0
1.50	66.7
2.00	50.0
2.50	40.0
3.00	33.3
3.50	28.6
4.00	25.0
4.50	22.3
5.00	20.0
6.00	16.7
6.50	15.4
7.00	14.3
7.50	13.3
8.00	12.5
8.50	11.8

(continued)

(continued)

Spacing (ft.)	Number Required per 100 Lin. Ft.
9.00	11.1
9.50	10.5
10.00	10.0

Example
If plants were required for one mile of screening at an 8 ft. spacing from the chart:

12.5 per 100 ft. × 52.8 = 660 plants per mile

5.13 PLANTS REQUIRED PER 100 SF: VARIOUS SPACINGS

Spacing (in. O.C.)	Plants per 100 SF	Spaces (in. O.C.)	Plants per 100 SF
4	900.00	18	45.00
6	400.00	24	25.00
8	225.00	30	16.00
9	178.00	36	11.11
10	144.00	48	6.25
12	100.00	72	2.78
16	56.00		

5.14 FLATS OF PLANTS REQUIRED PER 100 SF: VARIOUS SPACINGS AND VARIOUS QUANTITIES PER FLAT

Spacing (in. O.C.)	Quantities of Plants per Flat			
	50	64	81	100
4	18.00	14.06	11.11	9.00
6	8.00	6.25	4.94	4.00
8	4.50	3.51	2.78	2.25
9	3.55	2.78	2.20	1.78
10	2.88	2.25	1.78	1.44
12	2.00	1.56	1.23	1.00
16	1.12	0.87	0.69	0.56
18	0.90	0.70	0.56	0.45
24	0.50	0.39	0.31	0.25
30	0.32	0.25	0.20	0.16

5.15 NUMBER OF SHRUBS OR PLANTS FOR AN ACRE AT VARIOUS SPACINGS

Distance Apart	Number of Plants	Distance Apart	Number of Plants	Distance Apart	Number of Plants
3 × 3 in.	696,690	4 × 4 ft.	2,722	13 × 13 ft.	257
4 × 4 in.	392,040	4½ × 4½ ft.	2,151	14 × 14 ft.	222
6 × 6 in.	174,240	5 × 1 ft.	8,712	15 × 15 ft.	193
9 × 9 in.	77,440	5 × 2 ft.	4,356	16 × 16 ft.	170
1 × 1 ft.	43,560	5 × 3 ft.	2,904	16½ × 16½ ft.	160
1½ × 1½ ft.	19,360	5 × 4 ft.	2,178	17 × 17 ft.	150
2 × 1 ft.	21,780	5 × 5 ft.	1,742	18 × 18 ft.	134
2 × 2 ft.	10,890	5½ × 5½ ft.	1,440	19 × 19 ft.	120
2½ × 2½ ft.	6,970	6 × 6 ft.	1,210	20 × 20 ft.	108
3 × 1 ft.	14,520	6½ × 6½ ft.	1,031	25 × 25 ft.	69
3 × 2 ft.	7,260	7 × 7 ft.	889	30 × 30 ft.	48
3 × 3 ft.	4,480	8 × 8 ft.	680	33 × 33 ft.	40
3½ × 3½ ft.	3,556	9 × 9 ft.	537	40 × 40 ft.	27
4 × 1 ft.	10,890	10 × 10 ft.	435	50 × 50 ft.	17
4 × 2 ft.	5,445	11 × 11 ft.	360	60 × 60 ft.	12
4 × 3 ft.	3,630	12 × 12 ft.	302	66 × 66 ft.	10

5.16 PEAT MOSS COVERAGE: DEPTH IN INCHES PER SQUARE SURFACE FOOTAGE

Bale (Compressed), 5.6 Cu. Ft. When Loose,[a] Will Cover:		Bale (Compressed), 4.0 Cu. Ft. When Loose,[a] Will Cover:	
Depth (in.)	Coverage (SF)	Depth (in.)	Coverage (SF)
0.25	480	0.25	345.6
0.50	240	0.50	172.8
1.00	120	1.00	86.4
2.00	60	2.00	43.2
3.00	40	3.00	28.8
4.00	30	4.00	21.6
6.00	20	6.00	14.4
12.00	10	12.00	7.2

[a] Approximate expansion is 1.8 of compressed volume.

5.17 STEER MANURE COVERAGE: DEPTH IN INCHES PER SQUARE SURFACE FOOTAGE

One Sack (2.5 Cu. Ft.) Will Cover:		One Sack (2.0 Cu. Ft.) Will Cover:	
Depth (in.)	Coverage (SF)	Depth (in.)	Coverage (SF)
0.25	120	0.25	96.0
0.50	60	0.50	48.0
1.00	30	1.00	24.0
2.00	15	2.00	12.0
3.00	10	3.00	8.0
4.00	7.5	4.00	6.0
6.00	5.0	6.00	4.0
12.00	2.5	12.00	2.0

5.18 STEER MANURE: RATES PER ACRE BASED ON REQUIRED RATES PER 1,000 SF

Rate Per 1,000 SF (cu. yd.)	Equivalent Rate Per Acre (cu. yd.)
1.0	43.56
1.5	65.34
2.0	87.12
2.5	108.90
3.0	130.68
3.5	152.46
4.0	174.24
4.5	196.02
5.0	217.80

Example

3 cu. yds. required per 1,000 SF = $3 \times 43.56 = 130.68$ cu. yds./acre

5.19 NURSERY CONTAINER STOCK: APPROXIMATE BACKFILL VOLUME FOR VARIOUS CONTAINER STOCK: ROUND PLANT PITS, VERTICAL SIDES[a]

Container Stock	Plant Pit Volume (cu. ft.)		Container Stock Displacement (cu. ft.)		Backfill Necessary Cu. Ft.	Cu. Yd.
1 gal.	0.78		0.15		0.63	0.02
5 gal.	3.60		0.60		3.00	0.11
15 gal.	11.10		2.00		9.10	0.34
16" box	12.60		2.50		10.10	0.37
20" box	14.10		4.75		9.35	0.35
24" box	17.40	(minus)	6.70	(equals)	10.70	0.40
30" box	30.60		13.60		17.00	0.63
36" box	52.36		23.40		28.96	1.07
42" box	67.20		31.70		35.50	1.30
48" box	104.70		40.30		64.40	2.40
54" box	132.50		56.00		76.50	2.83
60" box	190.80		74.40		116.40	4.30

[a] Shrinkage factor not applied. See Table 1.4 for shrinkage.

5.20 NURSERY CONTAINER STOCK: VOLUME OF EXCAVATED SOIL RESULTING FROM MULTIPLE PLANTINGS: SQUARE PLANTING PITS

| Container Stock | Excavation Yardage per: | | |
	10 Plants	100 Plants	1,000 Plants
1 gal.	0.24	2.40	24.00
5 gal.	1.37	13.70	137.00
15 gal.	4.60	46.00	460.00
16" box	5.26	52.60	526.00
20" box	8.22	82.20	822.00
24" box	11.80	118.00	1,180.00
30" box	17.00	170.00	1,700.00
36" box	30.00	300.00	3,000.00
42" box	41.70	417.00	4,170.00
48" box	62.50	625.00	6,250.00
54" box	80.00	800.00	8,000.00
60" box	100.50	1,005.00	10,050.00

5.21 NURSERY CONTAINER STOCK: APPROXIMATE BACKFILL VOLUME FOR VARIOUS CONTAINER STOCK SQUARE PLANT PITS, VERTICAL SIDES[a]

Container Stock	Plant Pit Volume (cu. ft.)	Container Stock Displacement (cu. ft.)		Backfill Necessary Cu. Ft.	Cu. Yd.
1 gal.	0.7	0.15		0.55	0.02
5 gal.	3.7	0.60		3.10	0.115
15 gal.	12.5	2.00		10.50	0.04
16" box	14.2	2.50		11.70	0.43
20" box	22.2	4.75		17.45	0.64
24" box	32.0	6.70	(minus) (equals)	25.30	0.94
30" box	45.5	13.60		31.90	1.20
36" box	80.6	23.40		57.20	2.10
42" box	112.7	31.70		81.00	3.00
48" box	168.7	40.30		128.40	4.75
54" box	216.7	56.00		160.70	5.90
60" box	283.5	74.40		209.10	7.74

[a] Shrinkage factor not applied. See Table 1.4 for shrinkage.

5.22   NURSERY CONTAINER STOCK

Container	Approximate Weight (lb.)	Plant-Ball Volume (cu. ft.)
1 gal.	7	0.15
5 gal.	40	0.60
15 gal.	188	2.00
16" box	235	2.50
20" box	510	4.75
24" box	725	6.70
30" box	1,500	13.60
36" box	2,500	23.40
42" box	3,700	31.70
48" box	6,000	40.30
54" box	7,000	56.00
60" box	8,000	74.40

5.23   TREE PIT EXCAVATION FOR NURSERY CONTAINER STOCK SQUARE PITS, VERTICAL SIDES[a]

	Dimensions (in.)					Soil Volume
Container	Pit Length		Pit Width		Pit Depth	Displacement (cu. ft.)
1 gal.	10	x	10	x	12	0.7
5 gal.	18	x	18	x	20	3.7
15 gal.	30	x	30	x	24	12.5
16" box	32	x	32	x	24	14.2
20" box	40	x	40	x	24	22.2
24" box	48	x	48	x	24	32.0
30" box	54	x	54	x	27	45.5
36" box	66	x	66	x	32	80.6
42" box	78	x	78	x	32	112.7
48" box	90	x	90	x	36	168.7
54" box	102	x	102	x	36	216.7
60" box	108	x	108	x	42	283.5

[a] Swellage factor not applied. See Table 1.3 for swellage of various soil types.

5.24 TREE PIT EXCAVATION FOR NURSERY CONTAINER STOCK ROUND PITS, VERTICAL SIDES

Container Size	Pit Diameter (in.)	Pit Depth (in.)	Volume Displacement[a] (cu. ft.)
1 gal.	12	12	0.78
5 gal.	20	20	3.60
15 gal.	32	24	11.10
16" box	34	24	12.60
20" box	36	24	14.10
24" box	40	24	17.40
30" box	50	27	30.60
36" box	60	32	52.36
42" box	68	32	67.20
48" box	80	36	104.70
54" box	90	36	132.50
60" box	100	42	190.80

[a] Swellage factor not applied. See Table 1.3 for swellage of various soil types.

5.25 NURSERY CONTAINER STOCK: VOLUME OF EXCAVATED SOIL RESULTING FROM MULTIPLE PLANTINGS: ROUND PLANT PITS

Container Stock	Excavation Yardage[a] per:		
	10 Plants	100 Plants	1,000 Plants
1 gal.	0.29	2.9	29.00
5 gal.	1.33	13.3	133.30
15 gal.	4.10	41.0	410.00
16" box	4.67	46.70	467.00
20" box	5.21	52.10	521.00
24" box	6.44	64.40	644.00
30" box	11.33	113.33	1,133.30
36" box	19.40	194.00	1,940.00
42" box	24.90	249.00	2,490.00
48" box	38.70	387.00	3,870.00
54" box	49.00	490.00	4,900.00
60" box	70.66	706.60	7,066.00

[a] Swellage factor not applied. See Table 1.3 for swellage of various soil types.

5.26 SUGGESTED PLANTING DISTANCES FOR SOME FRUIT TREES

Variety	Feet Apart	Variety	Feet Apart
Almonds	22–30	Pears	20–35
Apples	30–35	Persimmons	16–20
Apples, Crab	20–25	Plums	18–25
Apricots	22–30	Pomegranates	15–20
Cherries-sweet	22–30	Prunes	18–25
Figs	30–40	Quince	15–20
Nectarines	20–25	Walnuts	40–60
Peaches	20–25		

5.27 TREE CABLING MATERIAL COMBINATIONS

Safe Loads for Dyamic Stresses Based on a Safety Factor of 4

	Anchor Units				Flexible Cable			
Safe Load (lb.)	Lag Screw Hooks (I) (in. diameter)	Beat Hook or Eye Bolts (in. diameter)	Drop-Forged Eye Bolts (in. diameter)	Amon Nuts (in. diameter)	7-Wire Galvanized (in. diameter)	7-Wire Copper Strand (in. diameter)	Single Wire Copper Covered Steel (No. A.W.G.)	3-Wire Copper Covered Steel Strands (No. A.W.G.)
100	¼				³⁄₁₆	³⁄₁₆	1 #12	
200	⁵⁄₁₆				³⁄₁₆	³⁄₁₆	1 #12	
300	⅜				³⁄₁₆	³⁄₁₆	1 #10	
500	½				¼	¼	1 #6	3 #12
600	½	½			⁵⁄₁₆	⁵⁄₁₆	1 #6	3 #10
900	⅝	⅝			⁵⁄₁₆	⅜		3 # 8
1,000		⅝			⁵⁄₁₆	⅜		3 # 8
1,200		⅝			⅜	⅜		3 # 8
1,400		¾	⅜		½			3 # 7
2,200			½		½			
3,000			⅝					
3,300				⅝				
3,700				⅞				

SOURCE: Courtesy of the United States Department of the Interior.

5.28 HYDROSEEDERS: AREA COVERAGE PER LOAD

To determine the coverage per load for any hydroseeder three questions must be answered prior to application. First, is the job to be done in "one step" (seed, fertilizer, and mulch applied proportionally per load) or in "two steps" (seed and fertilizer applied first, and then covered by mulch)? Second, at what rates (usually in pounds per 1,000 SF or pounds per acre) are the seeding materials to be applied? Finally, what are the loading capacities of the hydroseeder?

Application rates vary for different geographic locations, but in general, seed is applied at 6 to 10 lbs. per 1,000 SF; fertilizer is applied at a rate of approximately 4,000 lb. per acre. (*Note:* There are 43,560 SF per acre.) Fiber mulch is usually applied at 1,500 to 2,000 lb. per acre. Local agronomists, agricultural extension agents, or soil and water conservation officials should be contacted for more specific information on application rates for a given area.

The following tables show loading versus coverage rates for hydroseeders. Section 5.29 shows rates for "one-step" applications. The coverage area is determined by the fiber mulch capacity of the hydroseeder and the rate at which it is applied. Section 5.30 shows the area coverage when seeding only, where little or no mulch is applied.

5.29 HYDROSEEDER COVERAGE USING SEED, FERTILIZER, AND MULCH

| Size of Tank Gal | Amount of Material in Tank (lb.) | | | Coverage Area (SF) |
	Seed	Fertilizer	Mulch	
260	28	32	100–120	2,900–3,500
500	57	66	200–250	5,800–7,260
800	93	106.7	400	11,600
1,000	116.2	133.3	500	14,500
1,500	172.5	200	750	21,780
2,500	287.5	333.3	1,250	36,300
3,000	345	400	1,500	43,560

The table is based on rates per acre of 1,500 lb. mulch, 400 lb. fertilizer, 345 lb. (8 lb./1000 sq. ft.) seed.

Example

For a 500-gallon tank:

$$\frac{250 \text{ lb. mulch per tank}}{1,500 \text{ lb. mulch per acre}} = 0.167 \text{ acre}\left(\text{per load}\right)$$

400 lb. fertilizer per acre × 0.167 acre

= 66.8 lb. fertilizer per load

325 lb. seed per acre × 0.167 acre = 54.3 lb. seed per load

5.30 HYDROSEEDER COVERAGE USING SEED AND FERTILIZER ONLY

	Amount of Material in Tank (lb.)			Coverage Area	
Size of Tank Gal	Seed	Fertilizer	Total	SF	Acreage
260	260	300	560	32,670	0.75
500	522	600	1,122	65,340	1.50
800	784	900	1,684	98,010	2.25
1,000	1,045	1,200	2,245	130,680	3.00
1,500	1,742	2,000	3,742	217,800	5.00
2,500	3,136	3,600	6,736	392,040	9.00
3,000	3,485	4,000	7,485	435,600	10.00

Table is based on rates of 8 lb. seed and 9.2 lb. fertilizer/1,000 SF.

5.31 USAGE RATES FOR A COMBINATION OF ORGANIC AND SYNTHETIC PRODUCTS

Purpose	Conditions	Organic Tackifier (lb.)	Mulch (lb.)	Area (acre)
Erosion control	Severe conditions	120	3,000	1
	Moderate	90	2,250	1
Fiber mulch tackifier	Moderate	30	1,500	1

5.32   USAGE RATES FOR SYNTHETIC FIBER BOND

Tank Size (gal.)	Synthetic Fiber Bond (lb.)	Mulch (lb.)
500	3	200
800	5	400
1,000	7	500
1,500	10	750
2,500	15	1,250
3,000	20	1,500

5.33   USAGE RATES FOR GUM-BASED ORGANIC TACKIFIER

Fiber Mulch Tackifier[a]; to Tack 1 Acre Use:			Straw/Hay Mulch Binder[b]; to Bind 1 Acre Use:	
Organic Tackifier	Water	Mulch	Organic Tackifier	Water
40 lb.	3,000 gal.	1,500 lb.	30 lb.	80 gal.

Soil Stabilization[c]; to Stabilize 1 Acre Use:		Mulch Slurry Lubricant; to Increase Spraying Performance Use:	
Organic Tackifier	Water	Organic Tackifier	Mulch
40 lb.	1,500 gal.	1 lb.	Every 100 lb.

[a] For slopes > 2–1, increase to 60 lb.
[b] 150 lb. of fiber mulch may also be used as spray tracer.
[c] For slopes > 2–1, increase to 60 lb.

5.34 SEED FACTS FOR NATIVE PRAIRIE GRASSES AND LEGUMES: COOL SEASON

Native Grasses and Legumes	Seeds per Pound	Minimum Rainfall	Growing Season[a]	Seeding Rate (lb./acre)	Days to Germination	Preplant Treatment	Recommended Planting Dates
California brome	60,000	15"	C	8–30 PLS	14		Fall
Smooth brome	140,000	16"	C	7–40 PLS	14		Spring
Cresled wheat	200,000	9"	C	10–30 PLS	14		Fall/early spring
Western wheat	110,000	14"	C	10–40 PLS	28		Spring
Tuffled hairgrass	1,300,000	30"	C(Sun)	4–6 PLS	15		Fall
Indian rice	140,000	6"	C	7–30 PLS	42	None	Fall
Reed canary	530,000	Moist areas	C	3–10 PLS	21		Spring/fall
Blue wild rye	100,000	30"	C(Shade)	10–40 PLS	15		Fall
Alkaligrass	1,900,000	15"	C	20–80 PLS	28		Spring/fall
Junegrass	600,000	20"	C	5–10 PLS	28		Spring/fall

[a] W = Warm season; C = Cool season
[b] PLS = Pure live seed = Purity × Germination

5.35 SEED FACTS FOR NATIVE PRAIRIE GRASSES AND LEGUMES: WARM SEASON

Native Grasses and Legumes	Seeds per Pound	Minimum Rainfall	Growing Season[a]	Seeding Rate (lb./acre)	Days to Germination	Preplant Treatment	Recommended Planting Dates
Little bluestem	260,000	16"	W	5–40 PLS	14		Spring
Big bluestem	150,000	20"	W	10–60 PLS	14		Spring
Buffalograss	275,000	15"	W	15–80 PLS	28	0.5% KNO 24 hours	Spring
(Burr)	45,000						
Blue grama	800,000	10"	W	2–20 PLS	14		Spring
Sideoats grama	180,000	17"	W	5–40 PLS	14		Spring
Switchgrass	400,000	16"	W	4–30 PLS	14		
Weeping love	1,400,000	15"	W	4–40 PLS	14		
Jap millet	115,000	N/A	W	20–30 PLS	10		Fall/winter
Brownstep millet	142,000	N/A	W	10–20 PLS	7		Fall/winter

[a] W = Warm season; C = Cool season
[b] PLS = Pure live seed = Purity × Germination

5.36 SEED FACTS FOR LEGUMES

Legumes	Seeds per Pound	Minimum Rainfall	Growing Season[a]	Seeding Rate (lb./acre)	Days to Germination	Preplant Treatment	Recommended Planting Dates
Lespedeza sp.	90,000			25–60	21	Innoc.	Spring/fall
Crown vetch	110,000			20–25	14	Scarify & innoc.	Spring/fall
Birdsfoot trefoil	375,000			6–12	12	Scarify & innoc.	Spring/fall
Strawberry clover	300,000			5–10	7	Scarify & innoc.	Spring/fall
Crimson clover	140,000			15–30	7	Scarify & innoc.	Spring/fall
Red clover	275,000			6–12	7	Scarify & innoc.	Spring/fall
White clover	800,000			6–12	7	Scarify & innoc.	Spring/fall

5.37 SEED FACTS FOR COOL-SEASON TURFGRASSES

Turfgrass	Seeds per Pound	Minimum Preferred Percent Purity	Minimum Preferred Percent Germination	Seeding Rate per 1,000 SF	Days to Germination	Optimum Temperature (°F)	Recommended Planting Dates
Bent grass, colonial	5–8,000,000	98	85–90	½–1	14–18	59–86	Early spring/late summer
Bent grass, creeping	6–7,000,000	98	85–90	½–1	14–21	59–86	Early spring/late summer
Redtop	4,800,000	95–98	85–90	½–1	10–14	—	Early spring/late summer
Bluegrass, Kentucky	950,000–2,000,000	95–97	75–80	2–3	10–28	59–86	Early spring/late summer
Poa trivials	2,100,000	95–98	80–85	2	14–18	52–86	Early spring/late summer
Annual blue	2,250,000	95–97	75–80	3–7	14–18	68–86	Early spring/late summer
Fescue, red	550,000	95–97	80–85	4–5	18–21	59–77	Early spring/late summer
Fescue, hard	6–700,000	95–97	80–85	4–5	18–21	59–77	Early spring/late summer
Fescue, chewing	546,000	95–97	80–85	4–5	18–21	59–77	Early spring/late summer
Fescue, tall	220,000	95–98	85–90	8–10	14–21	68–86	Early spring/late summer
Ryegrass Annual	215,000	95–98	90–95	10–12	5–8	68–86	Early spring/late summer
Perennial	227–300,000	95–98	90–95	7–8	7–10	68–86	Early spring/late summer
Intermediate	205–215,000	95–98	90–95	7–8	6–10	68–86	Early spring/late summer

5.38 SEED FACTS FOR WARM-SEASON FORAGEGRASSES

Turfgrass	Seeds per Pound	Minimum Preferred Percent Purity	Minimum Preferred Percent Germination	Seeding Rate per 1,000 SF	Days to Germination	Optimum Temperature (°F)	Recommended Planting Dates
Bahiagrass	200,000	70-75	70-75	6-8	7-21	86-95	Mid spring through late summer
Bermuda grass							
Hulled	1,900,000	95-98	80-85	½-1	10-20	68-95	Early summer
Unhulled summer	1,400,000			2	14-21		
Carpetgrass	1,200,000	90-95	85-90	3	10-20	68-95	Early summer
Centipede grass	500,000	45-50	65-70	½	10-20	68-95	Mid spring through summer
Zoysia grass	1,000,000	90	45-50	1-2	14-35	68-95	Mid spring through summer

5.39   PLANT HARDINESS ZONES

Hardiness Zone	Average Minimum Temperature (°F)	Average Minimum Temperature (°C)
1	Below −50	Below −46
2a	−45	−43
2b	−40	−40
3a	−35	−37
3b	−30	−34
4a	−25	−31
4b	−20	−29
5a	−15	−26
5b	−10	−23
6a	−5	−20
6b	0	−18
7a	5	−15
7b	10	−12
8a	15	−9
8b	20	−7
9a	25	−4
9b	30	−1
10a	35	−2
10b	40	−4
11	Above 45	Above 7

5.40 USDA PLANT HARDINESS ZONE MAP

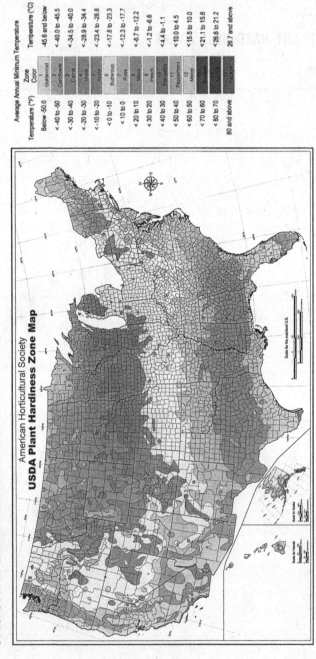

American Horticultural Society
USDA Plant Hardiness Zone Map

5.41 SHADE AND ORNAMENTAL TREES

Common Name	Botanical Name	Hardiness Zone	Height (ft.)	Spread (ft.)	Shape	Growth Rate	Foliage Texture	Full Foliage	Fruit	Bark	Unusual Characteristics
Almond, Halls	*Prunus amygdalus* "Halls"	4	15	12	Oval	Fast	Medium	None	Peachlike	Smooth	Nut fruit
Ash, Autumn Applause	*Fraxinus americana* "Autumn Applause"	3	50	30	Oval	Fast	Medium	Purple	Winged	Rough dark	Fall color
Ash, Autumn Purple	*Fraxinus americana* "Autumn Purple"	3	50	30	Oval	Fast	Medium	Red-purple	Seedless	Rough dark	Fall color
Ash, Black Hawk Mountain	*Sorbus aucaparia* "Black Hawk"	3	20	20	Oval	Medium	Medium	Gold	Orange	Gray brown	Orange fruit
Ash, Champaign County	*Fraxinus americana* "Champaign County"	3	50	30	Oval	Fast	Medium	Red	None	Dark brown	Glossy foliage
Ash, European Mountain	*Sorbus aucaparia* "European Mountain"	3	30	15	Oval	Medium	Fine	None	Orange	Brown	Orange fruit
Ash, Green	*Fraxinus pennsylvanica*	2	60	40	Oval	Fast	Medium	Yellow	Winged	Rough dark	Shiny green foliage
Ash, Marshall's Seedless	*Fraxinus pennsylvanica* "Marshall's Seedless"	3	60	40	Oval	Fast	Medium	Yellow	None	Rough dark	Seedless
Ash, Patmore	*Fraxinus pennsylvanica* "Patmore"	2	50	40	Oval	Fast	Medium	Yellow	None	Rough dark	Seedless
Ash, Rosehill	*Fraxinus americana* "Rosehill"	5	50	30	Oval	Fast	Medium	Bronze	None	Dark brown	Seedless

(continued)

5.41 SHADE AND ORNAMENTAL TREES (continued)

Common Name	Botanical Name	Hardiness Zone	Height (ft.)	Spread (ft.)	Shape	Growth Rate	Foliage Texture	Full Foliage	Fruit	Bark	Unusual Characteristics
Ash, Summit	Fraxinus pennsylvanica "Summit"	3	60	40	Oblong	Fast	Medium	Yellow	Very few	Rough dark	Straight growth habit
Bald Cypress	Taxodium destichum	5	70	30	Long	Medium	Medium-fine	Russet	None	Rough	Stately, hardy
Birch, Cutleaf Weeping	Betula pendula "Gracilis"	2	30	2.5	Oval	Medium	Fine	Golden	Conelike	Peeling	White bark
Birch, Clump European White	Betula pendula (Clump form)	2	30	2.5	Oval	Medium	Fine	Golden	Conelike	Peeling	White bark
Birch, European White	Betula pendula	2	50	2.5	Oval	Medium	Fine	Golden	Conelike	Peeling	White bark
Birch, Clump Paper	Betula papyrifera (Clump)	2	50	30	Oval	Medium	Fine	Yellow	Conelike	Peeling	White bark
Birch, Paper	Betula papyrifera	5	50	20	Oval	Medium	Fine	Yellow	Conelike	Peeling	White bark
Birch, Clump Monarch	Betula maximowicziana (Clump)	5	50	20	Oval	Medium	Fine	Yellow	Conelike	Peeling	Borer resistant
Birch, Monarch	Betula maximowicziana	5	50	20	Oval	Medium	Fine	Yellow	Conelike	Peeling	Borer resistant
Birch, Clump River	Betula nigra (Clump)	4	50	40	Oval	Medium	Fine	Yellow	Conelike	Red-brown Peeling	Borer resistant
Birch, River	Betula nigra	4	50	40"	Oval	Medium	Fine	Yellow	Conelike	Red-brown	Borer

Birch, Whitespire	Betula Platphylla japonica "Whitespire"	3	50	25	Oval	Medium	Fine	Golden	Conelike	Peeling	Very borer resistant
Catalpa, Northern	Catalpa speciosa	4	60	20	Irregular	Medium	Coarse	None	Pod	Gray brown	White blooms
Catalpa, Umbrella	Catalpa bungei	4	10	10	Umbrella	Fast	Coarse	Yellow	Pod	Rough dark	Umbrella form
Cherry, Flowering Akebono	Prurtus xyedoensis "Akebono"	5	20	20	Spreading	Medium	Medium	Red	Orange-red	Red-brown, smooth	Double flowers
Cherry, Flowering Canada Red	Prurtus virginia "Canada Red"	2	25	20	Oval	Medium	Medium	Red	None	Red-brown, smooth	Double flowers
Cherry, Flowering Sargent Columnar	Prurtus sargentii "Columnaris"	4	25	10	Vase	Medium	Medium	Amber	Red	Brown	Deep blush pink flowers, columnar habit
Cherry, Flowering Kwanzan	Prurtus serrulata "Kwanzan"	5	25	15	Vase	Medium	Medium	Red-gold	Small-yellow	Brown	White blooms
Cherry, Flowering Mt. Fuji	Prurtus serrulata "Mt. Fuji"	5	25	20	Drooping	Medium	Medium	Reddish	Orange-red	Dark	Drooping habit
Cherry, Flowering Sargent	Prurtus sargena	4	25	10	Round	Medium	Medium	Amber	Red	Dark	Pale pink flowers, vigorous
Cherry, Flowering Yoshino	Prurtus xyyeddensis "Yoshino"	5	20	20	Spreading	Medium	Medium	Red-gold	Orange-red	Red, smooth	Fragrant whitish pink blossoms
Cottonwood, Great Plains	Populus sargentii "Great Plains"	3	80	60	Vase	Rapid	Medium	Yellow-green	None	Light-yellow	Rapid growth

(continued)

5.41 SHADE AND ORNAMENTAL TREES *(continued)*

Common Name	Botanical Name	Hardiness Zone	Height (ft.)	Spread (ft.)	Shape	Growth Rate	Foliage Texture	Full Foliage	Fruit	Bark	Unusual Characteristics
Cottonwood, Narrowleaf	*Populus angustifolia* "Narrowleaf"	2	75	50	Round	Rapid	Medium	Golden	None	Light-brown	Drought tolerant
Cottonwood, Nor'Easter	*Populus x canadensis* "Nor'Easter"	2	60	40	Spreading	Rapid	Medium	Golden	None	Light-brown	Canker resistant
Cottonwood, Siouxland	*Populus deltoides* "Siouxland"	2	80	50	Spreading	Rapid	Medium	Golden	None	Light buff	Rust resistant
Crab, American Beauty	*Malus* "American Beauty"	4	15	12	Round	Medium	Fine	Diverse	Purple	Smooth	Double red blossoms
Crab, Brandywine	*Malus* "Brandywine"	4	25	20	Round	Fast	Fine	Deep-purple	Green	Smooth	Double rose blossoms
Crab, Candied Apple	*Malus* "Candied Apple"	4	15	12	Weeping	Medium	Fine	Diverse	Red	Smooth	Purple-pink blossoms
Crab, Centennial	*Malus* "Centennial"	3	18	18	Oval	Medium	Fine	Diverse	Red	Smooth	White blossoms
Crab, Centurian	*Malus* "Centurian"	4	25	20	Columnar	Fast	Fine	Diverse	Red	Smooth	Rosy red blossoms
Crab, Dolgo	*Malus* "Dolgo"	3	25	20	Round	Medium	Fine	Diverse	Crimson	Smooth	Large white blossoms
Crab, Eleyi	*Malus* "Eleyi Floribunda"	4	20	15	Upright	Medium	Fine	Diverse	Red	Smooth	Single pink blossoms
Crab, Floribunda	*Malus* "Floribunda"	4	25	20	Round	Medium	Fine	Diverse	Yellow-red	Smooth	Fragrant deep-pink blossoms

Crab, Halls Parkmani	*Malus* "Halls Parkmani"	4	15	12	Vase	Medium	Fine	Diverse	Red	Smooth	Double rose blossoms
Crab, Hopa	*Malus* "Hopa"	4	25	20	Upright	Medium	Fine	Diverse	Red	Smooth	Dense rose blossoms
Crab, Indian Magic	*Malus* "Indian Magic"	4	20	15	Round	Medium	Fine	Diverse	Red	Smooth	Pink blossoms
Crab, Liset	*Malus* "Liset"	4	15	15	Spreading	Medium	Fine	Diverse	Red-purple	Smooth	Purple blossoms
Crab, Makamik	*Malus* "Makamik"	4	15	15	Spreading	Medium	Fine	Diverse	Red	Smooth	Purple rose
Crab, Pink Perfection	*Malus* "Pink Perfection"	4	25	20	Spreading	Medium	Medium	Diverse	Red	Smooth	Double pink blossoms
Crab, Pink Spires	*Malus* "Pink Spires"	3	12	10	Pyramidal	Medium	Fine	Mahogany red	Maroon-red	Smooth	Light pink blossoms
Crab, Profusions	*Malus* "Profusions"	4	12	10	Shrub	Medium	Fine	Diverse	Oxblood red	Smooth	Reddish purple blossoms
Crab, Red Jade	*Malus* "Red Jade"	4	12	10	Weeping	Medium	Fine	Diverse	Red	Smooth	Blush white blossoms
Crab, Red Silver	*Malus* "Red Silver"	4	20	15	Upright	Medium	Fine	Diverse	Purple	Smooth	Fruit makes excellent jelly, rose blossoms
Crab, Red Splendor	*Malus* "Red Splendor"	3	18	12	Spreading	Medium	Fine	Diverse	Small red	Smooth	Rosy pink blossoms
Crab, Robinson	*Malus* "Robinson"	4	25	15	Upright	Medium	Fine	Diverse	Dark red	Smooth	Deep pink blossoms

(continued)

5.41 SHADE AND ORNAMENTAL TREES (continued)

Common Name	Botanical Name	Hardiness Zone	Height (ft.)	Spread (ft.)	Shape	Growth Rate	Foliage Texture	Full Foliage	Fruit	Bark	Unusual Characteristics
Crab, Royal Ruby	Malus "Royal Ruby"	4	15	10	Upright	Fast	Fine	Diverse	Red	Smooth	Double reddish pink blossoms
Crab, Royalty	Malus "Royalty"	4	15	10	Upright	Medium	Fine	Brilliant purple	Dark red	Smooth	Crimson-purple blossoms
Crab, Sargents	Malus "Sargenti"	4	8	6	Shrub	Medium	Fine	Diverse	Dark red	Smooth	White blossoms
Crab, Snowcloud	Malus "Snowcloud"	4	15	8	Upright	Fast	Fine	Diverse	Red	Smooth	Pure white blossoms
Crab, Snowdriff	Malus "Snowdriff"	3	20	15	Upright	Medium	Fine	Diverse	Orange-red	Smooth	White blossoms
Crab, Sparkler	Malus "Sparkler"	4	10	6	Horizontal	Medium	Fine	Diverse	Small red	Smooth	Bright pink blossoms
Crab, Spring Snow	Malus "Spring Snow"	4	20	15	Oval	Medium	Fine	Diverse	None	Smooth	Prolific white blossoms
Crab, Strathmore	Malus "Strathmore"	3	15	8	Pyramidal	Medium	Fine	Red	Purplish red	Smooth	Pink blossoms
Crab, Vanguard	Malus "Vanguard"	4	12	8	Vase	Medium	Fine	Diverse	Bright red	Smooth	Rose pink blossoms
Crab, White	Malus "White Angel"	4	15	8	Upright	Medium	Fine	Diverse	Glossy red	Smooth	Pure white blossoms

Common Name	Scientific Name										
Crab, White Candle	*Malus* "White Candle"	4	14	6	Columnar	Medium	Fine	Diverse	Few	Smooth	White with pink tint blossoms
Crab, Zumi	*Malus sieboldi* var. "Zumi Calocarpa"	5	15	8	Bushy	Medium	Fine	Yellow	Red-orange	Smooth	White blossoms
Dogwood Cherokee Chief	*Cornus florida* "Cherokee Chief"	5	20	12	Horizontal	Slow	Medium	Scarlet	Red	Tan smooth	Ruby red flowering
Dogwood, Cherokee Princess	*Cornus florida* "Cherokee Princess"	5	20	12	Horizontal	Slow	Medium	Scarlet	Red	Tan smooth	Light pink flowering
Dogwood, Cloud Nine	*Cornus florida* "Cloud Nine"	5	20	12	Horizontal	Slow	Medium	Scarlet	Red	Tan smooth	White flowering
Dogwood, Kouss	*Cornus florida* "Kouss"	5	20	12	Horizontal	Slow	Medium	Scarlet	Pinkish-red	Tan smooth	White flowering
Dogwood, Pink	*Cornus florida* "Pink"	5	20	12	Horizontal	Slow	Medium	Scarlet	Pinkish-red	Tan smooth	Pink flowering
Dogwood	*Cornus florida* "Rainbow"	5	20	12	Horizontal	Slow	Medium	Scarlet	Pinkish-red	Tan smooth	Varigated foliage, white flower
Dogwood, Red	*Cornus florida* "Red"	5	20	12	Horizontal	Slow	Medium	Scarlet	Pinkish-red	Tan smooth	Red flowering
Dogwood, White	*Cornus florida*	5	20	12	Horizontal	Slow	Medium	Scarlet	Pinkish-red	Tan smooth	White flowering
Golden Raintree	*Koelereutaria paniculata*	5	40	30	Spreading	Medium	Medium	Yellow	Brown	Light brown ridged	Yellow flowering in mid summer

(continued)

5.41 SHADE AND ORNAMENTAL TREES (continued)

Common Name	Botanical Name	Hardiness Zone	Height (ft.)	Spread (ft.)	Shape	Growth Rate	Foliage Texture	Full Foliage	Fruit	Bark	Unusual Characteristics
Hackberry	Celtis accidentalis	2	70	60	Oval	Medium-fast	Medium-coarse	Yellow-green	Orange-red to purple	Corky grayish brown	Hardy, tolerant tree
Hawthorn, Crumson Cloud P.P.#2679	Crantaegus laevigata "Crimson Cloud"	4	15	12	Oval	Medium	Medium	None	Scarlet	Brown	Red flowers
Hawthorn, Paul's Scarlet	Crartaegus laevigana "Paulii"	5	20	12	Oval	Slow	Fine	None	Scarlet	Flaky	Double scarlet flowers
Hawthorn, Toba	Crataegus "Toba"	3	15	12	Oval	Medium	Medium	None	Scarlet	Brown	Fragrant, double white flower
Hawthorn, Washington	Crataegus phaenopyrum "Washington"	4	30	25	Columnar	Medium-slow	Medium-fine	Red-purple	Glossy-red	Brown	White flowers in spring
Honey Locust, Thornless	Gleditsia triacanthos inermis "Thornless"	4	70	60	Spreading	Fast	Medium-fine	Yellow-green	Reddish-brown	Grayish brown	Fragrant white flowers
Honey Locust, Golden	Gleditsia triacanthos inermis "Golden"	4	60	40	Oval	Fast	Medium-fine	Yellow-green	None	Brown	Deligate golden foliage
Honey Locust, Green Glory	Gleditsia triacanthos inermis "Green Glory"	4	60	40	Pyramidal	Fast	Medium-fine	Yellow-green	None	Brown	Fragrant white flowers
Honey Locust, Moraine	Gleditsia triacanthos inermis "Moraine"	4	60	40	Spreading	Fast	Medium-fine	Yellow-green	None	Brown	Old favorite
Honey Locust, Rubylace	Gleditsia triacanthos inermis "Rubylace"	4	40	30	Spreading	Fast	Medium-fine	Yellow-green	None	Brown	Purplish bronze foliage

Common Name	Scientific Name		Height	Spread	Shape		Texture	Flower Color	Fruit	Bark	Notes
Honey Locust, Shadermaster	*Gleditsia triacanthos inermis* "Shadermaster"	4	60	40	Spreading	Fast	Medium-fine	Yellow-green	Note	Brown	Very dark green foliage
Honey Locust, Skyline	*Gleditsia triacanthos inermis* "Skyline"	4	60	60	Pyramidal	Fast	Medium-fine	Yellow-green	None	Brown	Very dark green foliage
Honey Locust, Sunburst	*Gleditsia triacanthos inermis* "Sunburst"	4	40	30	Pyramidal	Fast	Medium-fine	Yellow-green	None	Brown	Golden foliage on tips of branch
Linden, American	*Tilia americana*	3	75	50	Pyramidal	Medium	Medium	None	Nutlet	Rough-gray	Pyramidal habit
Linden. Glenleven	*Tilia cordata* "Glenleven"	3	50	35	Pyramidal	Medium	Medium	None	Nutlet	Rough-gray	Pyramidal habit
Linden, Greensire	*Tilia cordata* "Greenspire"	4	50	35	Pyramidal	Medium	Medium	None	Nutlet	Rough-gray	Spicy fragrant flowers
Linden, June Bride	*Tilia cordata* "June Bride"	3	50	35	Pyramidal	Medium	Medium	None	Nutlet	Rough-gray	Prolific flowering
Linden, Littleleaf	*Tilia cordata*	4	50	35	Oval	Medium	Medium	None	Nutlet	Rough-gray	Fragrant flowers
Linden, Redmood	*Tilia enchora* "Redmood"	4	40	30	Pyramidal	Medium	Medium	None	Nutlet	Rough-gray	Reddish bark in winter
Magnolia, Saucer	*Magnolia soulangiana*	5	30	25	Spreading	Medium	Coarse	None	Pod	Gray	Large white to pinkish flowers
Magnolia, Southern	*Magnolia grandiflora*	6	80	50	Pyramidal	Slow	Medium-coarse	None	Pod	Dark	Evergreen, large white fragrant flowers

(continued)

5.41 SHADE AND ORNAMENTAL TREES (continued)

Common Name	Botanical Name	Hardiness Zone	Height (ft.)	Spread (ft.)	Shape	Growth Rate	Foliage Texture	Full Foliage	Fruit	Bark	Unusual Characteristics
Magnolia, Star	Magnolia stellata	4	20	15	Oval	Slow	Medium	Yellow-brown	Pod	Brown	Double white fragrant flowers
Maple, Amur	Acer ginnala	2	20	20	½ round	Medium	Fine	Scarlet	Winged	Thin, dark	Extremely hardy
Maple, Autumn Flame	Acer rubrum "Autumn Flame"	3	50	35	Round	Fast	Medium	Scarlet	Winged	Gray	Fall color
Maple, Blair	Acer saccharinum "Blair"	4	80	50	Round	Fast	Medium	Pale yellow	Winged	Gray	Strong branching
Maple, Crimson, King	Acer platanoides "Crimson King"	4	40	30	Oval	Medium	Medium	None	Winged	Rough, dark	Purple foliage
Maple, Columnar Norway	Acer platanoides "Columnar"	4	50	20	Columnar	Medium	Medium	None	Winged	Rough, dark	Columnar
Maple, Emerald Lustre P.P. #4837	Acer platanoides "Pond"	3	50	30	Oval	Fast	Medium	Yellow	Winged	Rough, dark	Glossy foliage
Maple, Emerald Queen	Acer platanoides "Emerald Queen"	3	40	25	Ascending	Medium	Medium	Yellow	Winged	Rough, dark	Cutleaf
Maple, Japanese Dwarf	Acer palmatum	6	10	10	Round	Slow	Fine	Yellow	Winged	Reddish brown	Delicate appearance
Maple, Norway	Acer plananoides	3	50	30	Oval	Medium	Medium	None	Winged	Rough, dark	Dense foliage
Maple, Northwood Red	Acer rubrum "Northwood"	3	50	35	½ round	Fast	Medium	Scarlet	Winged	Gray	Fall foliage

Common name	Botanical name		Height	Shape	Growth	Texture	Flower	Fruit	Bark	Remarks
Maple, October Glory	*Acer rubrum* "October Glory"	4	50	½ round	Fast	Medium	Scarlet	Winged	Gray	Fall foliage
Maple Oregon Pride	*Acer platanoides* "Oregon Pride"	4	40	Oval	Medium	Medium	None	Winged	Rough, dark	Red foliage
Maple, Royal, Red	*Acer platanoides* "Royal Red"	4	40	Oval	Medium	Medium	None	Winged	Rough, dark	Dark red foliage
Maple, Red	*Acer rubrum*	4	60	½ round	Fast	Medium	Scarlet	Winged	Gray	Fall foliage
Maple, Red Sunset	*Acer rubrum* "Red Sunset"	3	50	½ round	Fast	Medium	Scarlet	Winged	Gray	Fall color
Maple, Schwedler	*Acer platanoides* "Schwedler"	4	60	Round	Medium	Medium	None	Winged	Rough, dark	Purplish red foliage
Maple, Silver	*Acer saccharinum*	4	100	Round	Rapid	Medium	Pale yellow	Winged	Gray	Fast growth
Maple, Silver Queen	*Acer saccharinum* "Silver Queen"	4	60	Round	Rapid	Medium	Pale yellow	Seedless	Gray	Seedless
Maple, Sugar	*Acer saccharinum*	3	75	Oval	Medium	Medium	Yellow-scarlet	Winged	Dark brown	Fall color
Maple, Bonfire Color	*Acer saccharinum* "Bonfire"	3	60	Oval	Medium	Medium	Brilliant red	Winged	Dark brown	Fall color
Maple, Sugar Green Mt.	*Acer saccharinum* "Mt. Green"	3	60	Oval	Medium	Medium	Orange-red	Winged	Dark brown	Fall color
Maple, Variegated Norway	*Acer platanoides* "Variegatum"	5	40	Oval	Medium	Fine	None	Winged	Rough, dark gray	Green leaves edged in white
Maple, Wiers	*Acer saccharinum* "Werri"	3	70	Weeping	Fast	Medium	Pale yellow	Winged	Gray	Cutleaf

(continued)

5.41 SHADE AND ORNAMENTAL TREES (continued)

Common Name	Botanical Name	Hardiness Zone	Height (ft.)	Spread (ft.)	Shape	Growth Rate	Foliage Texture	Full Foliage	Fruit	Bark	Unusual Characteristics
May Day	Prunus padus commutata	2	20	30	Round	Medium	Medium	None	Black	Dark brown	Large fragrant white flowers
Mimosa	Albizia julibrissin	6	30	20	Vase	Fast	Fine	None	Pod	Greenish	Foliage and brushlike pink flowering
Mt. Ash, Black Hawk	Sorbusri aucupana "Black Hawk"	3	20	25	Oval	Medium	Medium	Gold	Orange	Gray-brown	Orange fruit
Mt. Ash, Cardinal	Sarbusri aucuparia "Cardinal"	3	25	20	Ovate	Medium	Medium	Gold	Apricot color	Gray-brown	Apricot color fruit
Mt. Ash, Cardinal Royal	Sorbus aucupana "Card, royal"	3	25	20	Oval	Medium	Medium	Gold	Orange-red	Gray-brown	Orange-red fruit
Mt. Ash Cherokee	Sorbus aucuparia "Cherokee"	3	25	20	Oval	Medium	Medium	Gold	Red	Gray-brown	Red fruit
Mt. Ash, Decora	Sorbus decora	2	15	20	Oval	Medium	Medium	Gold	Red	Gray-brown	Hardy
Mulberry, Common	Morus alba	4	30	20	Rounded	Fast	Coarse	None	Flesh-purple	Brown	Fruiting
Mulberry, Fruitless	Morus alba "Fruitless"	6	50	50	Rounded	Fast	Coarse	None	None	Brown	Fruitless
Mulberry, Weeping	Morus alaba urabana	6	50	50	Rounded	Fast	Coarse	None	None	Brown	Weeping
Oak, Pin	Quercus palustra	4	70	40	Pyramidal	Slow	Medium	Scarlet	Acorn	Dark	Fast growth, fall color
Qak, Red	Quercus borealis	4	70	50	Oval	Slow	Medium	Dark red	Acorn	Rough, dark	Fast growth, fall color

Oak, Schumard	Quercus schumardi	6	80	60	Oval	Medium	Medium	Crimson	Acorn	Rough, dark	Fast growth, fall color
Oak, Water	Quercus nigra	6	60	40	Oval	Medium	Medium	None	Acorn	Smooth	Holds leaves into winter
Peach, Cardinal	Prunus persica "Cardinal"	5	15	15	Rounded	Medium	Medium	Yellow	Red blush	Red-brown	Double red pink flowers
Pear, Aristocrat	Pyrus calleryna "Aristocrat"	4	30	20	Oval	Medium	Medium	Purple-red	Small yellow	Brown	Fall color, white flowers
Pear, Bradford	Pyrus calleryna "Bradford"	4	30	20	Conical	Medium	Medium	Red-gold	None	Brown	Fall colors, white flowers
Pear, Capital	Pyrus calleryna "Capital"	5	50	15	Narrow upright	Medium	Medium	Red-purple	Few	Brown	Fall color, white flowers
Pear, Red Spires P.P.#3815	Pyrus calleryna "Red Spires"	5	30	20	Conical narrow	Medium	Medium	Crimson-purple	None	Brown	Fall color, white flowers
Pear, Whitehouse	Prunus calleryane "Whitehouse"	5	50	20	Columnar	Medium	Medium	Reddish purple	Few	Brown	Fall color, white flowers
Plum, Blireiana	Prunus x blireiana	5	15	10	Round	Medium	Medium	Purple	Dark red	Dark	Double fragrant flowers
Plum, K.V.	Prunus cerasifera "Krautet Vesuvius"	6	20	15	Oval	Fast	Medium	Purple	None	Dark red	Purple-black foliage
Plum, Newport	Prunus cerasifera "Newport"	3	15	15	½ round	Fast	Medium	Purple-red	Dark red	Rough, dark	Reddish purple
Plum, Thundercloud	Prunus cerasifera "Thundercloud"	4	15	15	½ round	Medium	Medium	Purple	Dark purple	Dark red	Purple foliage

(continued)

5.41 SHADE AND ORNAMENTAL TREES (continued)

Common Name	Botanical Name	Hardiness Zone	Height (ft.)	Spread (ft.)	Shape	Growth Rate	Foliage Texture	Full Foliage	Fruit	Bark	Unusual Characteristics
Poplar, Bolleana	*Populus alba* "Pyramidalis"	3	60	6	Columnar	Rapid	Medium	Golden	None	Smooth	Columnar
Poplar, Lombardy	*Populus rugra* "Thervestina"	3	60	6	Columnar	Rapid	Medium	None	None	Gray	Columnar
Poplar, Robusta	*Populus x canadensis* "Robusta"	2	50	15	Oval	Rapid	Medium	Golden	None	Light buff	Rapid growth
Poplar, Sliver	*Populus alba*	3	70	40	Rounded	Rapid	Medium-coarse	None	None	Gray-green	Silver-green foliage
Quaking Aspen	*Populus tremuloides*	2	60	30	Vase	Rapid	Medium	Yellow	None	Greenish white	Fall foliage
Red Bud, Eastern	Carcis canadersis	4	20	20	Round	Slow	Coarse	Yellow	Small pods	Dark gray	Rosy pink flowers
Red Bud, Forest Pansy	*Cercis canadensis* "Forest Pansy"	5	20	20	Round	Slow	Coarse	Yellow	Small pods	Dark gray	Purple foliage
Red Bud, Oklahoma	*Cersis canadensis* "Oklahoma"	6	20	20	Round	Slow	Coarse	yellow	Small pods	Dark gray	Glossy foliage
Red Bud, White	*Cersis canadensus alba*	6	20	20	Round	Slow	Coarse	Yellow	Small pods	Dark gray	White flowers
Russian Olive	*Elaeagnus angustifolia*	2	30	20	Spreading	Fast	Fine	None	Berries	Silver-brown	Silver-gray foliage
Sweet Gum	*Liquidambar styraciflua*	5	80	50	Pyramidal	Medium	Medium	Yellow-red	Winged	Gray-brown	Fall foliage

Common Name	Scientific Name										
Sycamore, American	*Platanus occidentalis*	4	100	100	Oval	Fast	Coarse	None	Balls	Brown-white	Adaptable
Sycamore, Bloodgood	*Platanus x acerifolia* "Bloodgood"	5	80	100	Oval	Fast	Coarse	None	Balls	Brown-white	Disease resistant
Smoke Tree	*Cotinus coggygra*	5	15	15	Spreading	Medium	Fine	Purple	Small pods	Brown-purple	Smoky pink flowers
Smoke Tree, Velvet Cloak	*Cotinus coggygria* "Velvet Cloak"	5	15	15	Spreading	Medium	Fine	Reddish purple	Small pods	Brown-purple	Purple foliage
Tulip Tree	*Luriodendron tulipfera*	5	90	50	Pyramidal	Fast	Coarse	Golden yellow	Winged	Brownish	Tulip-shaped flowers
Willow, Corkscrew	*Salix matsulana* "Tortoosa"	4	30	12	Upright	Rapid	Twisted	None	Cottony	Greenish	Oriental look
Willow, Globe	*Salix matsudana* "Globosa"	4	15	8	Round	Rapid	Fine	None	Cottony	Greenish	Globe shape
Willow, Niobe	*Salix alba tristis*	3	50	30	Weeping	Rapid	Fine	None	Cottony	Greenish	Golden color
Willow, Wisconsin Weeping	*Salix x blanda*	4	40	35	Weeping	Rapid	Fine	None	Cottony	Greenish branches	Pendulous
Wisteria Tree	*Wisteria sinensis*	4	30	15	Weeping	Fast	Fine	None	Pods	Grayish	Fragrant violet flowers

5.42 ORNAMENTAL SHRUBS

Common Name	Botanical Name	Hardiness Zone	Height (ft.)	Spread (ft.)	Shape	Growth Rate	Foliage Texture	Full Foliage	Fruit	Bark	Unusual Characteristics
Althea, Rose of Sharon	Hibiscus syriacus	5	10	5	Upright	Medium	Medium	None	Brown	Gray	Purple, red, white, pink flowers in summer
Almond, Pink Flowering	Prunus glandulosa "Rosea"	4	6	5	Upright	Medium	Medium	None	None	Dark	Pink, double flowers in early spring
Barberry, Crimson Pygmy	Berbiris thunbergii atropurpurea "Crimson Pygmy"	4	2	3	Mound	Slow	Fine	Red	None	Thorns	Extremely dwarf, deep crimson color
Barberry, Golden	Berberis thunbergii aurea	4	2	3	Mound	Slow	Fine	Gold	None	Thorns	Bright golden foliage
Barberry, Red Leaf	Berberis thunbergii aurea atropurpurea	4	5	5	Compact	Medium	Fine	Red	Red	Thorns	Mahogany red foliage
Barberry, Rosy Glow	Berberis thunbergii "Rosy Glow"	4	4	5	Compact	Medium	Fine	Red	None	Thorns	Red leaves with tinge of white and rose
Barberry, William Penn	Berberis x gladwynensis	5	1	3	Spreading	Medium	Medium	None	None	Thorns	Evergreen dwarf, bright yellow spring flowers
Burning Bush, Dwarf	Euonymusalatus "Compacta"	3	6	4	Compact	Slow	Medium	Red	Red	Winged	Brilliant red foliage in fall

(continued)

Common Name	Scientific Name				Habit	Rate	Texture				Remarks
Cotoneaster, Cranberry	Cotoneaster apiculatus	4	2	3	Mound	Slow	Medium	None	Red	None	Bright red berries in fall
Cotoneaster, Spreading	Cotoneaster divaricatus	4	2	3	Spreading	Medium	Medium	None	Red	None	Bright red berries hold into winter
Cotoneaster, Rock	Cotoneaster horizontalis	4	2	3	Spreading	Medium	Medium	Red	Red	None	Bright red berries in fall
Crape Myrtle	Lagerstroemia indica	6	25	15	Upright	Rapid	Medium	Yellow red	None	Smooth	Deciduous shrub or small tree, prolific cluster of flowers in a wide range of colors in summer
Dogwood, Bailey	Cornus stolonifera	3	8	6	Spreading	Medium	Coarse	Red	None	Red	Bright red twigs
Dogwood, Red Twig	Cornus sericea	3	8	6	Spreading	Medium	Coarse	Red	None	Red	Red twigs, small white flowers in summer
Dogwood, Variegated	Cornus alba "Elegantissima"	3	6	6	Upright	Slow	Coarse	None	White	None	Foliage is green with cream-colored fringe
Dogwood, Yellow Twig	Cornus stolonifera "Flaviramea"	3	8	5	Spreading	Medium	Coarse	None	White	Yellow	Bright yellow branches, white flowers in late spring
Forsythia, Lynnwood Gold	Forsythia x intermedia "Lynnwood Gold"	4	5	6	Spreading	Rapid	Medium	None	Capsule	Dull yellow	Brilliant yellow flowers in early spring

5.42 ORNAMENTAL SHRUBS (continued)

Common Name	Botanical Name	Hardiness Zone	Height (ft.)	Spread (ft.)	Shape	Growth Rate	Foliage Texture	Full Foliage	Fruit	Bark	Unusual Characteristics
Forsythia, Sunrise	Forsythia x Intermedia "Sunrise"	4	4	5	Spreading	Medium	Medium	None	None	Dull yellow	Shorter growing
Golden Elder	Sambucus canadensis "Aurea"	3	10	6	Open	Rapid	Coarse	Yellow	Red	Tan	Golden yellow leaves, clusters of tiny white flowers
Golden Nine Bark	Physocarpus opulifolius aureus	4	8	4	Round	Medium	Medium	None	Capsule	Orange-brown	Golden-colored leaves
Honeysuckle, Arnold's Red	Lonicera tatarica "Arnold Red"	3	8	8	Upright	Medium	Medium	None	Red	Gray	Red flowers in spring followed by red berries
Honeysuckle, Clavey's	Lonica xylosteum "Claveyi"	4	6	4	Compact	Slow	Medium	None	Red	Gray	Dwarf, fragrant creamy white flowers
Honeysuckle, Goldflame	Lonicera x beckrottii	4	2	10	Spreading	Fast	Soft	None	None	None	Brilliantly colored coral red flowers, yellow foliage
Honeysuckle, Hall's	Lonicera japonica "Halliana"	4	2	10	Spreading	Fast	Soft	None	None	None	Tubular pure white twig, yellow flowers, fragrant
Honeysuckle, Purpleleaf	Lonicera japonica "Purpurea"	5	2	10	Spreading	Fast	Soft	Purple	None	None	Foliage turns deep purple in winter
Honeysuckle, Zabel	Lonicera tatarica "Zabelii"	3	8	8	Upright	Medium	Medium	None	Red	Gray	Bright red blooms in spring

132

Hydrangea, P.G.	Hydrangea paniculata "Grandiflora"	3	8	5	Stiff	Medium	Coarse	None	Capsule	Buff	Profuse bloomers with large clusters of white flowers
Lilac, Old-Fashioned	Syringa vulgaris	3	12	10	Upright	Medium	None	None	Capsule	Drab	Clusters of lavender blossoms in spring
Lilac, Purple Persian	Syringa x chinensis†	3	8	6	Upright	Medium	Medium	None	Capsule	Dark gray	Profusion of lilac purple blooms in spring
Lilac, French Hybrid	Syringa French hybrids	3	15	8	Upright	Medium	Coarse	None	Capsule	Dark	Various attractive flowers in late spring
Mockorange, Minnesota Snowflake	Philadelphus x virginalis "Minnesota" Snowflake"	3	8	6	Upright	Medium	Medium	None	None	Gray	Double white fragrant flowers, very hardy
Mockorange, Sweet†	Philadelphus coronarius	3	10	6	Mound	Rapid	Medium	None	Capsule	Brown	Very fragrant single white flowers
Potentilla, Abbottswood	Potentilla fruticosa "Abbottswood"	3	3	3	Bushy	Medium	Medium	None	Capsule	Flaking	White flowers in summer
Potentilla, Coronation	Potentilla fruticosa "Coronation Triumph"	3	4	3	Compact	Medium	Fine	Golden	Seed pods	Brown	Brilliant yellow flowers in summer until frost
Potentilla, Goldfinger	Potentilla fruticosa "Goldfingers"	3	3	3	Round	Medium	Fine	None	Seed pods	Brown	Golden yellow flowers from spring until fall, very hardy

(continued)

5.42 ORNAMENTAL SHRUBS (continued)

Common Name	Botanical Name	Hardiness Zone	Height (ft.)	Spread (ft.)	Shape	Growth Rate	Foliage Texture	Full Foliage	Fruit	Bark	Unusual Characteristics
Potentilla, Jackmani	Potentilla fruticosa "Jackmani"	3	4	3	Upright	Slow	Fine	None	Seed pods	Brown	Dwarf with golden yellow flowers in mid summer
Potentilla, Katherine Dykes	Potentilla fruticosa "Katherine Dykes"	3	3	3	Spreading	Medium	Fine	None	Seed pods	Brown	Soft yellow flowers, very hardy
Privet, Golden Vicary	Ligustrum x vicaryi	5	6	5	Upright	Slow	Fine	None	Black	Gray	Bright yellow flowers, very hardy
Privet, AR North	Ligustrum ansurence	4	10	5	Upright	Medium	Medium	None	None	None	Very hardy hedge
Privet, AR South	Ligustrum sinence	6	10	5	Upright	Medium	Medium	None	None	None	Semi-evergreen hedge
Pussy Willow, Black	Salix discolor	4	15	8	Upright	Medium	Coarse	None	None	Brown	Pearly woolly catkins in spring, very hardy
Pussy Willow, French Pink	Salix caprea	4	15	12	Upright	Medium	Medium	None	None	Brown	Large pink catkins in spring, very hardy
Quince, Red Flowering	Chaenomeles japonica "Rubra"	5	6	4	Spreading	Medium	Medium	None	Green	Brown	Pinkish red, flowers in early spring
Sand Cherry Purpleleaf	Prunus x cistena	4	15	5	Bushy	Medium	Medium	Purple-red	Black	Rough	Deep purple-red foliage

Common Name	Botanical Name				Shape	Growth	Texture	Color		Bark	Flowers
Spirea, Anthony Waterer	Spires x bumalda "Anthony Waterer"	4	3	3	Mound	Slow	Medium	Red	None	Red-brown	Cluster of rosy blossoms summer/fall
Spirea, Billiardi	Spirea x billiardian	4	6	7	Upright	Medium	Medium	None	None	None	Bright pink flowers in July/August
Spirea, Blue Mist	Caryopteris incana	4	4	3	Mound	Medium	Medium	None	None	None	Powdery blue-fringed flowers in summer
Spirea, Froebeli	Spirea x bumalda "Froebeli"	4	4	4	Spreading	Medium	Medium	Red	None	Red-brown	Clusters of bright pink flowers in May/June
Spirea, Gold Flame	Sprirea x bumalda "Goldflame"	3	3	2	Round	Slow	Medium	Red	None	Brown	Foliage is golden red, clusters of flowers in summer
Spirea, Little Princess	Sprirea japonica "Little Princess"	4	3	2	Mound	Slow	Medium	None	None	None	Mint green foliage clusters of rose crimson flowers in summer
Spirea, Snowmound	Spirea ruponica "Snowmound"	3	4	3	Round	Medium	Fine	Yellow	None	Red-brown	Covered with white clusters in early spring
Spirea, Vanhoutte	Spirea x vanhouttei	4	10	8	Recurving	Medium	Fine	Orange-red	None	Red-brown	Arching branches covered with pure white blooms in late spring

(continued)

5.42 ORNAMENTAL SHRUBS *(continued)*

Common Name	Botanical Name	Hardiness Zone	Height (ft.)	Spread (ft.)	Shape	Growth Rate	Foliage Texture	Full Foliage	Fruit	Bark	Unusual Characteristics
Viburnum, Burkwood	*Viburnum x burkwoodii*	5	10	5	Upright	Medium	Coarse	None	Red to black	Hairy tan	Clusters of fragrant flowers in early spring
Viburnum, Dilatatum	*Viburnum dilatatum*	5	10	5	Upright	Medium	Medium	None	Red	None	Gray-green leaves, red fruit
Viburnum, Mariesi	*Viburnum plicatum totemtosum* "Mariesii"	5	8	6	Spreading	Medium	Coarse	None	Red to black	Brown	White flowers with flat heads in early spring
Viburnum, Snowball	*Viburnum opulus* "Sterilus"	3	10	15	Rounding	Medium	Medium	Purple	None	None	Pure white snowball-like flowers in summer
Viburnum, Pragense	*Viburnum x pragense*	4	8	6	Upright	Medium	Glossy	None	Red	None	Creamy white flowers in spring
Viburnum, Willowwood	*Viburnum rhytidophylloides* "Willowwood"	4	8	6	Upright	Medium	Glossy	None	Black-blue	Brown, smooth	Clusters of white flowers, foliage turns burgundy in fall
Weigela, Rosea	*Weigela florida*	5	6	6	Spreading	Medium	Medium	None	None	None	Showy rosy-pink trumpet-shaped flowers in May/June

Weigela, Vanicek	Weigela florida "Vanicek"	5	6	5	Spreading	Medium	Medium	None	None	Brown	Gray	Abundant red flowers in late May
Weigela, Variegated	Weigela florida "Variegata"	5	4	4	Compact	Medium	Medium	None	None	None	Gray	Pink ball-shaped flowers in spring, green/white leaves
Wisteria, Purple	Wisteria sinerisis	4	5	35	Spreading	Rapid	Medium	None	None	None	None	Vine with long grapelike clusters of blossoms in spring

5.43 FRUIT TREES

Variety	Fruit Color	Characterictics	Pollinator Required	Hardiness Zone
Apple, Anna	Greenish yellow	Low chill, good storage	Yes	7
Apple, Beverly Hills	Pale yellow	Low chill, cooking	Yes	8
Apple, Cortland	Greenish yellow	Large fruit, aromatic	Yes	5
Apple, Delicious Red	Red	Eating variety	Yes	5
Apple, Delicious Yellow	Yellow	Keeps well	Yes	5
Apple, Ein Shemer	Red striped	Low chill, juicy	Yes	8
Apple, Golden Delicious	Yellow	Large, crisp, sweet	No	5
Apple, Granny Smith	Green	Large fruit, keeps well	Yes	5
Apple, Grimes Golden	Yellow	Medium fruit, spicy	No	5
Apple, Gravestein	Green, red striped	Medium fruit, cooking	Yes	5
Apple, Haralson	Red striped	Keeps well	Yes	5
Apple, Honey Gold	Yellow	Hardy and sweet	Yes	4
Apple, Hyslop	Red	Tart, good jelly	Yes	5
Apple, Jonadel	Red	Eating variety, keeps well	Yes	5
Apple, Jonathan	Dark red	Crisp, keeps well	Yes	5
Apple, Keepsake	Red	Cooking variety	Yes	5
Apple, Lodi	Yellow	Large fruit, cooking variety	Yes	4
Apple, McIntosh	Red	Large fruit, aromatic	Yes	4
Apple, Red Baron	Red	Medium fruit, eating variety	Yes	5
Apple, Regent	Red	Dessert variety	Yes	3
Apple, Rome Beauty	Red	Cooking variety	Yes	4

Apple, State Fair	Red	Crisp, keeps well	3
Apple, Stayman Winesap	Red	Cooking and eating variety	5
Apple, Sweet Sixteen	Red striped	Medium fruit, crisp	3
Apple, Wealthy	Red	Tender dessert and cooking variety	5
Apple, Whitney Crab	Red	Juicy, sweet flavor	5
Apple, Winesap Crimson	Red	Wine flavored, cooking and keeps well	6
Apple, Yellow Transparent	Yellow	Medium fruit, cooking variety	5
Apricot, Chinese	Gold	Fuzzless, freestone	5
Apricot, Early Golden	Orange	Fuzzless	5
Apricot, Moongold	Gold	Early bearer	4
Apricot, Moonpark	Yellow	Large fruit, freestone	5
Apricot, Royal Blenheim	Orange-red	Large fruit, low chill	8
Apricot, Scout	Yellow	Hardy	3
Apricot, Sungold	Bright gold	Mild and sweet	4
Apricot, Tilton	Gold	Canning, drying variety	5
Cherry, Early Richmond	Red	Sour variety, highly productive	5
Cherry, Montmorency	Red	Sour variety, cooking	5
Cherry, Meteor	Red	Dwarf sour variety, cooking	4
Cherry, North Star	Red	Dwarf sour variety, large fruit, hardy	3
Cherry, Bing	Red	Good fresh or frozen, highly productive	6
Cherry, Black Tartarian	Purplish/black	Good pollinator	5
Cherry, Kansas Sweet	Red	Large fruit, highly productive	4
Cherry, Lambert	Black	Large fruit, fast grower	5

(continued)

5.43 FRUIT TREES (continued)

Variety	Fruit Color	Charactericrics	Pollinator Required	Hardiness Zone
Cherry, Rainier	Yellow	Keeps well, highly productive	Yes	5
Cherry, Royal Anne	Yellow	Large fruit, sweet and juicy	Yes	5
Cherry, Stella	Black	Good pollinator	No	6
Cherry, Windsor	Black	Highly productive, fast grower	Yes	5
Cherry, Van	Black	Large fruit, sweet and juicy	Yes	5
Cherry, Yellow Glass	Yellow	Hardiest of all sweet cherries	Yes	4
Fig, Brown Turkey	Mahogany brown	Few seeds, best eaten fresh	No	7
Fig, Texas Ever Bearing	Purple	Large fruit, sweet and juicy	No	7
Nectarine, Autumn Delight	Red	Large fruit, highly productive	No	5
Nectarine, Fantasia	Yellow	Large fruit, freestone	No	5
Nectarine, Flavortop	Red	Large fruit, freestone	No	5
Nectarine, Goldmine	Yellow blush	Large fruit, freestone, low chill	No	8
Nectarine, Golden Prolific	Yellow	True genetic dwarf, freestone	No	5
Nectarine, Redgold	Yellow	Juicy, highly productive	No	5
Nectarine, Sunred	Red	Semi-freestone, low chill	No	8
Nectarine, Surecrop	Yellow	Smooth skin, freestone	No	5
Peach, Bell of Georgia	White	Freestone	No	5
Peach, Bonanza	Red blush	True genetic dwarf, freestone	No	5
Peach, Desert Gold	Yellow	Highly productive, low chill	No	8
Peach, Early Elberta	Yellow	Freestone, good canning	No	5
Peach, Elberta	Yellow	Freestone, good canning	No	5
Peach, Elbertita	Yellow	True genetic dwarf, large fruit	No	5

Peach, Flordabelle	Yellow	Freestone, low chill	No	8
Peach, Flordasun	Yellow	Semi-freestone, low chill	No	8
Peach, Golden Jubilee	Yellow	Freestone, juicy	No	5
Peach, Golden Glory	Golden	True genetic dwarf, freestone	No	5
Peach, Halehaven	Yellow	Freestone	No	4
Peach, J.H. Hale	Yellow	Freestone, keeps well	Yes	5
Peach, Polly	White	Freestone, good canning	No	4
Peach, Redhaven	Yellow	Freestone, highly productive	No	5
Peach, Reliance	Yellow	Freestone, juicy	No	4
Peach, Richhaven	Red blush	Freestone, juicy	No	5
Pear, Bartlett	Golden red	Sweet, eating and canning variety	Yes	5
Pear, Bartlett Max Red	Red blush	Eating and canning variety	Yes	5
Pear, Clapp's Favorite	Red blush	Sweet eating variety	Yes	5
Pear, Dessert	Yellow	Large fruit, tart flavor	Yes	5
Pear, Douglas	Red blush	Blight resistant	Yes	5
Pear, Duchess	Green	Highly productive	No	5
Pear, Kieffer	Green	Blight resistant	No	4
Pear, Luscious	Yellow	Blight resistant	Yes	3
Pear, Moonglow	Yellow	Blight resistant	Yes	5
Pear, Parker	Red blush	Very prolific	yes	5
Pear, Patton	Yellow	Blight resistant	Yes	5
Pear, Red Sensation	Red	Sweet eating and canning variety	Yes	5
Pear, Seckel	Yellow-brown	Blight resistant, spicy flavor	Yes	5

(continued)

5.43 FRUIT TREES (continued)

Variety	Fruit Color	Characterictics	Pollinator Required	Hardiness Zone
Plum, Abundance	Red	Light yellow flesh	Yes	5
Plum, Blue Damson	Purple	Excellent preserves	Yes	5
Plum, Bruce	Wine red	Rich juicy flavor	Yes	5
Plum, Burbank	Red	Yellow flesh, canning variety	Yes	5
Plum, Green Gage	Green	Juicy, cooking variety	Yes	5
Plum, Hanska	Red-blue	Yellow flesh, extra sweet	Yes	5
Plum, Italian Prune	Purple	Large fruit, yellow flesh	Yes	5
Plum, La Crescent	Yellow	Extra juicy	Yes	4
Plum, Monitor	Red	The largest red plum	Yes	4
Plum, Mount Royal	Blue	Extra hardy	Yes	4
Plum, Ozark Premier	Red	Yellow flesh	No	4
Plum, Pipestone	Red	Large fruit, yellow flesh	Yes	4
Plum, Santa Rosa	Purple-red	Very prolific, compact growing	Yes	5
Plum, Sapa	Green	Small fruit, good sauce variety	Yes	4
Plum, Shropshire Damson	Purple	Yellow flesh, tart flavor	Yes	#
Plum, Stanley Prune	Purple	Freestone eating or canning variety	Yes	5
Plum, Superior	Red	Yellow flesh, very prolific	Yes	4
Plum, Toka	Red	Good pollinator, yellow flesh	Yes	4
Plum, Underwood	Red	Large fruit, freestone	Yes	4
Plum, Waneta	Red	Yellow flesh, extremely sweet	Yes	4

5.44 NUT TREES

Variety	Nut Size	Characterictics	Pollinator Required	Hardiness Zone
Pecan, Cherokee	Small to medium	Very vigorous and productive	Yes	6
Pecan, Cheyenne	Small to medium	Highly productive	Yes	6
Pecan, Choctaw	Medium	Unusually thin shelled	Yes	6
Pecan, Colby	Small to medium	Early maturity	Yes	5
Pecan, Camanche	Small	Vigorous grower	Yes	6
Pecan, Desirable	Small to medium	Very thin shelled, highly productive	Yes	6
Pecan, Mahan	Large	Disease resistant (largest of all pecan nuts)	Yes	6
Pecan, Mohawk	Large	Very thin shelled	Yes	6
Pecan, Major	Medium	Vigorous grower	No	5
Pecan, Peruque	Medium	Thin shelled	No	5
Pecan, Shawnee	Medium	Vigorous grower	Yes	6
Pecan, Stuart	Large	Scab resistant, highly productive	No	6
Pecan, Success	Large	Scab resistant, highly productive	No	6
Pecan, Western Chley	Large	Vigorous growth, highly productive	No	6
Pecan, Wichita	Large	Early maturity, highly productive	Yes	6
Walnut, Carpathian	Large	Vigorous growing, disease free	Yes	3
Walnut, Thomas Black	Medium	Vigorous growing, thin shelled	Yes	4

5.45 CONIFERS

Common Name	Botanical Name	Hardiness Zone	Height	Spread (ft)	Growth Shape	Foliage Rate	Texture	Unusual Characteristics
Arborvitae, Berckman's Gold	*Thuja orientalis* "Aurea Nana"	6	6–8'	4	Egg shape	Moderate	Soft	Dwarf egg-shaped evergreen, soft green foliage with bright golden tips
Arborvitae, Excelsa	*Thuja orientalis* "Excelsa"	6	10–12'	4	Pyramid	Rapid	Soft	Bright green foliage arranged in vertical planes, compact in habit
Arborvitae, Pyramidal	*Thuja occidentalis* "Pyramidalis"	3	20–30'	6	Pyramid	Moderate	Soft lacy	Soft lacy green foliage, minimal trimming required
Arborvitae, Emerald Green	*Thuja occidentalis* "Smaragd"	3	15–25'	4	Pyramid	Moderate	Soft lacy	Neat appearance, narrow pyramidal form, dense foliage, emerald green color
Arborvitae, Techni	*Thuja occidentalis* "Techney"	3	15–25'	6	Pyramid	Moderate	Soft lacy	Open pyramidal form, green foliage
Arborvitae, Woodwardi Globe	*Thuja occidentalis* "Woodwardii"	3	4–5'	4	Glove	Slow	Soft lacy	Small dense rounded evergreen, bright green foliage, slow growing
Juniper, Admiral	*Juniperus scopulorum* "Admiral"	4	20–25'	6	Pyramid	Moderate	Plumelike	Broad-based and bushy growth habit, gray-green color
Juniper, Ames	*Juniperus chinenses* "Ames"	4	4–6'	4	Pyramid	Slow	Medium	Dwarf dense pyramidal juniper, broad base, green color

Common Name	Botanical Name		Height		Habit	Growth Rate	Texture	Description
Juniper, Andorra	Juniperus horizontalis "Plumosa"	2	1–2'	10	Pyramid	Moderate	Plumelike	Low-growing, flat-top look, green foliage
Juniper, Andorra Youngstown	Juniperu horizontalis "Plumosa Youngstown"	2	1–2'	10	Low compact	Moderate	Plumelike	Densely branched prostrate grower, flattened branch structure
Juniper, Armstrong	Juniperus chinensis "Armstrong" Juniperus horizontalis	4	4'	4	Medium compact	Slow	Lacy	Medium-size, light green lacy juniper, dense branching habit
Juniper, Bar Harbor	Juniperus horizontalis "Bar Harbor"	3	1'			Moderate	Feathery	Feathery blue foliage, plum color in winter, good ground hugger
Juniper, Blue Chip	Juniperus horizontalis "Blue Chip"	4	1'	6	Ground cover	Moderate	Medium	Silver-blue low-mounding juniper, excellent ground cover
Juniper, Blue Danube	Juniperus sabina "Blue Danube"	3	2'	4	Low compact	Moderate	Soft	Semi-erect branching, greenish blue foliage, good spreader
Juniper, Blue Haven	Juniperus scopulorum "Blue Haven"	4	20'	5	Pyramid	Moderate	Medium	Neat compact narrow blue pyramid, likes exposed locations
Juniper, Blue Pacific	Juniperus scopulorum "Blue Pacific"	6	1–2'	6–8	Ground cover	Rapid	Medium	Blue foliage, superior plant, doesn't brown out during summer heat
Juniper, Blue Pfitzer	Juniperus chinensis "Pfitzertana Glauca"	4	5–6'	10	Upright spreader	Rapid	Medium	Spreading compact habit of growth, silvery blue foliage

(continued)

5.45 CONIFERS (cotninued)

Common Name	Botanical Name	Hardiness Zone	Height	Spread (ft)	Growth Shape	Foliage Rate	Texture	Unusual Characteristics
Juniper, Blue Point	*Juniperus chinensis* "Blue Point"	4	10–15'	4	Pyramid	Slow	Needlelike	Blue-gray foliage, dense pyramidal form, tolerant to heat and poor soil
Juniper, Blue Rug	*Juniperus horizontalis* "Wiltoni"	4	5'	10	Ground cover	Moderate	Medium	Dense short branchlets on long trailing branch, intense blue color
Juniper, Blue Sargent	*Juniperus chinensis* "Sargenti Glauca"	4	8–10'	6–8	Ground cover	Moderate	Medium	Broad-spreading, heavily branched, ground-bugging form, rich blue-green color
Juniper, Blue Star	*Juniperus squamata* "Blue Star"	4	3'	5	Irregular mounding	Slow	Needlelike	Dwarf form, irregular growth habit, mounding steel blue needlelike foliage
Juniper, Broadmoor	*Juniperus sabina* "Broadmoor"	3	1–2'	10	Mounding	Moderate	Soft	Low growing, attractive mounding, green foliage, takes full exposure
Juniper, Buffalo	*Juniperus sabina* "Buffalo"	3	1'	8	Ground cover	Moderate	Feathery	Very hardy, feathery branches create unusual spreading form, bright green color
Juniper, Burki	*Juniperus virginiana* "Burki"	4	10–20'	6–8	Pyramid	Moderate	Medium	Dense steel blue foliage, broad based, responds well to trimming, very hardy
Juniper, Calgary Carpet	*Juniperus sabina* "Calgary Carpet"	3	1'	6	Ground cover	Moderate	Soft	Attractive foliage is soft green, low-growth habit.

Juniper, Canaerti	Juniperus virginiana "Canaerti"	4	30'	8–10	Pyramid	Slow	Medium	Slow-growing, wide-based, rich dark green foliage with silver berries in the fall
Juniper, Cologreen	Juniperus scopulorum "Cologreen"	3	20'	6–8	Pyramid	Rapid	Medium	Compact upright grower, broad-based forest green foliage
Juniper, Columnaris	Juniperus chinensis "Hetzi Columnaris"	3	12–15'	4	Pyramid	Rapid	Medium	Rapid-growing, narrow-based pyramid, good green color
Juniper, Compact Irish	Juniperus communis "Hibernica"	6	12–20'	4–6	Pyramid	Rapid	Needlelike	Dark green foliage, very narrow base, erect grower
Juniper, Compact Pfitzer Nicks	Juniperus chinensis pfitzeriana "Compact Nicks"	4	2'	4–6	Medium compact	Moderate	Medium	Gray-green color, graceful branching, very hardy, compact in growth
Juniper, Dundee	Juniperus virginiana "Hilli"	4	10'	4–5	Pyramid	Slow	Soft	Blue-green foliage, compact habit of growth, changes color in fall
Juniper, Goldtip Pfitzer	Juniperus chinensis pfitzeriana "Aurea"	3	4'	4–6	Compact spreader	Rapid	Soft	Medium-size plant, greenish yellow foliage with golden yellow tips
Juniper, Gray Gleam	Juniperus scopulorum "Gray Gleam"	4	10–20'	4–6	Pyramid	Slow	Soft	Very dense columnar with gray-blue foliage
Juniper, Hetzi	Juniperus chinensis "Hetzi Glauca"	4	15'	15	Medium	Rapid	Medium	Rapid growing, medium to large plant, bright blue-green foliage

(continued)

5.45 CONIFERS (cotninued)

Common Name	Botanical Name	Hardiness Zone	Height	Spread (ft)	Growth Shape	Foliage Rate	Texture	Unusual Characteristics
Juniper, Hollywood	*Juniper chinensis* "Torulosa"	6	15'	4–5	Upright	Moderate	Medium	Rich green foliage, upright with twisted branching, fine accent plant
Juniper, Hughes	*Juniperus horizontalis* "Hughes"	4	2'	6–8	Medium compact	Moderate	Feathery	Blue-green foliage, requires little shearing, good for ground cover and slopes
Juniper, Keteleeri	*Juniperus chinensis* "Keteleeri"	5	15–20'	4–5	Pyramid	Moderate	Scalelike	Vigorous pyramid with compact ascending branches, bright green foliage
Juniper, Medora	*Juniperus scopulorum* "Medora"	3	20'	6–8	Pyramid	Moderate	Medium	Requires very little trimming, silver blue, upright
Juniper, Montana Green	*Juniperus scopulorum* "Montana Green"	3	10–15'	6	Pyramid	Moderate	Soft	Gray-green foliage, tends to have a wide base with age
Juniper, Mount Batten	*Juniperus scopulorum* "Moutana Green"	3	12'	6–8	Pyramid	Moderate	Needlelike	Gray-green in color, broad columnar with needlelike foliage
Juniper, Old Gold	*Juniperus chinensis* "Armstrong Aurea"	4	4'	4	Compact spreader	Slow	Medium	Rich gold coloring, dense compact growth, excellent mid-size plant
Juniper, Parsoni	*Juniperus squamata* "Parsoni"	4	1'	3–4	Ground cover	Rapid	Soft	Grows full and thick, hugging fashion, lovely gray-green foliage

Common Name	Botanical Name		Height		Form	Growth Rate	Texture	Description
Juniper, Pathfinder	*Juniperus scopulorum* "Pathfinder"	3	15–20'	3	Pyramid	Rapid	Medium	Showy blue-green foliage, likes full sun and low humidity, desirable for dry soil
Juniper, Pfitzer	*Junifer chinensis* "Pfitzeriana"	3	5'	12	Large spreader	Rapid	Medium	Rapid-growing, open-spreading shrub of medium height
Juniper, Prince of Wales	*Juniperus horizontalis* "Prince of Wales"	3	8'	8–10	Ground cover	Rapid	Soft	Bright green foliage snakes over ground or cascades over walls and ledges
Juniper, Procumbens Green Mound	*Juniperus procumbens* "Greenmound"	4	8'	6	Ground cover	Slow	Needlelike	Ground cover, light green foliage, short needlelike foliage, likes full sun, compact
Juniper, Procumbens	*Juniperus procumbens*	4	1–2'	6	Ground cover	Moderate	Needlelike	Gray-green in color, feathery foliage, on strong spreading branches
Juniper, Procumbens Dwarf	*Juniperus procumbens* "Nana"	4	1'	4–5	Ground cover	Slow	Needlelike	Low spreader with heavy compact ascending branches
Juniper, Variegated Procumbens	*Juniperus procumbens* "Variegata"	6	1–2'	6	Ground cover	Moderate	Needlelike	Low spreading with gray-green foliage, ascending branches
Juniper, Prostrate	*Juniperus communis* "Prostrata"	5	18'	6	Ground cover	Moderate	Scalelike	Stiff branching with dark green scalelike foliage

(continued)

5.45 CONIFERS (cotninued)

Common Name	Botanical Name	Hardiness Zone	Height	Spread (ft)	Growth Shape	Foliage Rate	Texture	Unusual Characteristics
Juniper, Pyramidalis Green	*Juniperus scopulorum* "Pyramidalis"	4	15–30'	6	Pyramid	Moderate	Needlelike	Foliage narrow pyramid, broadening with age
Juniper, San Jose	*Juniperus japonics* "San Jose"	4	1–2'	6–8	Ground cover	Moderate	Needlelike	Thickly branched full-growing, excellent ground cover and bonsai training
Juniper, Scandia	*Juniperus sabina* "Scandia"	3	18'	8	Low spreader	Rapid	Feathery	Good green color year-round, attractive low-growing spreader
Juniper, Sea Green	*Juniperus chinensis* "Sea Green"	4	4'	6	Vase	Moderate	Medium	Rich green coloring, arching branches, great eye-appeal plant
Juniper, Shore	*Juniperus conferta*	6	1'	6–8	Ground cover	Rapid	Medium	Excellent prostrate juniper, does well in poor, sandy soil, bluish green foliage
Juniper, Sneedi	*Juniperus scopulorum* "Sneedi"	4	15–20'	4–6	Pyramid	Rapid	Medium	Columnar and rather coarse growing foliage, compact gray-blue
Juniper, Spartan	*Juniperus chinensis densaerecia* "Spartan"	4	15–20'	6	Pyramid	Rapid	Needlelike	Rapid growth with a densely branched tail, pyramidal form, rich green foliage

Common Name	Botanical Name		Height	Shape	Growth Rate	Texture	Description
Juniper, Spiney Green	Juniperus excelsa stricta	5	10–15'	Pyramid	Rapid	Spiney	Short cone-shaped evergreen, dense gray-green foliage, erect branching
Juniper, springtime	Juniperus scopulorum "Springtime"	4	15–20'	Pyramid	Moderate	Soft	Good dark green color, very dense, pyramidal
Juniper, Sutherland	Juniperus scopulorum "Sutherland"	4	20'	Pyramid	Moderate	Medium	Strong-growing broad upright juniper, silver-green in color, requires light trimming
Juniper, Tamarix	Juniperus sabina "Tamariscifolia"	4	2'	Mounding	Moderate	Medium	Medium spreading ground cover, good green color year round
Juniper, Von Ehron	Juniperus sabina "Von Ehron"	3	6–8'	Vase	Rapid	Soft feathery	Rapid growing evergreen with loose upright branching, greenish color
Juniper, Welchi	Juniperus scopulorum "Welchi"	3	15–20'	Pyramid	Rapid	Soft	Dense growing, silver-green foliage, outstanding narrow upright
Juniper, Whicita Blue	Juniperus acopulorum "Wichita Blue"	4	20'	Pyramid	Moderate	Medium	Bright blue foliage, wide-based, upright
Cedar, Deodara	Cedrus deodara	7	40–60'	Upright	Moderate	Fine-needle	Dense-growth habit, widely spreading branches that droop at the tips, needles 1½" long, green to blue in color

(continued)

5.45 CONIFERS (cotninued)

Common Name	Botanical Name	Hardiness Zone	Height	Spread (ft)	Growth Shape	Foliage Rate	Texture	Unusual Characteristics
Pine, Austnan	*Pinus Nigra*	3	90'	2.5	Upright	Moderate	Needle	Grows practically anywhere, dark green needles 4–6" long, withstands adverse conditions, ideal for windbreak or as a specimen tree
Pine, Japanes Black	*Pinus Thunbergi*	5	80–100'	25	Upright	Moderate	Needle	Rugged look, irregular branching, good specimen tree, picturesque
Pine, Loblolly	*Pinus taeda*	6	80–100'	30	Upright	Rapid	Needle	Soft light green 6–9" needles, winter color is pale green to yellow brown
Pine, Mugho	*Pinus mugo mighus*	3	6–8'	10–15	Mounded	Slow	Needle	Bright green dwarf mounded pine, good in garden type landscape
Pine, Ponderosa	*Pinus ponderosa*	3	100–125'	40	Upright	Slow	Needle	Gray-green, 6–16" long needles, open growing, adapts to adverse conditions
Pine, Scotch	*Pinus sysvestris*	3	75'	25	Upright	Moderate	Needle	Irregular shaped, needles 2–3" long and twisted, orange-brown bark, resists drought

Common Name	Botanical Name	Zone	Height	Spread	Habit	Growth	Needle	Description
Pine, Slash	*Pinus elliotti*	7	80–100'	30–40	Upright	Rapid	Needle	Narrowly pyramidal when young, more oval with age, needles 10" long, medium green
Pine, White	*Pinus strobus*	3	80–100'	25–30	Upright	Moderate	Soft needle	Soft green graceful tree, when young it is pyramidal in habit, but as it grows older becomes open and picturesque
Spruce, Alberta Dwarf	*Picea glauca* "Conica"	3	6'	2	Cone	Slow	Small needle	Slow, dense conifer that grows in a cone shape without shearing, new growth is bright green, may take 10 years to reach 3" in height
Spruce, Black Hills	*Picea glauca densata*	3	60'	25	Upright	Slow	Needle	Slow-growing, short-needled tree, blue-green in color
Spruce, Colorado Blue	*Picea pungens glauca*	3	100'	30	Upright	Slow	Needle	Most popular spruce, often steel blue in color, desirable specimen
Spruce, Nest	*Piceas abies* "Nidiformis"	3	2'	2	Semi-globe	Slow	Needle	Main branches arch outward, leaving a central depression resembling a bird's nest, slow growing, ideal for rock gardens

(continued)

5.45 CONIFERS (cotninued)

Common Name	Botanical Name	Hardiness Zone	Height	Spread (ft)	Growth Shape	Foliage Rate	Texture	Unusual Characteristics
Yew, Dark Green Spreader	*Taxus x media* "Dark Green Spreader"	3	3-4'	4-5	Medium spreader	Slow	Needle	Hardiest variety grown, ¼" dark green needles, excellent foundation plant
Yew, Densiformis	*Taxus x media* "Densiforms"	3	3-4'	4-5	Medium spreader	Slow	Needle	Most versatile and popular yew, ¾" needles excellent foundation plant
Yew, Hicksi	*Taxus x media* "Hicksi"	3	6'	4	Upright	Slow	Needle	Upright growing yew, deep green needles, bright red berries in fall, extremely hardy in north, good for hedging
Yew, Browni	*Texus x media* "Brownii"	3	3'	4	Globe	Slow	Needle	Compact dense with medium dark green needles, can be sheared to a formal globe shape, very hardy

5.46 BROADLEAVES

Common Name	Botanical Name	Hardiness Zone	Height	Shape	Growth Rate	Foliage Texture	Fall Foliage	Fruit	Unusual Characteristics
Abelia, Grandiflora	Abelia grandiflora	6	4–8'	Compact	Rapid	Medium	None	None	Semigreen, dark purple flowers in spring
Aucuba, Gold Dust Plant	Aucuba japonica "Variegata"	6	6'	Compact	Rapid	Smooth	Berries	Red berries	Golden spotted, dark green glossy leaves
Aucuba, Picturata	Aucuba japonica "Picturata"	7	6'	Compact	Rapid	Smooth	None	None	Deep green leaves with golden yellow center
Azalea, Exbury	Azales exburry hybrids	Varies	Varies	Upright	Moderate	Medium-fine	None	None	Extremely winter hardy, deciduous
Azalea, Kurume	Azalea kurume	6	2–4'	Shrubby	Moderate	Leathery	None	None	Evergreen
Boxwood, Japanese	Buxus japonica	7	4'	Shrubby	Slow	Smooth	None	None	Very effective hedge
Boxwood, Korean	Buxus microphylla "Koreana"	5	18–24'	Compact	Moderate	Smooth	None	None	Excellent hedge plant
Boxwood, Winter Gem	Buxus microphylla asiaticum "Winter Gem"	6	2–3'	Oval	Moderate	Smooth	None	None	Excellent hedge plant
Elaeagnus, Fruitlandi (silverberry)	Eleagnus pungens "Fruitlandi"	7	10–12'	Upright	Rapid	Coarse	Fragrant flowers	None	Dainty fragrant flowers in fall
Euonumus, Boxleaf	Euonumus japonica "Microphylla"	6	2'	Compact	Moderate	Smooth	None	None	Excellent hedge or border

(continued)

5.46 BROADLEAVES (continued)

Common Name	Botanical Name	Hardiness Zone	Height	Shape	Growth Rate	Foliage Texture	Fall Foliage	Fruit	Unusual Characteristics
Euonumus, Coloratus	Euonumus fortunei "Colorata"	4	6'	Low mounding	Rapid	Coarse	Plum color	None	Good ground cover, plum color foliage in fall
Euonumus, Emerald Gaiety	Euonymus fortunei "Emerald Gaiety"	5	4'	Erect	Moderate	Smooth	Purple	None	White margin on green rounded foliage
Euonumus, Emerald 'N Gold	Euonymus fortunei "Emeralds 'N Gold"	5	3'	Mounding	Medium	Smooth	None	None	Deep green foliage with small green center
Euonumus, Golden	Euonumus japonica "Microphylia" "Aureo Marginata"	6	5'	Upright	Moderate to rapid	Smooth	None	None	Large golden foliage with small green center
Euonumus, Gold Spot	Euonumus japonica "Aureo Variegata"	6	3–4'	Upright	Moderate to rapid	Smooth	None	None	Large green foliage, gold in center
Euonumus, Green	Euonumus japonica	6	5"	Upright	Moderate to rapid	Smooth	None	None	Excellent hedge
Euonumus, Green Crusher	Euonumus fortunei "Green Crusher"	4	5'	Upright	Moderate to rapid	Coarse	None	None	Similar to sarcoxie in appearance, lighter colored foliage
Euonumus, Manhattan	Euonumus patens "Manhattan"	6	5–6'	Upright	Rapid	Glossy	None	None	One of the best semi-evergreen shrubs
Euonumus, Panlti	Euonumus patens "Pauli"	5	5–6'	Upright	Rapid	Smooth	None	None	Hardiest of patens varieties
Euonumus, Sarcoxie	Euonumus fortunei "Sarcoxie"	5	3'	Upright	Vigorous	Glossy	None	None	Hardy and useful as a foundation shrub

Common Name	Botanical Name	Zone	Height	Form	Growth	Texture	Feature	Fruit	Description
Euonumus, Sheridan Gold	*Euonumus fortunes* "Sarcoxie" "Sheridan Gold"	4	3'	Spreading	Moderate	Smooth	None	None	Deep green foliage with yellow color planted in full sun
Euonumus, Silver King	*Euonumus japonica* "Silver King"	6	8'	Upright	Rapid	Leathery	None	None	Large green foliage edged in creamy white
Euonumus, Vegetus	*Euonumus fortunei* "vegetus"	4	2'	Vine or semi-shrub	Moderate	Coarse	Orange fruits	None	Pink fruit capsules open in fall to reveal orange fruit
Grass, Mondo	*Ophiopogon japonicum*	6	10"	Clumpy	Fast	Slow	Grasslike	None	Used as bed edging and a ground cover
Grass, Pampas Pink	*Cortaderia selloana* "Rosia"	6	10'	Upright	Fast	Rapid	Serrated leaves	None	Produces beautiful pink plumes in fall
Grass, Pampas White	*Cortadena selloana*	6	10'	Upright	Fast	Rapid	Serrated leaves	None	Produces white plumes in fall
Hawthorn, Indian	*Raphiolepis indica*	7	5'	Upright	Moderate	Leathery	None	None	White to pink flowers in summer
Holly, Blue Prince P.P. #3517	*Iiex x meserveae* "Blue Prince"	5	15'	Pyramidal	Rapid	Crinkled	None	None	Male pollinator for the two female varieties
Holly, Blue Angel P.P. #3662	*Iiex x meserveae* "Blue Angel"	5	6–8'	Shrubby	Rapid	Crinkled	Berries	Red berries	Hardy, vigorous, dense branching
Holly, Blue Princess P.P. #3675	*Iiex x meserveae* "Blue Princess"	5	15'	Pyramidal	Rapid	Crinkled	Berries	Red berries	Excellent for a high, dense hedge
Holly, China Boy P.P. #4803	*Iiex x meserveae* "China Boy"	5	6–8'	Upright	Rapid	Glossy	None	None	Shears and shapes well
Holly, China Girl P.P. #4878	*Iiex x meserveae* "China Girl"	5	6–8'	Upright	Rapid	Glossy	None	Red berries	Shears and shapes well

(continued)

5.46 BROADLEAVES (continued)

Common Name	Botanical Name	Hardiness Zone	Height	Shape	Growth Rate	Foliage Texture	Fall Foliage	Fruit	Unusual Characteristics
Holly, Burford	*Ilex x cornuta* "Burfordi"	6	6'	Upright	Moderate	Glossy	None	Red berries	Deep glossy foliage
Holly, Dazzler	*Ilex cornuta* "Dazzler"	6	8'	Upright	Moderate	Spiney	None	Red berries	Good hedge or specimen
Holly, Dwarf Burford	*Ilex cornuta* "Burfordi Nana"	6	3'	Upright	Slow	Glossy	None	Red berries	Smaller leaves than Burford
Holly, Dwarf Chinese	*Ilex cornuta*	6	3–4'	Mounding	Medium	Spiney leaves	None	None	Mounding with spiney, glossy green foliage
Holly, Dwarf Yaupon	*Ilex vomitoria* "Nana"	6	2–3'	Compact	Slow	Smooth	None	None	Compact with small gray-green foliage
Holly, Fosteri	*Ilex opaca (attenuata)* "Fosteri"	6	20'	Pyramidal	Medium	Smooth	None	Large red berries	Fruits heavily with male
Holly, Green lustre	*Ilex crenata* "Green Lustre"	6	4'	Compact	Medium	Shiny	None	None	Ideal for specimen or foundation plantings
Holly, Helleri	*Ilex crenata* "Helleri"	6	3'	Compact	Slow	Smooth	None	None	Tiny dark green foliage tip ⅜" long
Holly, Japanese Compact	*Ilex crenata* "Compacta"	6	3–4'	Oval	Moderate	Glossy	None	Blue-black berries	Minutes shiny green foliage
Holly, Japanese Convex	*Ilex crenata* "Convexa"	6	4–6'	Mounding	Moderate	Smooth	None	None	Round leaves that cup downward
Holly, Japanese Hetzi	*Ilex crenata* "Hetzi"	6	5–6'	Upright	Vigorous	Glossy	Berries	Black berries	Dark green convex foliage

Common Name	Botanical Name		Height	Habit	Growth	Texture			Notes
Holly, Needlepoint	*Ilex cornuta* "Needlepoint"	6	10'	Upright	Moderate	Lustrous	Berries	Dark red berries	Long narrow dark green foliage
Holly, Nellie R. Stevens	*Ilex hybrid* "Nellie R. Stevens"	6	8–10'	Pyramidal	Rapid	Smooth	Berries	Red berries	Early fall berries
Holly, Rotundifolia	*Ilex crenata* "Rotundifolia"	6	3'	Upright	Moderate	Glossy	None	None	Round foliage to ¼" long
Ivy, English	*Hedera helix*	5		Spreading vine or mounding	Rapid	Leathery	None	None	Favorite ground cover
Jasmine, Carolina	*Gelsemium sempervirens*	7			Rapid	Smooth	None	None	Very showy, yellow fragrant flower in early spring
Jasmine, Nudiflorum	*Jasminum nudiflorum*	7		Viney	Moderate	Smooth	None	Yellow flowers	Use as a trailing or cascading planting
Ligustrum, Waxleaf	*Ligustrum texanum*	7	8–10'	Upright	Rapid	Glossy	None	None	White flowers in spring
Liriope, Green	*Liriope muscari*	6	18"	Clumpy	Fast	Moderate	Grasslike	None	Excellent for borders, purple flowers
Liriope, Variegated	*Liriope muscari* "Variegata"	6	18"	Clumpy	Fast	Moderate	Grasslike	None	Foliage is yellow-striped on outer margin, purple flowers
Nandina, Dwarf Purpurea	*Nandina domestica* "Nana Purpurea"	6	2'	Mounding	Slow	Smooth	Scarlet	None	Excellent for planter boxes
Nandina Domestica	*Nandina domestica*	6	5–6'	Compact	Rapid	Lacy	Red	Red berries	Foliage turns several shades of red in fall
Nandina, Harbour Dwarf	*Nandina domestica* "Harbour Dwarf"	6	18–20"	Mound	Moderate	Lacy	Red	Red berries	Red foliage in fall
Photinia, Freseri	*Photinia fraseri*	6	12–15'	Upright	Moderate	Glossy	Red	None	New foliage

(continued)

5.46 BROADLEAVES (continued)

Common Name	Botanical Name	Hardiness Zone	Height	Shape	Growth Rate	Foliage Texture	Fall Foliage	Fruit	Unusual Characteristics
Pittosporum, Tobira	Pittosporum tobira	8	6–10'	Upright	Moderate	Leathery	None	White to yellow flowers	Best for screens, massing or individual as free-standing small tree
Pittosporum, Variegated	Pittosporum tobira "Variegata"	8	4–5'	Upright	Moderate	Leathery	None	None	Makes an attractive trimmed hedge
Pittosporum, Wheeler's Dwarf	Pittosporum tobira "Wheeler's Dwarf"	8	4–5'	Compact	Moderate	Leathery	None	None	Fragrant white flowers
Pyracantha, Kasan	Pyracantha coccinea "Kasan"	5	6–10'	Upright	Rapid	Smooth	Berries	Orange-red Berries	Retain fruit for a long period, white flower in spring
Pyracantha, Lelandi	Pyracantha coccinea "Lelandi"	5	12–15'	Upright	Rapid	Smooth	Berries	Orange-red berries	Strong growing
Pyracantha, Mohave	Pyracantha, hybrid "Mohave"	6	8–10'	Upright	Rapid	Smooth	Berries	Orange-red berries	Produces orange-red berries in the fall
Pyracantha, Victory	Pyracantha koidzumni "Victory"	7	6–10'	Upright	Rapid	Smooth	Berries	Orange-red berries	Fire blight-resistant, produces orange-red berries in fall
Pyracantha, Wyatti	Pyracantha coccinea "Wyatti"	5	6'	Mounding	Moderate	Smooth	Berries	Orange berries	Foundation and hedge plantings, produces orange berries in fall
Rhododendron	Rhododendron	Varies	2–4'	Upright	Moderate	Leathery	None	None	Acid soil and good drainage are required
Yucca, Adam's Needle	Yucca filamentosa	4	12'	Upright	Moderate	Spiney leaves	None	None	Produces bold, stiff, swordlike rosettes

Section Six

SPORTS FIELDS

6.1   FIELD SPACES

Sports fields represent the largest area of land dedicated to recreation facilities. Issues surrounding sports field development are as follows:

- Appropriate available acreage.
- Lighting for playing field or security.
- Proximity to amenities, recreation center, concessions, bathrooms, and so on.
- Field design based on type and level of use.
- Size of field should always include a minimum of 10-yard safety buffer around perimeter. ·
- No streets, railroad tracks, waterways, trenches, settlement ponds, or storage yards should be located nearer than 100 yards to a facility or field.
- Excessive noise, smoke, odor, and dust should be avoided.
- Perimeter should be fenced or landscaped to keep players and spectators separated.
- Games of softball and baseball have three primary cautions (foul balls, home runs, and overthrows).

6.2   TOLERANCES FOR SELECTING THE BEST TURFGRASS FOR YOUR NEEDS (COOL AND WARM SEASON)

Selecting a turfgrass successfully requires knowing how the turf will be used, where it will be grown, and what appearance and maintenance level will be acceptable. Because each turfgrass species has good and bad features, one must learn the strengths and weaknesses of each of the species in order to choose the one best suited to a particular situation.

The following lists rank common turfgrass species according to important characteristics and requirements and their relation to each other. Within a category, a given grass may differ little from the one listed immediately above or below it; it may, however, differ greatly from one further up or down the list. The precise position of a turfgrass in a list may change slightly as more is learned about it or improved varieties are developed, but its location (high, low, or intermediate) is not likely to change and, therefore, can be usefully reviewed when preparing to plant.

The "warm-season" turfgrasses listed—Bermuda grass (common and hybrid), dychondra, kikuyu grass, seashore paspalum, St. Augustine, and zoysia grass—generally lose their green color and are dormant in winter if the average air temperature drops below 50 to 60°F (10 to 15.5°C). Some may die if exposed to subfreezing temperatures for extended periods.

The "cool-season" turfgrasses—bent grass, bluegrass, ryegrass, tall fescue, and weeping alkaligrass—ordinarily do not lose their green color unless the average air temperature drops below 32°F (0°C) for an extended period; they turn green again as soon as temperatures rise above freezing and are not usually damaged by subfreezing temperatures.

The following are the various criteria and factors used to evaluate the different turfgrasses:

Genetic color. Shades of color range from light green to dark green.

Leaf texture. Describes the width of the leaves. Coarser-textured grasses such as centipede grass, have a tougher "feel" and are not as thin or fine as Perennial Ryegrass or Fine Fescues.

Density. This measures the amount of grass plants in an area. The larger number of plants per unit area, the greater the density.

Disease/insect resistance. This relates to the turfgrass' ability to resist disease and insect attacks. Some plants are more resistant to these pests because of their genetic makeup due to naturally occurring fungi known as endeophytes. They live within certain tall fescue, fine fescue, and perennial ryegrass varieties, helping these grasses ward off insects. Some of these endophytes can be harmful to grazing animals; therefore, be sure to inquire as to the use of open areas.

Drought resistance. This is the ability of the turfgrass to survive and/or thrive during drought conditions. The most desirable grass will maintain its green color and good quality during prolonged drought. However, for basic survival during drought, grasses often lose their green color and go dormant.

Heat and cold tolerance. This is the observation of the turfgrass ability to survive extreme winter and summer temperatures. Cold tolerance ratings can be misleading for it is often the sudden change in temperature that causes winterkill not the actual temperature.

Rate of establishment. This is the evaluation of how quickly the turfgrass produces 100% ground cover. This is especially important in resisting weed invasion, controlling erosion, and recovering from disease or insect damage.

Shade tolerance. Trees in the landscape area are inevitable; subsequently, we need species that can survive and thrive in shaded areas.

Traffic and wear tolerance. This characteristic is important in parks, athletic fields, golf courses, playgrounds, and home lawns.

Thatch production. Some grass types produce thatch, or dead roots and stems (not from leaf clippings), faster than soil microorganisms can decompose the thatch. Thatch is a place for diseases and insects to thrive and it prevents water from reaching the turfgrass' roots.

Grasses are divided into warm- and cool-season species. Cool-season grasses are generally grown in areas where frosts and freezing temperatures are routine for portions of the year, while warm-season grasses are grown in regions where mild to hot weather predominates. Buffalo grass, zoysia grass, and creeping bent grass are exceptions in that they are often grown successfully beyond traditional boundaries.

Cool-Season Species

Kentucky bluegrass is the most popular cool-season grass with its medium-textured, green to dark green turf of good density. The aggressive sod-forming habit of bluegrass is attributable to its strong rhizome development, lateral spreading potential, and excellent recovery potential. It has fair high-temperature tolerance and good-to-excellent cold-temperature tolerance. Some cultivars are pest susceptible.

Turf type tall fescue has the coarsest texture of any cool-season grass. The recuperative potential is quite low, as it is a bunch grass and does not spread laterally. Tall fescue has a very extensive root system that is used to draw on soil moisture reserves and resist insect feeding damage. Thus, it is considered pest resistant and drought tolerant. It has fair cold-temperature tolerance and good-to-excellent heat tolerance.

Fine-leaf fescue is made up of the species Hard Fescue, Sheep Fescue, Creeping Red Fescue, and Chewings Fescue. Although there are minor differences, they are grouped together for practicality. Except for Creeping Red Fescue, these are bunch grasses and do not spread significantly. They are medium dark green and exhibit good-to-excellent shade tolerance, and are predominately mixed with shade-tolerant cultivars of bluegrass for use in turf areas that receive three to six hours of sun per day.

Perennial ryegrasses are also commonly mixed with Kentucky bluegrass with wear tolerance and quick establishment as the desired results. Ryegrasses are shiny, medium to dark green, and fine to medium in texture. They germinate rapidly, making them useful in sports turfs. They have fair cold and warm temperature tolerance. Like the fescues, they are bunchgrasses and do not spread laterally.

Creeping bent grass is a very low, very fine textured grass. It has good cold-temperature tolerance and fair heat tolerance. It spreads readily through rhizomes. Bent grasses are susceptible to a wide range of fungal diseases, and along with a requirement for very frequent mowing, it is considered a high-maintenance grass.

Warm-Season Species

Bermuda grass is one of the most widely used warm-season species due to its many uses. Common Bermuda grass is a bit coarse, having medium texture, while improved seeded types of Bermuda grass are medium fine. Cold-temperature tolerance is poor to moderate, while heat tolerance is good. All Bermuda grasses are aggressive spreaders, giving it excellent recuperative potential and rapid establishment.

Zoysia grass is similar to Bermuda grass in that it has rhizomes and stolons, but it differs with regard to cold tolerance, stiffness, and growth rate. Some zoysia species have good cold-temperature tolerance, and all have good-to-excellent heat tolerance. The growth rate is slower, and is quite stiff and tough compared to other grasses. The texture is medium, and the color is medium green in color. Zoysia grass has a deep root system, allowing it to avoid drought stress in many situations. Zoysia grass is low growing and tolerates a low mowing height.

Centipede grass is a low-maintenance, medium-coarse grass, which spreads by short leafy stolons and forms a mat of low-growing stems and leaves. Centipede grass is light to medium green color. The cold-temperature hardiness is quite poor, while the heat tolerance is good. The shade tolerance is intermediate but better than Bermuda or zoysia grass. The recuperative potential is poor, due to its slow growth rate and spreading ability.

Buffalo grass is one of the few turfgrasses native to the United States; it is the low-maintenance warm-season turf, adapted to the central part of the United States from Texas to Minnesota, and Colorado to Illinois. It is gray-green in color and possesses a medium-fine texture. Improved cultivars provide turf type medium density. The cold and heat tolerance is good to excellent, while pest resistance is excellent. It requires no irrigation once established and has a very low fertility requirement. It is slow growing and slow to establish, which limits its ability to recuperate from stress or injury.

Seashore paspalum (SeaIsle 1) is a warm-season variety developed for coastal areas that encounter high levels of salt either from water or soils.

Two major injuries can occur at high soil or water salinity level: drought stress brought on by the plant's inability to uptake water, and injury to root and shoot tissues due to high salt levels. Based on tests on shoot and root growth responses to increasing salinity levels, SeaIsle 1 showed a minimal decline in rooting and maintained a high level of growth when compared to Adalayd and Tiftway 419. SeaIsle 1 is the most salt tolerant of the warm-season grasses.

Drought resistance involves the ability to produce extensive, deep root systems. A 1997 study of seven warm-season turfgrasses comparing responses to surface soil drought stress and rewatering demonstrated the SeaIsle1 and TifBlair centipede had superior drought resistance. Both exhibited enhanced root growth, rapid root-water uptake at deep soil levels, excellent soil surface root viability, and excellent root regeneration after rewatering.

SeaIsle 1 does quite well in the southern transition zone (between 30 to 35 degrees N-S latitudes) and has a cold tolerance similar to hybrid Bermudas. Fine-textured paspalums like SeaIsle 1 are generally the last warm-season grasses to go off color in the fall by two or three weeks, and it normally takes temperatures of at least 28°F (–2.2°C) for them to go into full winter dormancy. They are also less responsive to mild mid-winter and early-spring temperature swings.

On golf courses, wear injuries usually result from physical abrasions and torn plant tissue caused by maintenance vehicles, golf carts, and incoming balls. On athletic fields, compaction, as well as cleat and spike injuries, are the most common problems. A 1998 wear simulation study involving seven paspalums and three Bermudas revealed that, in general, the fine-textured paspalums had less or similar injury from wear than Bermudas of similar texture.

SeaIsle 1, like most other warm-season grasses, does not tolerate tree shade very well and should not be planted in areas with thick canopies or heavy shade. However, in rainy, low-light environments or cloudy, foggy, or smoggy conditions, SeaIsle 1 performs exceptionally well and maintains a dark green color and good turf quality versus elongated, spindly leaves that are typical of Bermuda grass response to the same reduced light conditions.

Paspalums in general will tolerate pH ranges from a very acidic pH of 3.5 to a highly alkaline pH of 10.2, and they root equally well in sands, heavy clays, silts, and mucks. In acid soil field trials, SeaIsle 1 placed in the "high-tolerance" group, while SeaIsle 2000, the fine-bladed greens-type paspalum cultivar, scored in the "very high tolerance" group. In extremely alkaline soils, paspalums are usually the only turf type grass species that can survive.

Paspalums have a history of tolerating complete ocean water inundations and the low-oxygen problems associated with waterlogged or wet and boggy environments. That's why paspalums are so ideally suited for low-lying or poorly drained areas that tend to stay wet for long periods of time. In fact, paspalums may actually help firm up such problem areas through water extraction, so that sports and other recreational activities can take place.

SeaIsle 1 should be maintained at ½ in. for fairways and ¾ in. for tees. At these lower mowing heights, it develops higher shoot densities, is more competitive against broadleaf and annual grass weed invasions, and maintains better turf quality. Maintaining the cutting heights at or below ½ in. is more important than N fertility levels in terms of low-temperature color retention in the fall as well as grass color during spring green up.

Paspalums do not have a wide variety of disease problems that tend to plague other warm-season grasses. More than likely this is because the grass evolved in a wet, humid ecosystem where it developed resistance. When irrigating with brackish or straight seawater, disease problems are negligible, since most diseases do not function well at high salt levels.

The paspalums developed and thrived in stressful ocean-exposed ecosystems. This helps explain how paspalum developed such efficient nutrient uptake and utilization mechanisms. Paspalum will grow when the availability of nutrients is quite low, as well as in the situations with severe nutrient imbalances. Because paspalum is so efficient in its uptake of nitrogen, less than 5 pounds of N per 1,000 SF is recommended via spoon-feeding on an annual basis.

Most weeds lack the necessary level of salt tolerance to compete with paspalum in salt-affected turfgrass environments. A close mowing height of less than 13 mm (½ in.) will also provide a tight, dense canopy that deters weed growth. The exceptions include those grassy weeds like kikuyu grass, torpedo grass, and kyllinga, which are extremely aggressive in warm coastal venues, wet soil conditions, and aquatic waste sites.

Turfgrass Characteristics

Warm-Season Turfgrasses

> Optimum temperature: 85 to 90°F
>
> Planting: May, June, July
>
> Mowing: May to September
>
> Watering: Summer
>
> Fertilizing: May to October

Cool-Season Turfgrasses

> Optimum temperature: 60 to 75°F
>
> Planting: September
>
> Mowing: April to November
>
> Watering: Spring to Fall
>
> Fertilizing: May to November

		Culture		
	Mowing Requirement	Irrigation Requirement	Nitrogen Requirement	Thatch
Bermuda grass	Medium	Low	Medium	Moderate
Buffalo grass	Low	Low	Low	Light
Zoysia grass	Low	Low	Low	Light
Creeping bent grass	High	High	High	Heavy
Kentucky bluegrass	Medium	High	Medium	Moderate
Perinnial ryegrass	Medium	High	Medium	Light
Tall fescue	High	Medium	Medium	Light

	Soil pH	
	Tolerance Range	*Preferred*
Bermuda grass	5.5–7.5	6.0–7.0
Buffalo grass	6.0–8.0	6.0–7.0
Zoysia grass	5.0–7.8	6.0–6.5
Annual ryegrass	6.0–7.5	6.0–7.0
Creeping bent grass	5.5–7.2	5.5–6.5
Kentucky bluegrass	6.0–7.2	6.0–7.0
Perennial ryegrass	6.0–7.5	6.0–7.0
Tall fescue	4.7–8.5	5.5–6.5
Fine fescue	5.5–7.0	5.5–6.5

General Turf Adaptations and Tolerances

Texture (leaf blade width)
Coarse

Dichondra
St. Augustine
Kikuyugrass
Tall Fescue
Common Bermuda
Zoysiagrass
Kentucky Bluegrass
Perennial Ryegrass
Seashore Paspalum
Highland Bentgrass
Weeping Alkaligrass
Colonial Bentgrass
Hybrid Bermudagrass
Creeping Bentgrass
Red Fescue

Fine

Heat Tolerance
High

Zoysiagrass
Hybrid Bermuda
Common Bermuda
Seashore Paspalum
St. Augustine
Kikuyugrass
Tall Fescue
Dichondra
Creeping Bentgrass
Kentucky Bluegrass
Highland Bentgrass
Perennial Ryegrass
Colonial Bentgrass
Weeping alkaligrass
Red Fescue

Low

Cold Tolerance
High

Creeping Bentgrass
Kentucky Bluegrass
Red Fescue
Colonial Bentgrass
Highland Bentgrass
Perennial Ryegrass
Tall Fescue
Weeping alkaligrass
Dichondra
Zoysiagrass
Common Bermuda
Hybrid Bermuda
Kikuyugrass
Seashore Paspalum
St. Augustine

Low

Drought Tolerance
High

Hybrid Bermuda
Zoysiagrass
Common Bermuda
Seashore Paspalum
St. Augustine
Kikuyugrass
Tall Fescue
Red Fescue
Kentucky Bluegrass
Perennial Ryegrass
Highland Bentgrass
Creeping Bentgrass
Colonial Bentgrass
Weeping alkaligrass
Dichondra

Low

Recovery from Moderate Wear
Fast

Hybrid Bermuda
Kikuyugrass
Common Bermuda
Seashore Paspalum
Tall Fescue
Perennial Ryegrass
St. Augustine
Kentucky Bluegrass
Dichondra
Highland Bentgrass
Creeping Bentgrass
Red Fescue
Weeping alkaligrass
Zoysiagrass
Colonial Bentgrass

Slow

Mowing Height Adaptation
High

Tall Fescue
Red Fescue
Kentucky Bluegrass
Perennial Ryegrass
Weeping alkiligrass
St. Augustine
Common Bermuda
Dichondra
Kikuyugrass
Colonial Bentgrass
Highland Bentgrass
Zoysiagrass
Seashore Paspalum
Hybrid Bermuda

Low

Disease Incidence

High

Dichondra
Creeping Bentgrass
Weeping Alkiligrass
Colonial Bentgrass
Highland Bentgrass
Kentucky Bluegrass
Red Fescue
Perennial Ryegrass
St. Augustine
Seashore Paspalum
Hybrid Bermuda
Tall Fescue
Zoysiagrass
Common Bermuda
Kikuyugrass

Low

Recovery from Severe Injury

Common

Hybrid Bermuda
Kikuyugrass
Common Bermuda
Seashore Paspalum
Zoysiagrass
Creeping Bentgrass
Highland Bentgrass
Kentucky Bluegrass
Dichondra
St. Augustine
Tall Fescue
Perennial Ryegrass
Red Fescue
Colonial Bentgrass
Weeping Alkiligrass

Partial

Nitrogen Requirement

High

Creeping Bentgrass
Hybrid Bermuda
Dichondra
Perennial Ryegrass
Kentucky Bluegrass
Seashore Paspalum
Colonial Bentgrass
Highland Bentgrass
Weeping Alkiligrass
Common Bermuda
St. Augustine
Tall Fescue
Red Fescue
Zoysiagrass
Kikuyugrass

Low

Shade Tolerance

High

Red Fescue
St. Augustine
Zoysiagrass
Seashore Paspalum
Dichondra
Kikuyugrass
Creeping Bentgrass
Colonial Bentgrass
Highland Bentgrass
Tall Fescue
Kentucky Bluegrass
Perennial Ryegrass
Weeping Alkiligrass
Hybrid Bermuda
Common Bermuda

Low

Establishment Rate

Fast

Perennial Ryegrass
Tall Fescue
Common Bermuda
Dichondra
Red Fescue
Highland Bentgrass
Colonial Bentgrass
Creeping Bentgrass
Kentucky Bluegrass
Weeping Alkiligrass
Hybrid Bermuda
Kikuyugrass
Seashore Paspalum
St. Augustine
Zoysiagrass

Slow

Salinity Tolerance

High

Seashore Paspalum
Weeping Alkiligrass
Hybrid Bermuda
Zoysiagrass
St. Augustine
Common Bermuda
Kikuyugrass
Creeping Bentgrass
Tall Fescue
Perennial Ryegrass
Kentucky Bluegrass
Red Fescue
Highland Bentgrass
Colonial Bentgrass
Dichondra

Low

Wear Resistance

High

Zoysiagrass
Kikuyugrass
Hybrid Bermuda
Tall Fescue
Common Bermuda
Seashore Paspalum
Perennial Ryegrass
Kentucky Bluegrass
Red Fescue
St. Augustine
Highland Bentgrass
Colonial Bentgrass
Creeping Bentgrass
Weeping Alkiligrass
Dichondra

Low

Maintenance Cost and Effort

High

Creeping Bentgrass
Dichondra
Hybrid Bermuda
Kentucky Bluegrass
Colonial Bentgrass
Seashore Paspalum
Perennial Ryegrass
St. Augustine
Highland Bentgrass
Zoysiagrass
Tall Fescue
Common Bermuda
Kikuyugrass

Low

General Turf Adaptations and Tolerances

Turf Species	Heat	Cold	Drought	Shade	Salinity	Wear/Traffic	Temperature Adaptation	Planting Method
Bermuda grass	High	Low	High	Low	High	High	Warm-Season	Seed, sod, stolons, plugs
Seashore paspalum	High	Low	High	Low	Very High	High	Warm-Season	Sod, stolons, plugs
Kentucky bluegrass	Low	High	Low	Moderate	Low	Moderate	Cool-Season	Seed, sod
Perennial ryegrass	Low	High	Low	Low	Moderate	High	Cool-Season	Seed, sod
Red fescue	Low	High	Moderate	High	Low	Moderate	Cool-Season	Seed, sod
St. Augustine	High	Low	Moderate	High	High	Moderate	Cool-Season	Sod, stolons, plugs
Tall fescue	Mod.–High	Mod-erate	Moderate	Moderate	Moderate	Mod.–High	Cool-Season	Seed, sod

6.3   TYPES OF FIELD TURF

1. **Synthetic surface.** Rubber, polymer, pigment, PVC, thermset, or thermoplastic. They require a substrate that usually consists of crushed gravel or sand. Synthetic fields have a higher initial cost but eliminate most of the maintenance costs, water, mowing, aerating, herbicides, and replacement of worn sod associated with a natural field. Also, games do not have to be canceled because of wet field conditions. Current life expectancy is around 15 years. Earlier concerns about increased injuries on synthetic surfaces have been greatly reduced in recent years with improvements in both field materials and construction.

2. **Natural.** Sports fields are special facilities. When natural turf is chosen, a sports turf specialist is needed to oversee the development of a total field management program. A comprehensive program should include the following:

   a. Selecting an adapted grass for the locality

   b. Mowing this selected grass at proper height and frequency

   c. Fertilizing at the proper time and rate

   d. Irrigating as needed to encourage establishment and to reduce stress

   e. Aerifying to relieve compaction and to dethatch according to the amount of play

   f. Using the appropriate pre- and postemergence herbicides to ensure you have a vigorous turf to compete with the weeds.

6.4   LIGHTING

Lighting is a controversial issue for most recreational facilities and sports fields. It increases the initial cost but offers opportunities to increase field usage and generate more revenue. Residents near a proposed field should be involved in the planning process to reduce conflicts later.

Multipurpose sports fields are more complicated but more cost-effective to light than single-purpose fields. However, most new field construction combines activities to maximize the use and minimize the acreage needed. There should be no shadows, glare, or irregular bright patches on the playing field. All lighting poles should be located outside the playing area.

Your planning for lighting should include the following:

- Determine the layout of the field and all its potential uses. The most common multipurpose fields combine football with soccer, or football, soccer, softball, or rugby.

- Determine the quantity (level) as well as the quality needed. The Illuminating Engineering Society of North America (IESNA) publishes the "Sports and Recreational Area Lighting Guidelines."

- Determine the number of luminaire assemblies (luminaire assemblies consist of lamp reflector, ballast mounting, cross arm, mounting hardware, and poles required to light the playing surface while avoiding spill and glare).

- Decide on the type of poles to be used: wood, concrete, and steel are standard options. Each has advantages and disadvantages.

- Consider all aspects of safety. The lighting system must comply with the national electric code as well as state and local codes and use illuminaire assemblies that have Underwriter Laboratory approval.

- Establish switching controls that allow for maximum flexibility and efficiency.

- Recalling that some activities require more lighting than others, switches should afford higher and lower levels of illumination. Switching capacity becomes even more important with overlapping fields.

Illumination Levels

For baseball and softball. 20 foot-candles (outfield) and 30 foot-candles (infield)

For soccer, rugby, football, and lacrosse. 30 foot-candles

6.5 DRAINAGE

The ability of a field to drain quickly reduces the number of games that have to be canceled due to wet conditions. Playing on wet turf will damage the sod and increase operating costs due to replacement. Underground drainage is the most desirable, but it is not always an option because of cost. If it is not an option, the playing field should be crowned (3/1 to 5/1 slope) toward the sideline to allow runoff.

Typical Drainage (Rectangular Fields)

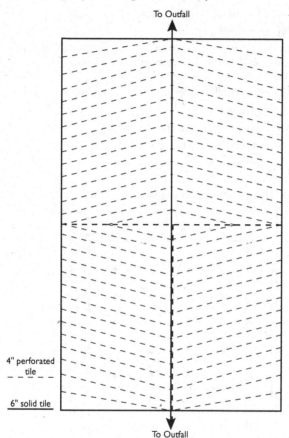

To Outfall

4" perforated
tile

6" solid tile

To Outfall

6.6 IRRIGATION

There are three types of irrigation systems:

- Underground with sprinkler heads throughout the field (preferred)
- Underground with sprinkler heads on the perimeter of the field
- Above ground with portable piping and sprinkler heads and hoses

Note: Underground systems should be designed as a "blocked" system where there is no water pressure on the field proper until watering commences. This will minimize damage in case of a leak or blowout.

The planners of the irrigation system should consider the following:

1. The safety of the participant
2. Types of sprinkler heads
3. The watering pattern layout (i.e., the number of overlapping zones based on the available water pressure, to reach all areas of the field evenly)
4. The source (wells or lake with a pumping system, city water, or other)
5. The timing system
6. A winterization plan in colder climates
7. Tie-ins for toilet facilities, and so on
8. The possibility for fertigation

Typical Irrigation (Rectagular Fields)

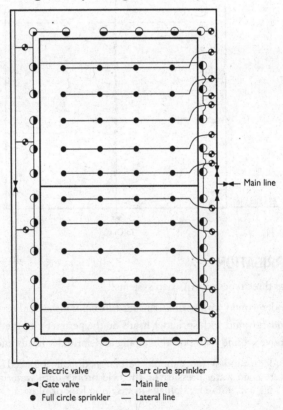

6.7 SPORTS FIELD SPACE REQUIREMENTS (GENERAL)

Activity	Category	Dimensions/ Play Area	Dimensions/ Total	Area/SF
Archery	Min	50'×300'	110'×40'	44,000
Badminton	Singles	17'×44'	27'×54'	1,458
	Doubles	20'×44'	30'×54'	1,620
Baseball	Official	350'×350'	400'×400'	160,000
	Pony	250'×250'	300'×300'	90,000
	Little League	200'×200'	250'×250'	62,500
Basketball	Official	50'×94'	62'×114'	7,068
	High school	50'×84'	62'×104'	6,448
	Recreation	40'×70'	52'×90'	4,680
Croquet		35'×70'	45'×80'	3,600
Deck Tennis	Singles	12'×40'	20'×50'	1,000
	Doubles	18'×40'	26'×50'	1,300
Field Hockey		180'×300'	200'×320'	64,000
Football		160'×360'	180'×380'	68,400
LaCrosse- Men	Min	160'×330'	180'×350'	63,000
	Max	180'×330'	200'×350'	70,000
LaCrosse- Women	Min	150'×360'	170'×380'	64,600
	Max	150'×420'	170'×440'	74,800
Soccer—Men	Min	195'×330'	115'×350'	40,250
	Max	225'×360'	245'×380'	93,100
Soccer—Women	Min	120'×240'	140'×260'	36,400
	Max	180'×300'	200'×320'	64,000
Softball	Fast pitch	225'×225'	250'×250'	62,500
	Slow pitch—men	275'×275'	295'×295'	87,025
	Slow pitch—women	250'×250'	270'×270'	72,900
Tennis	Singles	27'×78'	51'×120'	6,120
	Doubles	36'×78'	60'×120'	7,200
Volleyball		30'×60'	50'×80'	4,000

6.8 ARCHERY

Archery is the art of shooting arrows from a bow at a target. Although it is essentially an individual pursuit, it is possible to participate as a member of a team.

1. **Target archery** consists of shooting arrows from various distances at a target of standard size.

The target

Two sizes of target face are used for shooting under international rules, each at a different shooting distance. A 1,220 mm (48") target face is used for distances at 90 m (295'), 70 m (230'), and 60 m (197') and an 800 mm (32") diameter face is used for distances of 50 m (164'), and 30 m (98').

Both these faces are divided into 10 zones, each color of the standard five-zone target face being halved, the scoring range then ranging from 10 for the central gold down to 1 for the outermost ring of white.

The target is erected on a wooden stand of a height such that its exact center, the pin hole, is 1,300 mm (52") vertically above the ground.

The range

The ideal archery range consists of a level area of closely mown grass in a reasonably sheltered position. The ground should be laid out in a north-south axis with the targets set up at the southern end. There should be a safety zone of at least 22.86 m (75') behind the targets. Archers stand astride a clearly defined shooting line or mark to shoot, with a waiting line at least 4.572 m (15') behind them. The various distances being shot are clearly defined on the ground by means of white lines, ropes, or spots, and the targets, which are at least 3.658 m (12') apart for communal shooting, are usually numbered from left to right for identification purposes.

An area 150 m (492') × 26 m (85') will allow for six shooters in a 90 m (295') range.

2. **Field archery** consists of shooting at black and white faces, shot over varying distances from 5 m (16') to 60 m (197') set out in a bush setting.

3. **Clout shooting** requires shooting arrows high in the air to fall on a target marked out on the ground at much greater distances than those used in normal target archery.

Distances shot are from 145 m (475') (women) to 185 m (607') (men). The overall field length including safety zones is approximately 250 m (820') with 8 m (26') clear of overhead cables and other obstructions.

A flag marks the center of the target which measures 7.315 m (24')
in diameter and has scoring rings at radii of 0.457 m (18"), 0.914
m (3'), 1.829 m (6'), 2.734 m (9'), and 3.658 m (12'), respectively.

4. **Flight shooting** is a highly specialized pursuit with the sole object
of reaching great distances.

5. **Popinjay shooting** consists of a series of wooden cylinders with
feathers attached, variously called cocks, hens, and chicks, set on a
frame called a roost that is mounted on a mast 25.908 m (85')
high. The object is to dislodge these birds, each of which has a
score value, by shooting blunt-headed arrows vertically from the
base of the mast.

Archery Range

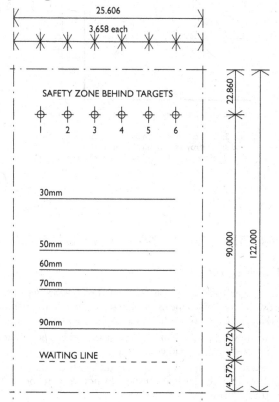

Indoor Archery

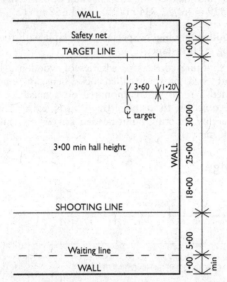

Indoor Archery. The official distances for competition are
Round 1: 30 arrows shot from 18 m at a target face of 400 mm.
Round 2: 30 arrows shot from 25 m at a target face of 800 mm
Round 3: 30 arrows shot from 30 m at a target face of 800 mm.
The center of the target face is 1300 mm above the floor level.

6.9 ATHLETICS (TRACK AND FIELD) FIELD EVENTS

1. Throwing events

The danger zone sector for hammer throw is 85 degrees. As most
throwers are right-handed, it is better to locate the hammer circle
so that the danger of the hammer landing on the track is minimized.

There should be a javelin runway at the other end of the track to
allow for different wind conditions.

Some venues locate the steeple water jump on the outside of the
track so that the steeple hurdles do not have to be moved onto
the track during the event.

When large numbers of athletes have to be catered for, and additional high jump, long jump, javelin, and shot put facilities may be required.

2. High jump

The runway length is unlimited; however the minimum length is 15 m. Where possible the recommended minimum length should be 18 m.

The maximum overall incline of the runway and takeoff area is 1:250 in the direction of running.

The takeoff area should measure not less than 5 m long by 3 m wide.

3. Pole vault

The runway shall have a minimum width of 1.22 m. The length of the runway is unlimited; however, the minimum length is 40 m. Where possible, the recommended length should be 45 m.

The maximum allowance for the lateral incline of the runway is 1:100 and in the running direction 1:1000.

No marks are placed on the runway, but a competitor may place marks (supplied by the organizing committee) alongside the runway. No marks are placed on any pit or landing area.

The landing area should measure not less than 5 m by 5 m.

The pole vault box is constructed of some suitable rigid material, sunk level with the ground and 1.0 m in length, measured along the inside of the bottom of the box, 600 mm in width at the front end and tapering to 150 mm in width at the bottom on the stop board. The length of the box at ground level and the depth of the stop board will depend upon the angle formed between the base and stop board which shall be 105 deg.

The base of the box slopes from ground level at the front end to a vertical distance below ground level of 200 mm at the point where it meets the stop board. The box should be constructed in such a manner that the sides slope outward and end next to the stop board at an angle of approximately 120 deg. to the base.

If the box is constructed of wood, the bottom is lined with 2.5 mm sheet metal for a distance of 800 mm from the front of the box.

4. Long jump

The runway shall have a minimum width of 1.22 m. The length of the runway is unlimited; however, the minimum length is 40 m. Where possible the recommended minimum length is 45 m.

The maximum allowance for lateral inclination of the runway is 1:100 and in the running direction 1:1000. No marks are placed on the runway, but a competitor may place marks (supplied by the organizing committee) alongside the runway. No marks are placed in the pit.

The takeoff is marked by a board sunk level with the runway and the surface of the landing area. The edge of the board that is nearer to the landing area is called the takeoff line. Immediately beyond the takeoff line there is placed a board of plasticine or other suitable material for recording the athlete's footprint when he or she has foot-faulted.

The takeoff board is made of wood or some other suitable rigid material and measures 1.21 m to 1.22 m long, 198 mm to 202 mm wide and maximum l00 mm deep. It is painted white.

The plasticine indicator board consists of a rigid board, 98 mm to 102 mm wide and 1.21 m to 1.22 m long. It is mounted in a recess or shelf in the runway, on the side of the takeoff board nearer the landing area. When mounted in this recess, the whole assembly must be sufficiently rigid to accept the full force of the athlete's foot.

The surface of the board beneath the plasticine is of a material that the spikes of an athlete's shoes will grip and not skid.

The landing area is a minimum width of 2.75 m and 3 m maximum and, if possible, should be so placed that the middle of the runway when extended coincides with the middle of the landing area.

The distance between the takeoff board and the end of the landing area must be at least 10 m.

The takeoff board must not be less than 1.0 m from the edge of the landing area.

5. Triple jump

The runway is the same as for the long jump.

The takeoff board is the same as for the long jump; however, placed at least 13 m from the landing area.

The landing area is the same as for the long jump.

6. Putting the shot

For a valid trial, the shot must fall so that the point of impact is within the inner edges of lines 50 mm wide, marking a sector of 40 degrees set out on the ground so that the radii lines cross at the center of the circle. (The 40-degree sector may be laid out accurately and conveniently by making the distance between the two points on the sector lines 20 m from the center of the circle exactly 13.68 m apart.)

The ends of the lines marking the sector should be marked with sector flags.

The maximum allowance for the downward inclination in the throwing direction of the putting is 1:1000.

The circle is made of band iron, steel, or other suitable material, the top of which is flush with the ground outside. The inside diameter of the circle measures 2.135 m (plus-or-minus 5 mm). The rim is at least 6 mm in thickness and painted white.

The interior of the circle may be constructed of concrete, asphalt, or some other form but not slippery material. The surface of this interior is level and 20 mm (plus-or-minus 6 mm) lower than the upper edge of the rim of the circle.

A portable circle meeting these specifications is permissible.

The stop board is made of wood or some other suitable material in the shape of an arc so that the inner edge coincides with the inner edge of the circle, and also so made that it can be firmly fixed to the ground.

The board measures 1.21 m to 1.23 m long on the inside, 112 mm to 116 mm wide, and 98 mm to 102 mm high in relation to the level of the inside of the circle. This is also painted white.

7. Discus

The discus is thrown from a circle having an outside diameter of 2.5 m (plus-or-minus 5 mm). The competitor must commence the throw from a stationary position and is allowed to touch the inside edges of the circle.

For a valid trial, the implement must fall so that the point of impact is within the inner edges of lines 50 mm wide marking a sector of 40 degrees set out on the ground so that the radii lines cross at the center of the circle.

The maximum allowance for the inclination in the throwing direction of the throwing field shall not exceed 1:1000.

For construction of the circle, see "Putting the Shot."

Discus Throwing Cage

All discus throws are made from an enclosure or cage to ensure the safety of spectators, officials, and competitors. The cage specified in this note is intended for use in a major stadium with spectators all the way around the outside of the arena and with other events taking place in the arena. Where this does not apply, and especially in training areas, a much simpler construction may be satisfactory.

The cage should be u-shaped in plan, consisting of a minimum of six panels of netting 3.17 m wide. The width of the mouth is 6 m, positioned 5 m in front of the center of the throwing circle. The minimum height of the netting panels should be at least 4 m.

Provisions must be made in the design of the cage to prevent a discus forcing its way through any joints in the cage or the netting or underneath the netting panels.

Note: The hammer throwing cage may also be used for discus throwing, either by installing a 2.135 m and 2.5 m concentric circle or by using an extended version of that cage with a second discus circle. (Refer to "Hammer throwing.")

The maximum danger sector for discus throws from this cage is approximately 98 degrees, including both right- and left-handed throwers. The position and alignment of the cage in the arena is, therefore, critical for its safe use.

8. Hammer throwing

The hammer must be thrown from a circle and the competitor must commence the throw from a stationary position.

For a valid trial, the hammer must fall so that the point of impact is within the inner edges of lines 50 mm inside marking a sector of 40 degrees set out on the ground so that the radii lines cross at the center of the circle.

The circle is the same as for the discus and shot put, but has an inside diameter of 2.135 m (plus-or-minus 5 mm).

The throwing field has a maximum allowance for the inclination in the throwing direction not exceeding 1:1000.

Hammer Throwing Cage

All hammer throws are made from an enclosure or cage to ensure the safety of spectators, officials, and competitors. The cage specified in this note is intended for use in a major stadium, with spectators all the way around the outside of the arena, and other events beside hammer throwing taking place.

The cage should be u-shaped in plan, consisting of a minimum of seven panels of netting, each 2.74 m wide, as shown on the diagram. The width of the mouth is 6 m, positioned 4.2 m in front of the center of the throwing circle. The minimum height of the netting panels is at least 7 m.

Two moveable netting panels 2 m in width are provided at the front of the cage, only one of which will be operative at a time. The minimum height of the panels is 9 m.

Combined Hammer and Discus Cages

Where it is desired to use the same cage for discus throwing, the installation can be adapted in two alternative ways. Most simply, a 2.135 m and 2.5 m concentric circle may be fitted, but this involves using the same surface in the circle for hammer and discus throwing.

Where it is desired to have separate circles for hammer and discus, the two circles must be placed one behind the other with the centers 2.37 m apart on the centerline of the throwing sector and with the hammer circle at the front. The shape of the rear of the cage must then be enlarged using a minimum of eight fixed panels 2.83 m wide and two moveable panels 2 m wide, as shown on the following diagram. The minimum height of the panels, both fixed and moveable, for this enlarged cage shall be exactly the same as for the standard cage.

Note: The safety of the hammer cage installation is dependent on the position and alignment in the arena. The maximum danger sector for all throws, including for both left- and right-handed throwers, is approximately 85 degrees.

9. Javelin

The length of the runway must not be more than 36.5 m, but not less than 30 m, and is marked by two parallel lines 50 mm in width and 4 m apart. (It is recommended where possible that the runway not be less than 33.5 m.)

The throw is made from behind an arc of a circle drawn with a radius of 8 m and consists of a strip made of painted wood or metal 70 mm in width, painted white and flush with the ground. Lines are drawn from the extremities of the arc at right angles to the parallel lines marking the runway. These lines are 7.5 m in length and 70 mm in width.

The maximum allowance for lateral inclination of the runway is 1:100 and the inclination in the running direction 1:1000.

One Suggested Field Event Layout

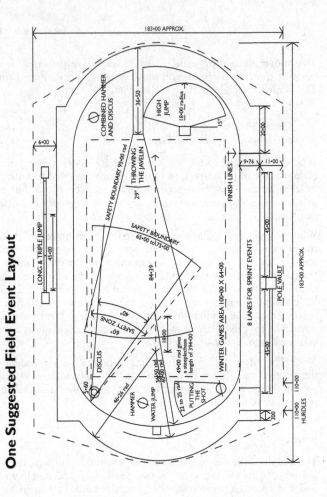

Discus Throwing Circle

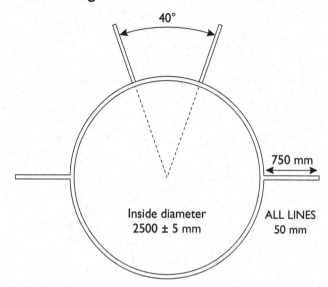

40°

750 mm

Inside diameter
2500 ± 5 mm

ALL LINES
50 mm

Track and Field Areas and Distances

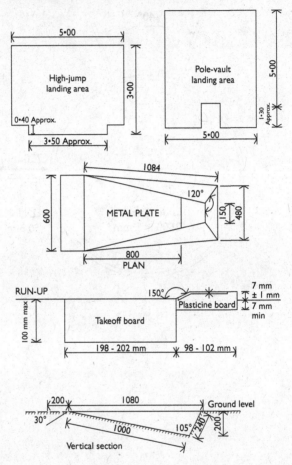

High-jump landing area

5·00

3·00

0·40 Approx.

3·50 Approx.

Pole-vault landing area

5·00

1·30 Approx.

5·00

METAL PLATE

1084

120°

600

150

480

800

PLAN

RUN-UP

150°

Plasticine board

Takeoff board

100 mm max

7 mm ± 1 mm

7 mm min

198 - 202 mm

98 - 102 mm

200

1080

Ground level

30°

1000

105°

220

200

Vertical section

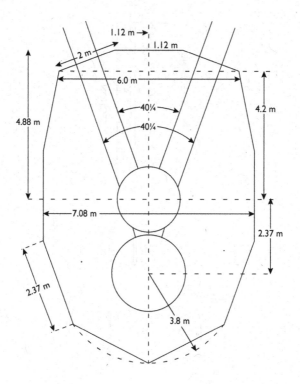

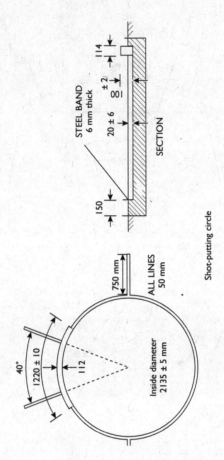

STEEL BAND
6 mm thick

114

100 ± 2

20 ± 6

150

SECTION

750 mm

ALL LINES
50 mm

40°

1220 ± 10

112

Inside diameter
2135 ± 5 mm

Shot-putting circle

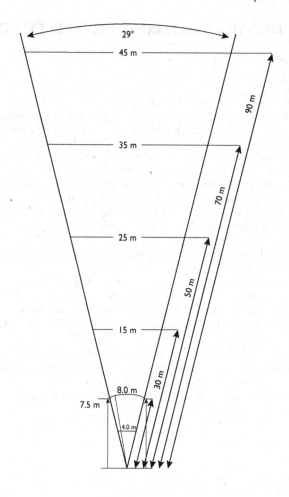

6.10   ATHLETICS (TRACK AND FIELD) TRACK EVENTS

1. Track events

400 Meters

Track (refer to the figure on page 194)

The running track should not be less than 400 m in length and not less than 7.32 m wide. If possible, the track should be bordered on the inside with a curb of concrete or other suitable material, approximately 50 mm in height and a minimum 50 mm in width. Where it is not possible for the inner edge of the running track to be a raised border, the inner edge is marked with lines 50 mm in width.

For grassed tracks the inside line should be flagged at intervals of 5 m. Flags must be so placed on the line as to prevent any competitor running on the line, and are placed at an angle of 60 degrees with the ground away from the track. Flags approximately 250 mm by 200 mm in size, mounted on a staff 450 mm long, are the most suitable for the purpose.

Although the preferred track length is 400 m, there is no standard measurement for the length of straights or the radius of the curves. Both measurements can be varied according to the area of land available, provided the track length remains 400 m. However, the recommended radius is between 35 and 38 m.

The actual track measurement is taken 300 mm outward from the inner border of the track, or where no border exists, 200 mm from the line marking the inside of the track.

The formula for measuring track lengths is as follows:

Track length (400 m) = $(2 \times l) + 2 \text{ pi } R$

where l is the length of the straights curves.

Given that the track length is 400 m and the dimension of one of the two variables has been determined, the dimension of the other can be derived using the above formula.

Remaining lanes shall be measured 200 mm from the outer edges of the lines.

Example

If the length of the straights is 84.389 m then the radius of the curves r is

$$r = \frac{\text{Track length} - 2\,(l)}{2 \text{ pi}}$$

$$= \frac{400 - 168.778}{2 \text{ pi}}$$

$$= 36.8 \text{ m}$$

The radius of the track border, which is 300 mm inside the actual track measurement (refer to the first paragraph), will be 36.5 m. The radius of a line marking the inside of the track, which is 200 mm inside the actual track measurement (refer to first paragraph), will be 36.6 m.

Alternatively, if the radius of the curve has been set at 36.8 m, then the length of the straights (l) is as follows:

$$
\begin{aligned}
l &= \frac{\text{Track length} - 2pi\ r}{2} \\
&= \frac{400 - 36.8\,(2pi)}{2} \\
&= 84.389\ \text{m}
\end{aligned}
$$

2. Preferred track layout

The shape of the track can be varied to suit the shape of the land available.

Lanes

In international meetings the track should allow for at least six lanes, preferably eight lanes, particularly for major international events. All lanes must be of an identical width, a minimum of 1.22 m and a maximum of 1.25 m marked by lines 50 mm in width.

The inner lane is measured as stated above, but the remaining lanes are measured 200 mm from the outer edges of the lines. The line on the right hand only of each lane is included in the measurement of the width of each lane.

An eight-lane athletic track requires an area of 180 m × 95 m. (This allows for approximately 1.5 m clearance.)

The direction for running is left-hand inside.

The maximum allowance for lateral inclination of tracks is 1:100 and the inclination in the running is 1:1000.

3. 100 m start

The line is measured 100 m behind the finish line so that the event is not run around a curve.

4. 200 m start and 4 × 100 m relay

For the 4 × 100 m relay event, the first leg of competitors start from the 400 m staggered starting positions. The first takeover marks are

100 m in advance of each of the relevant 400 m staggered starting positions. The third takeover marks are 100 m from the finish line. At each stage the takeover zone will be within two lines set out 10 m either side of the actual takeover mark in each lane.

5. 4 × 200 m and 4 × 400 m relays

In the 4 × 200 m and 4 × 400 m relays, competitors run the first full lap in lanes.

The second-stage runners in the 4 × 400 m relay and the third-stage runners in the 200 m relay remain in their respective lanes until they enter the back straight.

The following method should be adopted to determine the staggered starting positions for the first runner in each lane to ensure all teams run the prescribed distance of 4 × 200 m or 4 × 400 m.

Inside lane S1 is identical with A1 (refer to the diagram on p. 196).

Lane S2 = A2 plus normal stagger for 400 m.

Lane S3 = A3 plus normal stagger for 400 m.

Lane S4 = A4 plus normal stagger for 400 m.

Note: The positions of A2, A3 . . . A8 allow for the 200 m stagger, plus the compensatory adjustments as prescribed for the 800 m event.

The central lines of the first takeover zones are determined by advancing the normal starting stagger in each lane for a 200 m race by the distance the points B2, B3 . . . B8 are in advance of the line B1Y and are identical to the 800 m staggered starting positions.

The takeover zone is within two lines set out 10 m either side of the central line in each lane. The takeover zones for the second and last takeovers (4 × 400 m) will be the normal 10 m lines either side of the start/finish line AA.

The arc across the track at the entry to the back straight showing the positions at which the second-stage runners (4 × 400 m) and third-stage runners (4 × 200 m) are permitted to leave their respective lanes is identical to the arc for the 800 m event.

6. Hurdle races

The following are the standard distances for hurdle races. There are different distance sprint hurdle races for school-age athletes. These are not uniform throughout the world.

Men: 110 m and 400 m.

Women: 100 m and 400 m.

There are 10 flights of hurdles in each lane, set out in accordance with the following table:

Distance of Race	Height of Hurdle	Distance from Start Line to Finish Line	Distance between Hurdles	Distance from Last Hurdle to Finish
Men				
110 m	1.067 m	13.72 m	9.14 m	14.02 m
400 m	0.914 m	45 m	35 m	40 m
Women				
100 m	0.840 m	13 m	8.5 m	10.5 m
400 m	0.762 m	45 m	35 m	40 m

The Start and Finish

The start and finish of the race is denoted by a line 50 mm in width at right angles to the inner edge of the track. The distance of a selected race is measured from the edge of the starting line farther from the finish to the edge of the finish line nearer to the start.

7. Steeplechase

The standard distances are 2,000 m and 3,000 m (2,000 m for juniors only). The water jump, including the hurdles, is 3.66 m (plus-or-minus 2 cm) in length for men, 3.06 m (plus-or-minus 2 cm) for women, and 3.66 m (plus-or-minus 2 cm) for both sexes. The bottom is 700 mm deep in front of the hurdle and slopes to the level of the ground at the approach end. The hurdle at the water jump is firmly fixed in front of the water and is of the same height as the others in the competition. The women's steeplechase hurdles are 0.762 m (plus-or-minus 3 mm) and the men's are 0.914 m (plus-or-minus 3 mm). A horizontally and vertically adjustable hurdle for the water jump should be provided to allow for the women's steeplechase.

To ensure safe landing of the competitors, the bottom of the water jump is covered at the approach end with suitable matting at least 3.66 m wide and 2.5 m long, the thickness of which should not exceed 25 mm.

400-Meter Track

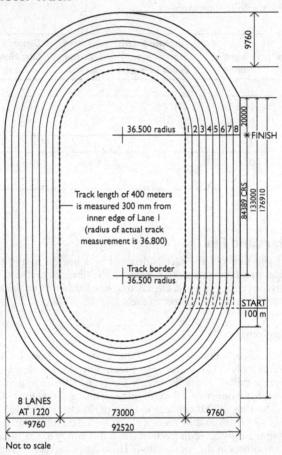

36.500 radius | 1 2 3 4 5 6 7 8

FINISH

20000

84389 CRS
133000
176910

Track length of 400 meters
is measured 300 mm from
inner edge of Lane 1
(radius of actual track
measurement is 36.800)

Track border
36.500 radius

START
100 m

8 LANES
AT 1220
*9760

73000

9760

92520

9760

Not to scale

Starting and Takeover Positions 4 x 100 m Relay

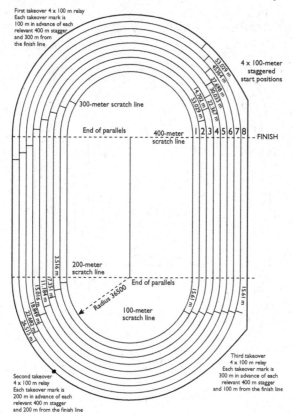

First takeover 4 x 100 m relay
Each takeover mark is
100 m in advance of each
relevant 400 m stagger
and 300 m from
the finish line

4 x 100-meter
staggered
start positions

53,029 m
43,364 m
37,698 m
30,033 m
22,367 m
14,701 m
53,029 m

300-meter scratch line

End of parallels

400-meter
scratch line

1 2 3 4 5 6 7 8

— FINISH

200-meter
scratch line

3,516 m
17,351 m
15,016 m
11,184 m
18,849 m
22,682 m
26,515 m

End of parallels

15,61 m

Radius 36,500

100-meter
scratch line

Third takeover
4 x 100 m relay
Each takeover mark is
300 m in advance of each
relevant 400 m stagger
and 100 m from the finish line

Second takeover
4 x 100 m relay
Each takeover mark is
200 m in advance of each
relevant 400 m stagger
and 200 m from the finish line

Not to scale

Starting and Takeover Positions

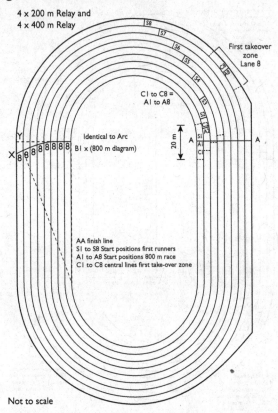

4 x 200 m Relay and
4 x 400 m Relay

First takeover zone Lane 8

C1 to C8 = A1 to A8

Identical to Arc
B1 x (800 m diagram)

20 m

AA finish line
S1 to S8 Start positions first runners
A1 to A8 Start positions 800 m race
C1 to C8 central lines first take-over zone

Not to scale

Starting Positions for 800 m

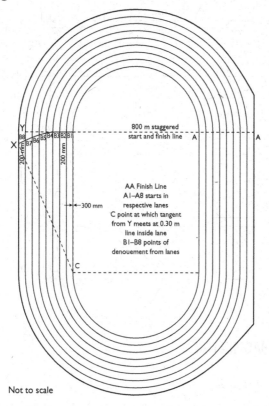

800 m staggered start and finish line

AA Finish Line
A1–A8 starts in respective lanes
C point at which tangent from Y meets at 0.30 m line inside lane
B1–B8 points of denouement from lanes

200 mm

300 mm

Not to scale

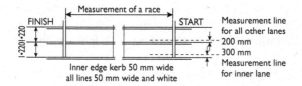

Measurement of a race

FINISH | START

Measurement line for all other lanes
200 mm
300 mm
Measurement line for inner lane

Inner edge kerb 50 mm wide
all lines 50 mm wide and white

Water Jump for Men's Event

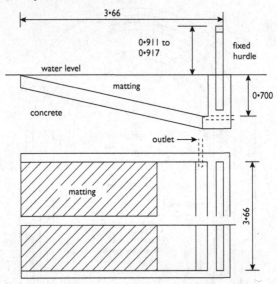

6.11   BADMINTON

General Specifications

- **Court orientation.** For yearlong play the axis should be northwest to southeast 22 degrees off true north. For April to October play, the long axis should be north to south.

- **Slope.** A slope of 0.05 to 0.15% is acceptable depending on the type of surface. Courts should be sloped side to side or end to end. Under-court drainage and a flat surface is preferred, but the above is acceptable if this is not present.

- **Dimensions.** Singles—44 ft. long × 17 ft. wide; doubles court—44 ft. × 20 ft.

- **Court markings.** 1.5 in. wide.

- **Safety distance.** 8 ft. unobstructed behind the back boundary line and 5 ft. on each sideline or between courts.

How to Set Up a Badminton Court

1. Measure space for a doubles-play standard court, defined by the International Badminton Federation as a rectangle 44' long and 20' wide.

2. Narrow the court width from 20' to 17' for singles-only play.

3. Mark the borders, or sidelines, of the court with paint or chalk. Lines should be about 1½" wide.

4. Use cones or natural boundaries such as trees to mark the perimeter for a casual setup.

5. Tie the ends of the net to posts positioned at the sidelines. The net should divide the length of the court evenly.

6. Hang the net so its top is approximately 5' from the playing surface at the center of the court and 5' 1" above the surface at the sidelines.

7. Measure and mark short service lines for singles play. These lines should be parallel to the net, approximately 15½' from the back boundary lines.

8. Measure and mark the long service lines for doubles play. These lines are parallel to the net, approximately 2½' from the boundary lines.

9. Measure and mark centerlines that extend from both short service lines to both back boundary lines.

Since competitive badminton is strictly an indoor game, the ceiling area is important. The ceiling should be at least 30' high with no overhangs, girders, or other obstructions, and indirect lighting is preferred.

On an ideal badminton court, the background (walls and ceiling) should be dark to provide contrast to the shuttle. There should be no windows or skylights. This may not always be possible in all facilities, and sometimes light backgrounds can be fixed by draping black plastic over the walls and windows.

Usually, three badminton courts will fit on a basketball court. There should be at least 4 feet of clear floor space surrounding each court and between any two courts.

6.12 BASEBALL

Baseball Field Layout and Construction (provided by the University of Florida)

Properly laid out constructed fields are paramount to the game. Whether you are a parks and recreation type, work for a local school system, or just want your own regulation backyard baseball field, knowing a few basics is necessary before you build your own field. The following instructions are designed to help set up a field from a relatively level, open area of ground. In addition to the field setup requirements, keep in mind that to have a quality turfgrass playing surface, sports fields must have the following:

1. Adequate drainage
2. Properly designed, installed, and maintained irrigation systems
3. A sound maintenance program to address turf and soil conditions
4. The necessary field equipment (maintenance, bases, pitching rubber, home plate, etc.) and surrounding structures such as fences

Baseball and softball are the only major sports that are played on fields that have both turf and exposed soil for a playing surface. Since about 66% of the game is played on the infield, "skinned" areas should receive as much attention as the turf areas. The concept of clay management is similar to turf management in that it is difficult to write a maintenance program for all infield skinned areas due to the diversity among infield soils. One thing that does not change, though, is the basic layout.

The following is a basic 13-step program for laying out a baseball field. If you can follow these 13 basic steps, you can build your own field of dreams. In addition to the steps, a few tips and suggestions are also included. A few basic tools such as shovels, rakes, a couple of measuring tapes, a small sledge hammer, a tamp or roller, as well as some supplies such as stakes, string, paint (inverted aerosol cans), pitching rubber, bases, and home plate are needed to complete this project. Power tools and some extra hands will make the project go much faster.

Steps

1. Start with a flat, open area. If some elevation is on-site, it should be in the infield area. Ideally, the open area has a good, dense stand of turf, or with a little help, one can rejuvenate it. If that is not the case, plan a turf management program to coincide with the construction of your ball field. It is helpful to mark out the components of an infield with paint as outlined below to visualize the field before you actually start removing turf.

2. Placement of home plate determines the layout of the field. Be sure to plan for some type of backstop to contain stray pitches and to protect fans from tipped balls. If it is truly a backyard field, and fans are not likely, planting shrubs about 60 feet (minimum for high school and college fields) behind home plate may prevent balls from rolling too far from the field.

3. Using the apex of home plate (back corner), cut out turf in a 13' radius.

4. The next step is to locate second base. Measure from the back tip of home plate to a distance of 127' 3⅜" (see the table on page 203 for the distance between bases for various leagues). Mark with a wooden stake. When installing the base pads, this will be the center of second base.

5. With the tape measure still in place, it is easiest to go ahead and mark the location of the pitching rubber. The placement can be marked by measuring from the back tip of home plate along a string stretched to second base at a distance of 60' 6".

6. The easiest way to find first and third base is to use two tape measures. Stretch one tape from second base toward the first-base line, and the second tape from the back tip of home plate toward the first-base area. The point where the two tapes cross at the 90' mark is the back corner of the bases. Repeat this step to find third base. A baseball diamond is actually a 90' square.

7. First and third base fit within the square, but second base is measured to the center of the bag. An improperly placed second base is one of the most common mistakes made when a baseball field is set up.

8. To make a "slide area" around the bases, cut out turf around bases measuring a 13' radius within the 90' square. You can leave the base paths grassed if you like, or you can turn them into "skinned" base paths.

9. Next, turn your attention to the pitcher's mound. The diameter of the pitcher's mound clay is 18', with 10' from the front of the rubber, toward home plate and 8' from the back of the rubber.

10. The top of the mound is a plateau that is 5' wide.

11. A regulation pitcher's mound is 10½" high (compared to the surface level of home plate). Miscalculation of the pitcher's mound height is

probably the second most common error in setting up a baseball field. A transit or field level is best for setting the height, but in a pinch, other methods may also work. Some have been known to peer through a cheap scope clamped to a carpenter's level on a makeshift tripod. Another option is to use your stakes with a taut string and a ruler. A standard pitching rubber is 24" × 6".

12. Building a pitcher's mound is as much an art as it is science. Build the mound from the ground up 1" at a time, keeping in mind the mound's slope. As you add each layer, tamp or roll the soil.

13. Beginning 12" in front of the pitcher's rubber and measuring toward home plate, for every foot of distance the slope will fall 1" (until the slope hits ground level).

The mix used to build the pitcher's landing area (and often the batter's and catcher's box) should have a significant concentration of clay to provide the necessary stability to resist degradation from increased traffic. A good material will be about 40% sand, 20% silt, and 40% clay. If necessary, you can mix individual components together. Just be sure that individual components are evenly distributed throughout the material.

A quality infield material will have a lower concentration of clay than the pitcher's mound. The infield skin should be moist and firm, not hard and baked dry. To achieve firmness, an infield mix should not be too sandy. An infield mix with greater than 75% sand causes unstable footing for ball players and increases infield skin maintenance problems.

A sandy infield will create low spots more quickly and is more likely to create lips at the skin/turf interface. Ideally, the infield mix should be between 50 and 75% sand and 25 to 50% clay and silt. A combination that has been successfully used is a 60% sand, 20% silt, and 20% clay base mix (sandy clay loam to sandy loam). The silt and clay give the firmness. If the mix contains too much silt and clay, compaction and hardness become a problem.

You may want to now consider your outfield fence. The distance varies with the level of play. Confer with league officials for data listed and recommended placement of outfield fences. Refer to the following table for a summary of base, pitching rubber, and outfield wall distances.

Infield Mix Quantities (Note: 1.3 tons = 1 cubic yard)

Field Type and Size	Depth – 1"	2"	3"	4"	5"	6"
90' Bases with –95' Arc to Center Field						
Grass infield—skinned area 11,550 SF	36 yards	71	107	143	178	214
Skinned infield—skinned area 18,300 SF	57 yards	113	169	226	282	339
80' Bases with 80' Arc—315' to Center Field						
Grass infield area 8,400 SF	26 yards	52	78	104	130	156
Skinned infield area 13,650 SF	42 yards	84	126	168	210	252
70' Bases with 70' Arc—275' to Center Field						
Grass infield area 6,800 SF	21 yards	42	63	84	105	126
Skinned infield area 10,700 SF	33 yards	66	99	132	165	198
60' Bases with –50' Arc—215' to Center Field						
Grass infield area SF 3,850 SF	12 yards	24	36	48	60	72
Skinned infield area 6,700 SF	21 yards	41	62	82	103	124

Distance between Bases, from Pitching Rubber to Home Plate, and from Outfield Wall to Home Plate for Various Levels of Play

Field Use	Base to Base	Pitching Rubber to Home Plate	1st to 3rd Home to 2nd	Home to Outfield Wall
Baseball	90'	60'6"	127' 3 ⅜"	Varies
Little League	60'	46'	84' 10 ¼"	180' radius
Pony League	75'	54'	106' ¾"	250' radius
Babe Ruth	90'	60'6"	127' 3 ⅜"	300' radius

Baseball Field

Confer with League Officials for placement of Outfield Fence and verification of data listed.

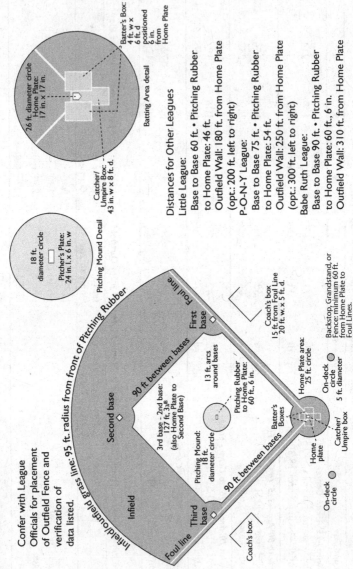

Infield/outfield grass line: 95 ft. radius from front of Pitching Rubber

Pitching Mound Detail
18 ft. diameter circle
Pitcher's Plate: 24 in. l. x 6 in. w

Batting Area detail
26 ft. diameter circle
Home Plate: 17 in. x 17 in.
Catcher/Umpire Box: 43 in. w x 8 ft. d.
Batter's Box: 4 ft. w x 6 ft. d positioned 6 in. from Home Plate

First base

Second base

Third base

90 ft. between bases

3rd base - 2nd base: 127 ft. 3⅜ (also Home Plate to Second Base)

13 ft. arcs around bases

Pitching Rubber to Home Plate: 60 ft., 6 in.

Pitching Mound: 18 ft. diameter circle

Batter's Boxes

Home plate

Home Plate area: 25 ft. circle

Catcher/Umpire box

On-deck circle 5 ft. diameter

On-deck circle

Foul line

Coach's box 15 ft. from Foul Line 20 ft. w. x 5 ft. d.

Coach's box

Backstop, Grandstand, or Fence: minimum 60 ft. from Home Plate to Foul Lines.

Distances for Other Leagues
Little League:
Base to Base 60 ft. • Pitching Rubber to Home Plate: 46 ft.
Outfield Wall: 180 ft. from Home Plate (opt.: 200 ft. left to right)

P-O-N-Y League:
Base to Base 75 ft. • Pitching Rubber to Home Plate: 54 ft.
Outfield Wall: 250 ft. from Home Plate (opt.: 300 ft. left to right)

Babe Ruth League:
Base to Base 90 ft. • Pitching Rubber to Home Plate: 60 ft, 6 in.
Outfield Wall: 310 ft. from Home Plate

Baseball Layouts

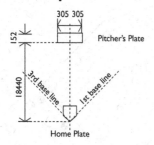

Pitcher's Plate

Home Plate

Playing Field from Home Base
12190 Radius

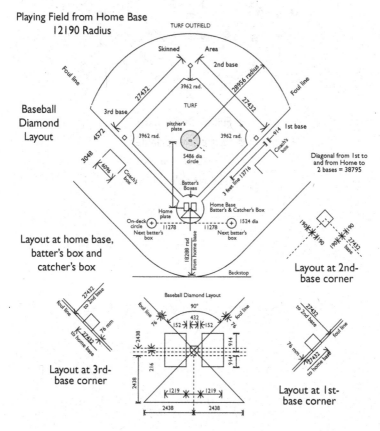

Baseball
Diamond
Layout

Layout at home base,
batter's box and
catcher's box

Layout at 3rd-
base corner

Layout at 2nd-
base corner

Layout at 1st-
base corner

Typical Irrigation of Baseball and Softball Fields

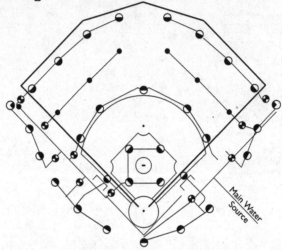

- • Full circle sprinkler
- ⊖ Part circle sprinkler
- ⊘ Adjustable sprinkler
- ⊕ Electric solenoid valve

Typical Drainage for Baseball and Softball Fields

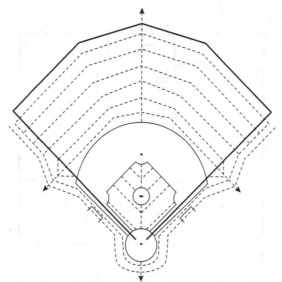

6.13 BASKETBALL

Outdoor (Recreational) Court Specifications

General Specifications

- **Court orientation.** Year-round play to have long axis northwest to southeast 22 degrees off true north. April to October play to have long axis north to south.
- **Slope.** A slope of 0.05 to .15% is acceptable depending on the type of surface. Courts should be sloped side to side or end to end.
- **Dimensions (Junior high, high school, and recreation play).** 84' ×× 50'.
- **Court markings.** 2" wide
- **Safety distance.** 10' minimum or 8' unobstructed behind the back boundary line and a minimum of 6' on each sideline or between courts.
- **Three-point arch.** 19'9" from the center of the basket.
- **In-ground pole.** Should be padded, off the playing court, and extended at least 4' onto the court.
- **Fence height.** If used, should be 10'.

High School and College Basketball Court

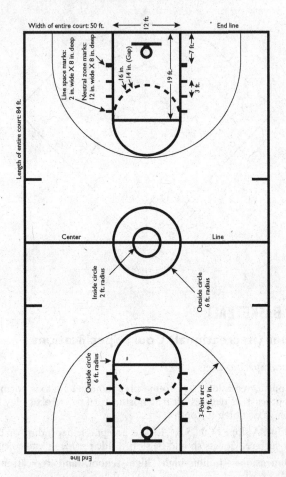

High School:

Court Size: 50 ft. × 84 ft.

Free-Throw Line: 15 ft. from backboard

Free-Throw Lane: 12 ft. wide

3-Point Arc: 19 ft. 9 in.

Professional Basketball Court

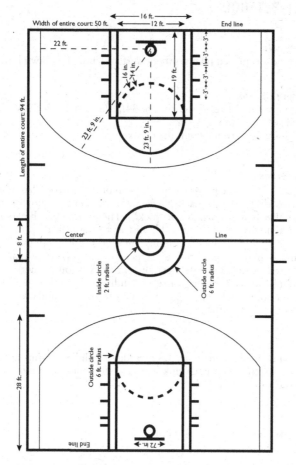

Professional Basketball:
Court Size: 50 ft. × 94 ft.
Free-Throw Line: 15 ft. from backboard
Free-Throw Lane: 16 ft. wide

6.14   INDOOR AND LAWN BOWLS AND BOULES (BOCCE/PETANQUE)

Indoor Bowls
Indoor bowls is played on a synthetic surface simulating lawn bowls.

Dimension of Rink
The recommended size of the rink is 34.75 m × 4.27 m. However, the standard size is 32 m × 4.27 m.
Rinks are usually grouped in fours or sixes.

Lawn Bowls
Lawn bowls is a game played on a flat lawn or turf surrounded by a ditch and enclosed by a bank. Each player or team delivers a bowl toward a white smaller ball (known as the "jack"). The object of each person is to position as many bowls as possible nearer to the jack than the nearest opposing bowls. Each bowl fulfilling that object scores a point.

A match may be person against person (singles), two against two (pairs), three against three (triples), or four against four (fours).

In contests it is customary for several individual matches to take place, usually simultaneously. Six rinks of fours is the most popular inter-group formula.

Rinks
The green is divided into rinks 5.8 m wide (for club play, rinks may be a minimum 4.4 m wide). Each rink is to have its centerline marked with a peg on the bank.

Mat
The mat is 600 mm long × 360 mm wide.

Lawn Bowls Layout

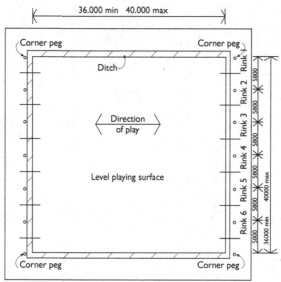

Field of Play

Boules (Bocce/Petanque)

Boules or bocce/petanque is played between two players or teams. Players seek to place their boules nearer to the target jack than their opponent's boules, or to displace their opponents and to so improve the position of their own boules in relation to the jack.

Dimensions of the Pitch

The standard pitch length is 27.5 m (it may be reduced to a minimum of 24.5 m). Width of pitch is 2.5 m to 4 m. For international matches, the minimum pitch width is 3 m, and there must be banks at least 20 mm high.

The Pitch

Any surface may be used; however, it is usually clay or gravel.

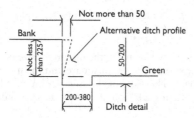

Boules Layout

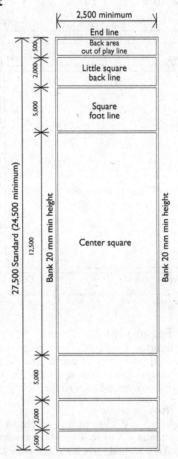

6.15   BOXING (INTERNATIONAL)

Boxing is fist-fighting with gloves worn by two people in a roped area.

Dimensions of the Ring

For international competition, the minimum size between the inside line of the ropes must be 4.9 m square and the maximum size must be 6.1 m square. The floor of the ring shall not be less than 914 m and not more than 1,219 mm above the surrounding floor.

The floor of the ring must project not less than 457 mm beyond the line of the ropes.

Three ropes a minimum 300 mm and maximum 50 mm in diameter are tightly drawn from the four corner posts.

The heights of the three ropes from the ring floor are 400 mm, 800 mm, and 1,300 mm, respectively. These ropes are covered with a soft or smooth material.

For club play or recreational purposes, a ring of 3.65 m × 3.65 m may be used.

Clear Space about the Ring

A minimum of 2.0 m should be allowed for around the ring. An area of 10 m to 11 m square is needed.

Boxing Ring

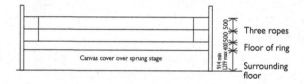

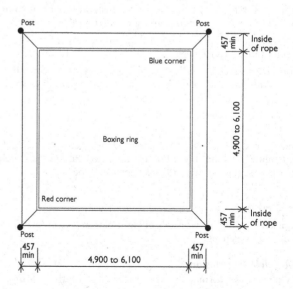

6.16   CRICKET (OUTDOOR AND INDOOR)

Outdoor Cricket

Indoor cricket is a match played between two sides consisting of 11 players on each side, using a bat and a ball.

Dimensions of the Pitch

The pitch is 20.12 m long between the centerline stumps, and 1.52 m wide each side of the center stump.

Outfield

International. Suggested 64 m to 68.6 m radius requiring an area of 1.5 ha. approximately.

Pennant. 68.68 m radius from center of pitch.

Junior. 40 m to 50 m radius.

Additional Notes

Pitch to be 100 mm thick concrete slab reinforced with F62 mesh.

Run-up to be 100 mm thick unreinforced concrete.

Pitch surrounds to be 75 mm thick unreinforced concrete.

Individual pitches to be poured first. Install fence prior to installation of pitch and prior to pouring of surrounds.

Concrete surface to be "wood float finish" except for "transverse broom finish" to run-ups.

Concrete surface to be flush with adjacent grass surface. Pitches to be provided with maximum 30 mm longitudinal fall as shown.

Fence to be 3,000 mm high, black, PVC-coated mesh with top and bottom rails. Bottom rail to be installed 25 mm above finished concrete surface.

Backdrop attached fence to be conveyor belt (secondhand) 3,000 mm wide and 2,000 mm high. Provide intermediate rail for attachment. Synthetic surfacing should be of approved type.

Indoor Cricket

Indoor cricket consists of two sides of eight players each playing within a pitch bounded by netting.

Playing Area/Net Dimensions

The playing area is formed by the dimensions of the netting, which should conform to the following specifications:

	Min	Max
Length	28 m	30 m
Width	10.5 m	12 m
Height	4.0 m	4.5 m

Dimensions of Pitch

The pitch is 20.12 m long between the centerline of the stumps and 1.52 m wide each side of the centerline.

Clear Space above Pitch

Minimum clear height above the playing area is 5 m.

Cricket Layout

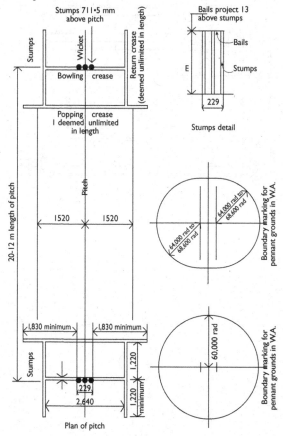

Plan of pitch

Practice Wickets

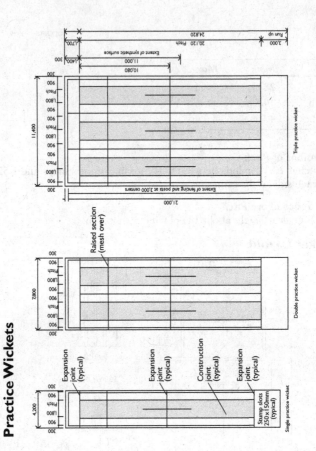

Single practice wicket

Double practice wicket

Triple practice wicket

Stump slots 250×150mm (typical)

Expansion joint (typical)

Construction joint (typical)

Raised section (mesh over)

Extent of synthetic surface

Extent of fencing and posts at 3,000 centers

6.17 CROQUET

Croquet is a lawn game played with balls and a mallet. The object is to score through all the hoops in the correct order.

Dimensions of the Green (International)
The standard court is 32 m × 25.6 m to the inside edge of the boundaries. The standard setting is shown on the diagram on page 219.

Hoops
Hoops should be 305 mm above the ground, measured to the top of the hoop.

Peg
The peg should be 38 mm in diameter and extend 457 mm above the ground, measured to the top of the peg.

Specifications (English)

Recommended area. 3,000 SF

Size and dimension. Playing area is 35' × 70', plus a minimum 2' to 6' on each end and side.

Orientation. Orientation is not critical, and may be adjusted to suit local topographical conditions.

Surface and drainage. Playing surface is to be turf closely cropped and rolled with a maximum 2% slope (preferably level) with adequate under drainage.

Special considerations:

Boundary lines are marked with strong cotton twine held by corner staples.

Arches are ½" diameter steel rod—3⅜" wide and 9" above the ground when in place.

Playing lines may either be imaginary, marked with white chalk, or with smaller twine wired close to the ground.

Stakes shall be made of steel and shall be firmly anchored. They shall be 11" high and set 1½" outside the playing line halfway between the end corners.

Croquet Layout

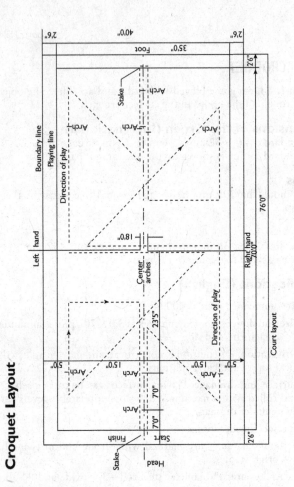

Court layout

Croquet Standard Setting

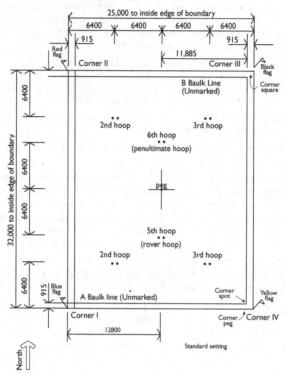

Standard setting

6.18 EQUESTRIAN

Equestrian events are competitive events for horse and rider designed to test the horse's development and training, his jumping ability, and the all-round ability of horse and rider.

1. Dressage competitions test the harmonious development of the horse's physique and ability, and demand a high degree of understanding between horse and rider.

Competitors carry out official tests incorporating a variety of paces, halts, changes of directions, movements, and figures. All tests are judged in an area measuring 60 m × 20 m. There should be 15 m clear space between the arena and the public, and for indoor arenas a minimum distance of 3 m.

2. **Show jumping** competitions are those in which the horse's jumping ability and the rider's skill is tested under various conditions over a course of obstacles. The arena is enclosed, and while a horse is jumping, all exits are shut. The maximum length of the course is usually the number of obstacles multiplied by 60 m.

The obstacles vary in number, height, and width, and this depends on the level of competition. However, for a high level of competition (e.g., the final at the Olympic Games), the course length does not exceed 700 m, with between 10 and 12 obstacles, varying in height from 1.4 to 1.6 m and between 1.5 to 2.2 m wide. The water jump is at least 4.3 m wide.

Dimensions of indoor arenas are 70 m × 30 m (minimum) and for outdoor 80 m × 100 m minimum.

3. **The three-day event** is designed to test the all-round ability of horse and rider. The events consist of three distinct phases that take place on separate days. The first phase is dressage. The second phase is designed to test the speed, endurance, and jumping ability of the true cross-country horse. The length of the course varies according to the importance of the competition—the Olympic Games course is a maximum distance of 30.13 km.

The course consists of roads and tracks totaling 15.4 to 18.7 km, steeplechase 3.105 to 3.45 km, and cross-country 7.41 to 7.98 km.

The third phase is the show-jumping event, which is not an ordinary show-jumping competition nor a test of style or endurance. The object is to prove that, on the day after a severe test of endurance, horses are still able to continue in service.

6.19 FENCING

Fencing is a form of sword fighting that has a long history with its roots in the tradition of chivalry. Fencing competitions take place on a piste using a range of three weapons: the foil, the epee, and the saber.

Dimensions of the Piste

14 m long × 1.8 to 2 m wide

Space about the Piste

No space about piste is required for competitors, but it is necessary for instructor, officials, and others who wish to watch.

Clearance space at both ends 1.5 to 2 m for all standards of play.

Clearance space at both sides.

Clearance space at both sides (including match officials table space one side) 2.25 to 3 m for Olympic Games and 1.25 to 2 m for others.

Overall sizes 17 m × 4.3 m minimum for club and preferred 18 m × 6 m for international events.

Fencing Dimensions

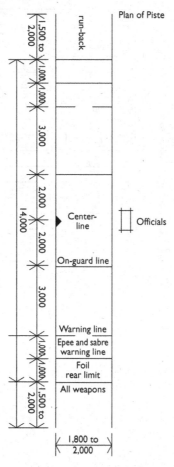

Plan of Piste

6.20 FOOTBALL (AMERICAN-CANADIAN)

The following diagrams illustrate the proper setup for various levels of North American (American and Canadian) football.

High School Football

Additional Specs—High School
Hash Marks - Recommended 24' long and 4" wide.
Yard Markers - Recommended 24' long and 4" wide.
End Lines and Side Lines - Recommended 4" wide.
Yard Line Numbers - Recommended 3' high.
Placement should be 27 ft. from sideline to top of Yard Line Number.

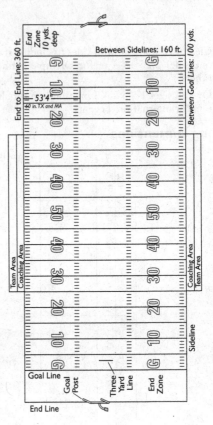

Uprights: 20' high

Width: 23'4" Basic 18'6" high

End to End Line: 360 ft.

End Zone 10 yds. deep

Between Sidelines: 160 ft.

53'4"
40 in. TX and MA

Between Goal Lines: 100 yds.

Team Area
Coaching Area

Coaching Area
Team Area

Sideline

Goal Line

Goal Post

Three Yard Line

End Zone

End Line

College Football

Coaching Box: 6' from field and marked with diagonal lines. Coaching Box and Team Area: 6 x 50 yds. Entire field surrounded by 12' wide "Limit Line." All other lines: 4" wide.

Additional Specs—College
Same as High School (above), but Yard line numbers are recommended 6' high.

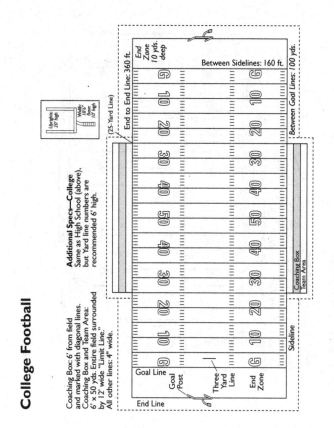

Uprights: 20' high
Width: 18'6"
Base: 10' high

(25-Yard Line)

End Zone 10 yds. deep

End to End Line: 360 ft.

Between Sidelines: 160 ft.

Between Goal Lines: 100 yds.

Goal Line

Goal Post

Three-Yard Line

End Zone

Sideline

Coaching Box / Team Area

End Line

223

6-8-9 Man Football

Additional Specs—High School

Hash Marks: Recommended 24' long and 4" wide.
Yard Markers: Recommended 24' long and 4" wide.
End lines and side lines must be 4" continuous white line.
Yard Line Numbers: Recommended 3' high.
Placement should be 27 ft. from sideline to top of yard line number.

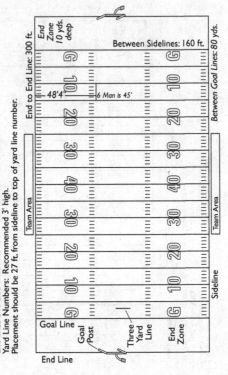

Canadian Football
The Canadian Field

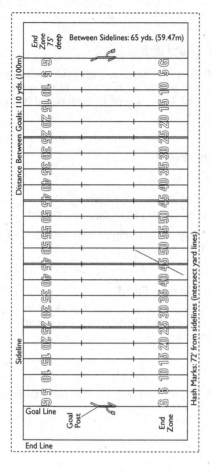

Distance Between Goals: 110 yds. (100m)

Between Sidelines: 65 yds. (59.47m)

End Zone 75' deep

Sideline

Goal Line

End Line

Goal Post

End Zone

Hash Marks: 72' from sidelines (intersect yard lines)

The Canadian Field
The area between the goal lines and sidelines is the field of play. It is divided between the goal lines by parallel lines 5 yds. apart. These lines are intersected by short lines 24 yds. from the sidelines (hash marks). The boundary lines are outside the field of play.

6.21 FOOTBALL (AUSSIE RULES)

Aussie rules football is a fast-moving, high-goal-scoring game. High marking and long kicking are the main features of the sport. A match is played by two teams, each consisting of 18 players and two interchange reserves.

Dimensions of the Field

The shape of the field is oval.

Maximum is 185 m long × 155 m wide.

Minimum is 135 m long × 110 m wide.

The National Football Association recommends 165 m long × 135 m wide as the ideal size.

Space about the Playing Field

There should be a recommended minimum 4 m between boundary line and fence. For a 165 m × 135 m field including 4 m free space, an area of 173 m × 143 m is required.

Goalposts

Goalposts are 6 m high. Behind posts are 3 m high. Distance between inside edge of posts is 6.4 m.

U7-U10

The pitch (oval) is 82.5 m × 65 m, approximately.

Goalposts are 3 m and 6 m high.

Aussie Rules Football Field

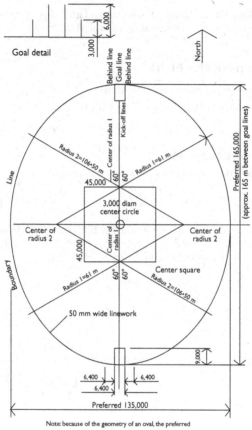

Goal detail

Note: because of the geometry of an oval, the preferred
dimensions stated are approximate only.

6.22   FOOTBALL (GAELIC)

Gaelic football is a 15-players-a-side ball and goal game that superficially looks like a compromise between Aussie rules, rugby, and soccer.

Dimensions of the Field

Maximum is 146 m long × 91 m wide.
Minimum is 128 m long × 77 m wide.

Space about the Field

The recommended minimum is 3 m around the field.

Goals

Goalposts are 5 m high and 6.4 m apart with a crossbar 2.44 m above the ground. A semicircular arc of 13 m radius centered at the midpoint of the 20 m line shall be marked outside of each 20 m line.

Gaelic Football Field

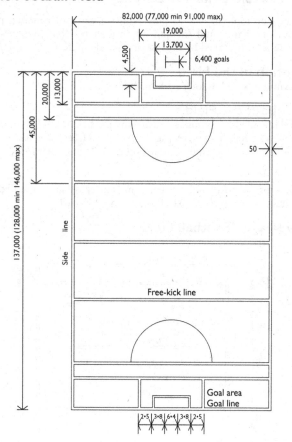

6.23 FOOTBALL (RUGBY LEAGUE)

Rugby League football is a 13-players-a-side game of running, passing from hand to hand, and kicking an oval-shaped ball. Derived from Rugby Union, it is played by professionals and amateurs.

Dimension of the Field
The maximum is 100 m long × 68 m wide.

Space about the Field
The recommended minimum is 3 m around the field (with 6 m preferred). For a 100 m × 68 m field including the preferred free-space and dead-ball area, 134 m × 80 m is required based on maximum dimensions.

Goals
The height of the posts must exceed 4 m. The crossbar is 3 m high from the ground to the top of the bar. Posts are 5.5 m apart.

Rugby League Football Field

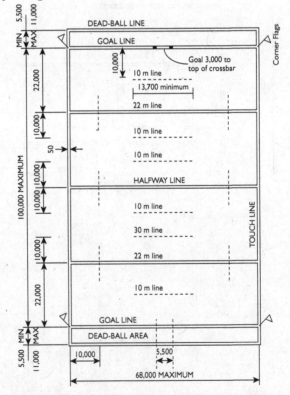

6.24   FOOTBALL (RUGBY UNION)

Rugby Union football is a body contact sport in which the ball may be handled as well as kicked. The game has different rules than Rugby League, and teams consist of 15 a side.

Dimensions of the Field
The maximum is 10 0 m long × 70 m wide.

Space about the Field
Recommended space is 3 m around the field (preferred is 6 m). For a 100 m × 70 m field, including preferred free-space and dead-ball area, 156 m × 81 m is required based on maximum dimensions.

Goals
The height of the posts must exceed 3.4 m. The height from the ground to the top of crossbar is 3 m. Posts are 5.6 m apart.

Modified Rules

Walla Rugby
The pitch size is 70 m × 50 m.

Rugby Union Football Field

UNION RUGBY
(shown) is the version of Rugby played by high schools, colleges, and other nonprofessional organizations.

LEAGUE RUGBY
(not shown) is the professional version of the game.

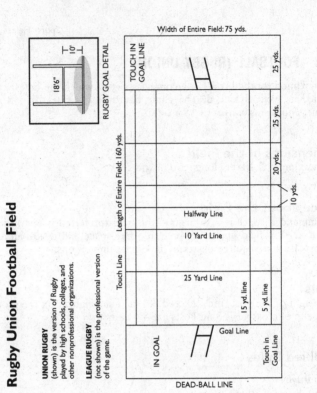

Rugby Dimensions

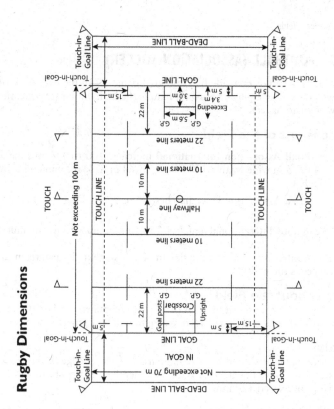

6.25   FOOTBALL (ASSOCIATION SOCCER)

Soccer is an 11-players-a-side ball and goal game. With the exception of the two goalkeepers, players may not handle the ball, but must propel the ball with head or feet.

Dimensions of the Field

Football Association International matches. 100 to 110 m long × 64 to 75 m wide, with a 100 m long by 64 m wide recommended by the F.A.

Football Association U.K. matches. 90 to 120 m long × 45 to 90 m wide.

Schoolboy International matches. 75 m minimum × 55 m minimum.

Note: The pitch must be rectangular in shape so that the length in all cases exceeds the width.

Space about the Field

End margin of 9 m and side margins of 6 m are recommended. For 100 m × 64 m pitch, an area of 118 m × 76 m is required (8,968 square meters).

Goals

The inside opening should be 7.32 m apart. The lower edge of the crossbar is 2.44 m above the ground. Width and depth of goalposts and crossbar shall not exceed 120 mm.

Modified Soccer

U8, U9

Size of pitch. 50 m × 25 m maximum.

Penalty area. 20 m × 10 m.

Goals. 1.52 m × 4.57 m wide.

Number of players. Seven per side.

Soccer Field Size (United States)	
Age (% of adult size)	**Field Size (in yards)**
U-14 (100%)	60×100
U-12 (80%)	50×80
U-10 (70%)	40×70
U-8 (50%)	25×50
U-6 (25%)	15×30

Soccer Field

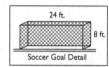

24 ft.

8 ft.

Soccer Goal Detail

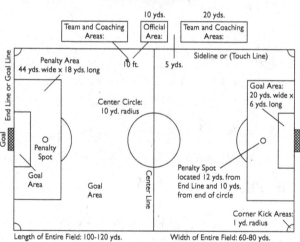

10 yds.   20 yds.

Team and Coaching Areas: Official Area: Team and Coaching Areas:

Penalty Area 44 yds. wide x 18 yds. long

Sideline or (Touch Line)

10 ft.   5 yds.

End Line or Goal Line

Goal

Center Circle: 10 yd. radius

Goal Area: 20 yds. wide x 6 yds. long

Penalty Spot

Goal Area

Goal Area

Center Line

Penalty Spot located 12 yds. from End Line and 10 yds. from end of circle

Corner Kick Areas: 1 yd. radius

Length of Entire Field: 100-120 yds.   Width of Entire Field: 60-80 yds.

Line Court Marking

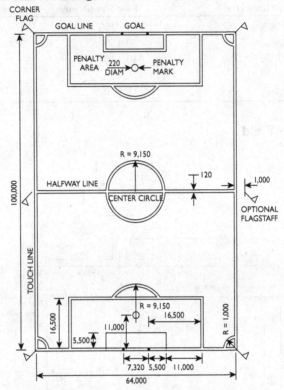

6.26 HANDBALL

Handball is a noncontact game between two 11-players-a-side teams, played out-of-doors. It is played with one or two hands by catching, interpassing, and throwing the ball.

Dimensions of the Field

The maximum is 110 m × 65 m.

The minimum is 100 m × 60 m.

Space about the Field

For a 100 m × 60 m field including 3 m free space, about an area of 0.7 is required.

Goals

Goals are the same as for soccer goals. Goalposts are 2.44 m in height to underside of crossbar. The crossbar is 7.3 m in length to the inside of the goalposts. The woodwork is 127 in cross section, and the side facing the field is painted in alternating black and white segments.

Handball Dimensions

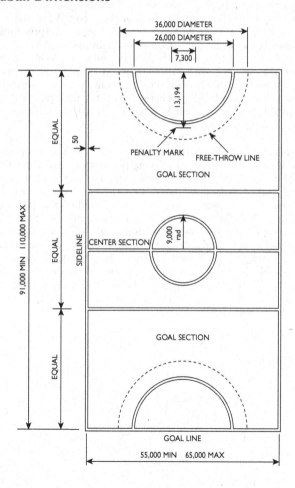

6.27 HORSESHOES

Court Construction

Whether you, your club, or your parks and recreation department plan to build or remodel horseshoe courts, good planning should be done in advance.

Site selection and garding should call for reasonably level ground with suitable drainage away from the pits and walkways.

The actual court "playing area" measures from the back of the pitching platform to the back of the opposite pitching platform at the other end of the court and the 6' width of the pitcher's box. Additional suitable space is necessary for safety. Space courts at least 10' (preferably 12') apart measured stake to stake. Construct chain-link fencing across both ends of the courts at least 8' (preferably 10') measured from the stakes.

Backstops are often used for containment of loose pit material, but they also stop shoes that may have hit and bent the chain-link fence.

Fence gates should be located between courts and, if practical, located every second court to give pitchers a safe, direct exit from their court without crossing adjacent courts.

Surrounding yard areas should be graded level with or slope away from walkways to eliminate ledges and steps that may cause ankle injuries. If practical, the paved pitching platforms should extend the full length of the court as continuous walkways. The surrounding areas may be gravel packed, seeded, or paved. Construction joints in paved walkways should be accurately positioned to act as foul lines or pitching platform dimensions; otherwise, painted lines will be necessary.

"In-ground" courts should have top of pit fill material level with pitching platforms. "Portable" (aka "temporary" and "raised") courts have a maximum height limit of 7" above the pitching platform surface. Some portable courts have a slight adjustment in pitching distance to compensate for the raised height of the pit, and some portable courts eliminate much of the front edge of the "box," which otherwise blocks a view of the pit surface.

Horseshoes Court Layout

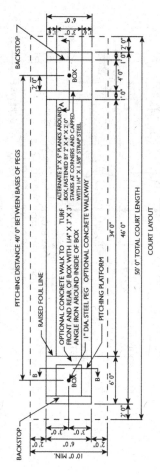

CourtLayout

BACKSTOP

PITCHING DISTANCE 40' 0" BETWEEN BASES OF PEGS

RAISED FOUL LINE

OPTIONAL CONCRETE WALK TO
FRONT AND REAR OF BOX WITH 1/4" X 3" X 3"
ANGLE IRON AROUND INSIDE OF BOX

1" DIA. STEEL PEG

PITCHING PLATFORM

TURF   ALTERNATE 2" X 5" PLANKS AROUND
BOX, FASTENED BY 2" X 4" X 2' 0"
STAKES AT CORNERS AND CAPPED
WITH 1/4" X 1 5/8" STRAP STEEL

OPTIONAL CONCRETE WALKWAY

BOX

BOX

BACKSTOP

BACKSTOP

6' 0"

1' 6"   3' 0"   1' 6"

2' 0"

1' 0" 2' 0"

4' 0"

1' 0"

2' 0"

34' 0"

46' 0"

50' 0" TOTAL COURT LENGTH

6' 0"

2' 0"

2' 0"   3' 0"   3' 0"   2' 0"

6' 0"

10' 0" MIN.

239

Horseshoes Peg Layout

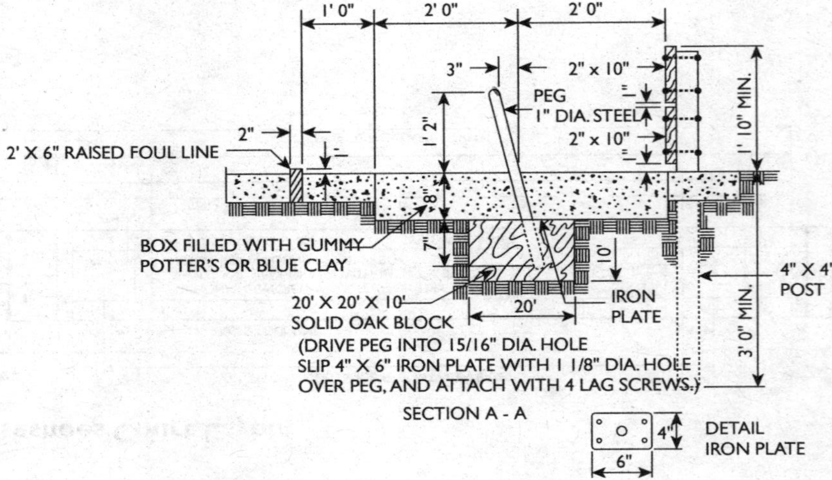

1' 10" MIN.

2' 0"

2" x 10"

PEG

2" x 10"

1" DIA. STEEL

3"

2' 0"

2'1"

8"

7"

1' 0"

1"

2"

4" X 4" POST

3' 0" MIN.

IRON PLATE

10"

20'

2' X 6" RAISED FOUL LINE

BOX FILLED WITH GUMMY POTTER'S OR BLUE CLAY

20' X 20' X 10' SOLID OAK BLOCK

(DRIVE PEG INTO 15/16" DIA. HOLE SLIP 4" X 6" IRON PLATE WITH 1 1/8" DIA. HOLE OVER PEG, AND ATTACH WITH 4 LAG SCREWS.)

SECTION A - A

DETAIL IRON PLATE

4"

6"

240

Lighting Guidelines

Horseshoe court lighting should be uniform over the playing surface, and for a few feet outside the sidelines and backstops. Outdoor lighting offers no reflective background, so all light must be direct from the fixture. For club/tournament play, 25 to 35 foot-candles is recommended, and for nontournament play, 15 to 25 foot-candles is sufficient. All the surface area should be evenly lighted. Lighting fixtures should be placed outside the courts so that the beams are generally aimed across the courts. No light sources should be located directly behind the courts or at the back corners of courts if possible. Quartz, metal halide, fluorescent, and high-pressure sodium lamps are all applicable to horseshoe court lighting. Poles should be 30' to 35' long, have cross arms at the top to which fixtures are attached, and be capable of withstanding at least 100 mph winds. Fluorescent fixtures should be mounted 14' to 16' above the court surface, outside the lines, and tilted inward. Pole-mounted fluorescent fixtures may be located 22' above the court surface. Poles should be primed and painted a dark color. All wiring should be installed underground and outside the court area. Basic design techniques are the same for almost any type of field.

Horseshoe pitching tournaments normally have "class" sizes of 6, 8, 10, 12, and 16 pitchers. Each class pitches a "round-robin" schedule where each contestant pitches against another in the same class. Thus, a six-man class would need three courts, an eight-man class would need four courts, and so on. Accordingly, court installations should be constructed with enough courts to handle the largest anticipated tournament. For example, if 40 entries were expected, they could be divided into five 8-man classes, requiring 20 courts. Alternately, two 8-man morning classes and three 8-man afternoon classes would require no more than 12 courts (with a suggested minimum of 18 to 24).

6.28   HOCKEY (FIELD)

Field hockey is a stick-and-ball game between two teams of 11 persons per side. The sport is played by men and women.

Dimensions of the Field (International)

Length should be 91.44 m.

Width should be between 50 m and 55 m. (See the diagrams on the following pages for both English and metric field measurements.)

Goal Dimensions

Posts. 3.66 m in length to inside of posts.

Crossbar. 2.14 m in height to underside of crossbar.

Goalposts and crossbar must be rectangular, 50 mm wide and 75 mm deep, and painted white. A backboard 3.66 m long and not exceeding 460 mm high is placed at the foot of and inside goal nets.

Side boards. 1.2 m long and not exceeding 460 mm high. Should be are placed at right angles to the goal line. The front of the goalposts must touch the outer edge of the goal line.

Netting. Should be 25 mm to 38 mm and attached firmly to the goalposts at intervals of not more than 150 mm.

Space about the Field

A minimum of 3 m is recommended for the side margin and 4.5 m for the end margin. For maximum size field, an area of 99.44 m × 61 m is required.

Allow a minimum of 6 m clear space between adjoining hockey fields.

Field Hockey (English Dimensions)

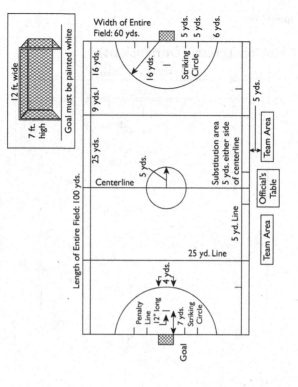

12 ft. wide

7 ft. high

Goal must be painted white

Width of Entire Field: 60 yds.

16 yds.

9 yds.

16 yds.

5 yds.

5 yds.

6 yds.

Striking Circle

25 yds.

Centerline

5 yds.

Length of Entire Field: 100 yds.

Substitution area 5 yds. either side of centerline

5 yd. Line

5 yds.

Team Area

Official's Table

Team Area

25 yd. Line

Penalty Line

12" long

7 yds.

4 yds.

Striking Circle

Goal

Field Hockey (Metric Dimensions)

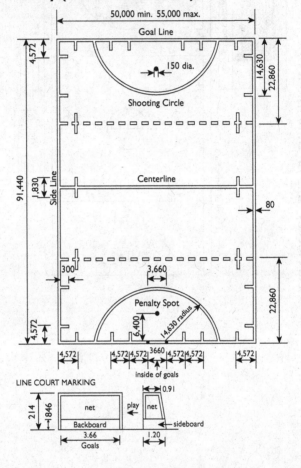

6.29   HOCKEY (INDOOR)

Indoor hockey is a game adapted from field hockey for two teams of six players each. Although usually played indoors, the game can take place outdoors if the surface is hard and flat.

Dimensions of the Playing Area

The maximum is 44 m long × 22 m wide.

The minimum is 36 m long × 18 m wide.

International size court is 40 m long × 20 m wide.

Goals

Posts. 2 m high to underside of crossbar.

Crossbar. 3 m long between inside edge of goalposts.

Nets. Attached firmly to goalposts at intervals of not less than 150 mm.

Space about the Court

Clearance behind goals. Minimum 1.5 m, preferred 3 m.

Clearance outside sideboards. Minimum 1.5 m, 2.7 m wide on one side to accommodate official's area (1.2 m wide).

Area required for 36 m × 18 m court. 42 m × 22.2 m.

Area required for 44 m × 22 m court. 50 m × 26.2 m.

Indoor Hockey Field

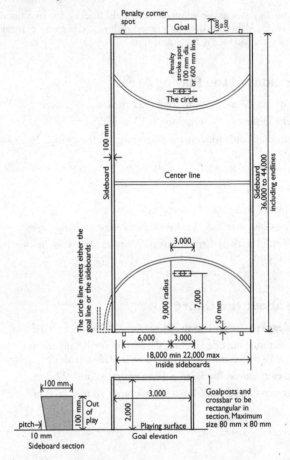

6.30 HURLING (MEN'S)

Hurling is arguably the oldest and fastest field sport in the world. It evolved as a method of training Irish warriors for battle or in lieu of battle to settle disputes over property claims for example. Today hurling is one of Ireland's national sports and is growing in popularity in the United States.

Teams consist of 15 players including a goalkeeper, three fullbacks, three halfbacks, two midfielders, three half forwards, and three full forwards. Teams are allowed three substitutes per game. Players may switch positions on the field as much as they wish.

Officials for a game are composed of a referee, two linesmen, and four umpires.

Dimensions of the Field

150 yards long × 100 yards wide

137 m long × 82 m wide

Space about the Field

A minimum of 3 m (10') around the field is recommended, but 6 m (20') is preferred.

Goals

The goalposts are the same shape as on a rugby pitch. The uprights are 7 m high, and the crossbar is 6.5 m wide from inside to inside. The height of the crossbar is 2.5 m, with a net attached similar to soccer. The outside dimension of the goalie box is 19 m wide × 13 m deep. The inside box is 14 m wide × 4.5 meters deep.

Hurling (Men's Field)

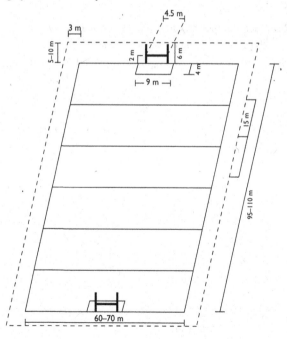

6.31 LACROSSE (MEN'S)

Men's lacrosse is a robust ball-and-goal game played with a netted stick. International men's lacrosse consists of a 10-person-a-side team plus up to 9 substitutes. National competition consists of 10-persons-a-side teams plus up to 9 substitutes.

Dimensions of the Field

Pitch. 100 m long × 55 m wide

Distance between goals. 72 m

Space about the Playing Area

A minimum of 3 m is recommended about the playing field. For a 100 m × 55 m pitch, an area of 106 m × 61 m is required.

Goals

Posts. 1.83 m high to underside of crossbar.

Crossbar. 1.83 m long to inside of posts.

Net. Shall be triangular in plan and extend to a point 2 m behind goal.

Men's Lacrosse Field

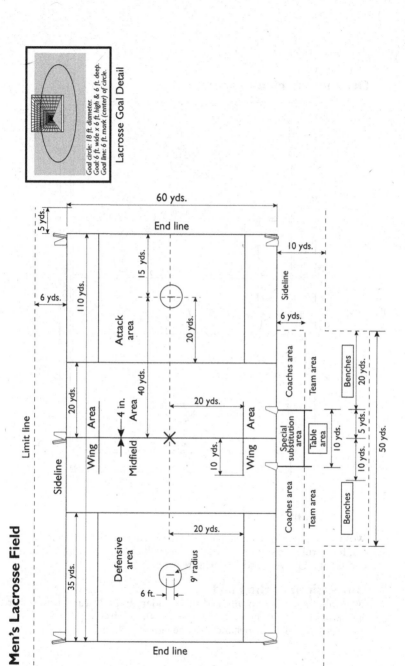

Lacrosse Goal Detail

Goal circle: 18 ft. diameter.
Goal: 6 ft. wide x 6 ft. high & 6 ft. deep.
Goal line: 6 ft. mark (center) of circle.

60 yds.

5 yds.

End line

10 yds.

6 yds.

110 yds.

15 yds.

Sideline

Attack area

20 yds.

6 yds.

Coaches area

Team area

Benches

20 yds.

20 yds.

Wing Area

40 yds.

Midfield Area

4 in.

20 yds.

Wing Area

Special substitution area

Table area

10 yds.

5 yds.

50 yds.

Limit line

Sideline

10 yds.

10 yds.

Coaches area

Team area

Benches

Defensive area

35 yds.

20 yds.

9' radius

6 ft.

End line

249

Distances in Men's Lacrosse

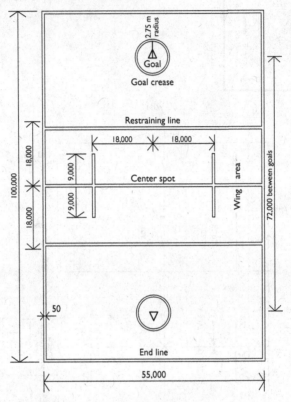

6.32 LACROSSE (WOMEN'S)

Women's lacrosse differs from men's lacrosse in that there are more players, fewer rules, and no body contact or stick checking is allowed. Teams consist of 12 players and 1 substitute per side.

Dimensions of the Field

Women's pitches are not marked by boundary lines. Natural features or spectators mark the boundary. However, the desirable pitch is 110 m long × 60 m wide. The distance between goals is 92 m.

Space about the Playing Field

Recommended minimum of 3 m about the playing field. For a 110 m ×
60 m pitch an area of 116 m × 66 m is required.

Goals

Posts. 1.83 m high to underside of crossbar.

Crossbar. 1.83 m long to inside of posts.

Women's Lacrosse Field

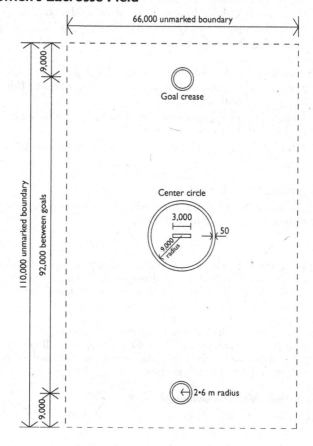

6.33 NETBALL

Netball is a seven-person-a-side game. Points are scored by sending the ball through a ring at one's own end of the court.

By rule, netball is played on a firm surface. The game can be enjoyed as much indoors as outdoors.

Dimensions of the Court

The dimensions should be 30.5 m long × 15.25 m wide. The longer sides shall be called sidelines and the shorter sides goal lines.

The court shall be divided into three equal parts: a center third, and two goal thirds, by traverse lines drawn parallel to the goal lines.

A semicircle with a radius of 4.98 m and with its center at the midpoint of the goal line shall be drawn in each goal third. This shall be called the goal circle.

A circle 0.9 m in diameter shall mark the center of the court. This shall be called the center circle.

All lines are part of the court and shall not be more than 50 mm wide.

Space about the Court

The recommended minimum is 3.7 m all round. The court including free space requires an area 37.9 m × 22.65 m.

Goals

A goalpost, which shall be vertical and 3.05 m high, shall be placed at the midpoint of each goal line. A metal ring with an internal diameter of 380 mm shall project horizontally 150 mm from the top of the post, the attachment to allow 150 mm between the post and the near side ring. The ring shall be of steel rod 15 mm in diameter, fitted with a net clearly visible and open at both ends. Both ring and net are considered to be part of the goalpost. If padding is used on the goalpost, it shall not be more than 25 mm thick and shall start at the base of the goalpost and extend between 2 m and 2.4 m up the goalpost.

Netta Netball (Modified Game—Under 9)

Court size as for adults. Goalpost height 2.4 m. Players may use a smaller ball.

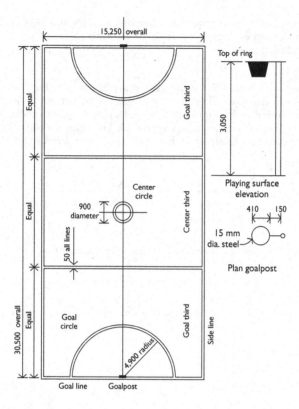

6.34 POLO

Polo is a four-person-a-side game whose players are mounted on horses and use wooden mallets, or sticks, to strike a wooden ball in an attempt to score goals.

Dimensions of the Field

A full-size ground is 275 m long and 180 m wide if unboarded, or 146 m wide if boarded. This full-size ground dimension includes a safety area of about 10 m at the sides and 30 m at the ends. If boards are used, they should not exceed 270 mm high and are placed on the sidelines.

An area of 5 ha. is required. All perimeter lines and lines at 27.5 m and 55 m from each goal line are marked. Marks are also placed at the center of each goal.

Goals

Goalposts are placed 7.3 m apart and are about 3 m high and about 254 mm in diameter. They are not of solid construction but usually of wicker covered with a cloth fixed around a central pole of about 38 mm diameter that fits into a slot in the ground. This allows a player or horse to collide with the goalpost without the certainty of serious injury.

Polo Field

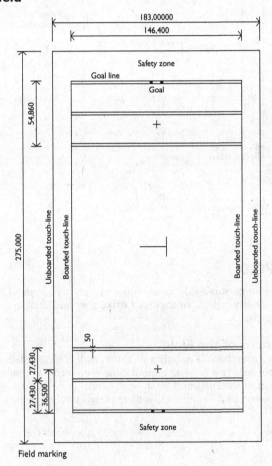

Field marking

6.35   SHUFFLEBOARD

Recommended area. 312 SF minimum.

Surface and drainage. Surface is to be concrete with a burnished finish. Court surface is to be level with drainage away from the playing surface on all sides.

General Specifications

Court orientation. Long axis is north to south.

Slope. A slope of 0.05 to 0.15% is acceptable depending on the type of surface. Courts should be sloped side to side.

Court dimensions. 39' long × 6' wide. Total dimensions 52' × 10'.

Safety area. Should be 6' at both ends of the court and 2' along the sidelines.

Playing lines. Should be 1" wide and brightly colored.

Triangle. 10 off area is 3" at the base.

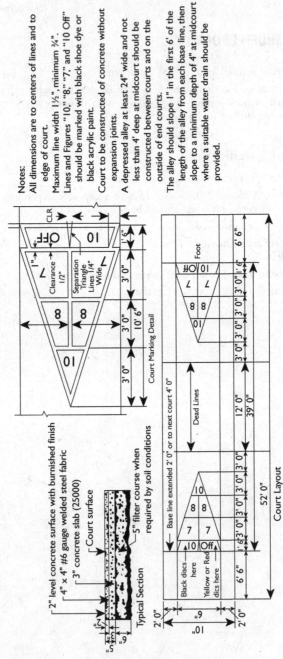

Notes:

All dimensions are to centers of lines and to edge of court.

Maximum line width 1½", minimum ¾".

Lines and Figures "10," "8," "7," and "10 Off" should be marked with black shoe dye or black acrylic paint.

Court to be constructed of concrete without expansion joints.

A depressed alley at least 24" wide and not less than 4' deep at midcourt should be constructed between courts and on the outside of end courts.

The alley should slope 1" in the first 6' of the length of the alley from each base line, then slope to a minimum depth of 4" at midcourt where a suitable water drain should be provided.

Court Marking Detail

CLR

OFF

10

1"

Clearance
1/2"

Separation
Triangle
Lines 1/4"
Wide

7

8   8

10

1' 6"

3' 0"

3' 0"

10' 6"

3' 0"

Typical Section

2" level concrete surface with burnished finish

4" x 4" #6 gauge welded steel fabric

3" concrete slab (25000)

Court surface

5" filter course when required by soil conditions

Court Layout

Base line extended 2' 0" or to next court 4' 0"

Dead Lines

Foot

Black discs here

Yellow or Red discs here

10   Off

7   7

8   8

10

2' 0"

6' 0"

9"

10"

6' 6"

1' 6"   3' 0"   3' 0"   3' 0"

12' 0"

3' 0"   3' 0"   3' 0"   1' 6"

6' 6"

39' 0"

52' 0"

256

6.36 SOFTBALL

The care and setup procedures for a softball field are the same as for baseball. Use the same 13-step procedure outlined in Section 6.12 according to the dimensions set forth in the softball field diagram.

Infield Mix Quantities

	1"	2"	3"	4"	5"	6"
60' bases—60' arch—200' to center field						
Skinned infield area 8,350 SF	26 yards	51	77	103	128	154
60' bases—65' arch—275' to center field						
Skinned infield area 9,300 SF	29 yards	57	86	114	143	172

Softball Field

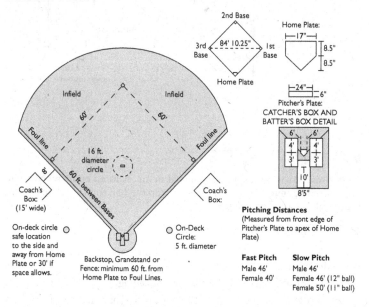

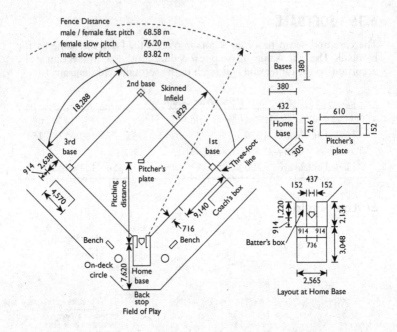

6.37 SQUASH RACKETS

Squash rackets is a racket game played in an enclosed court by two participants. The object of the game is to keep the ball in play while making it difficult for the opponent to do so.

Dimensions of the Court

The court is 9.754 m long × 6.4 m wide internal face-to-face dimensions.

Height of front wall to underside of ceiling:

International: 5.8 m. Also, for an international court, the ceiling height is 6.4 m high at the distance of 3.5 m from the front wall.

Championship: 5.5 m. Minimum is 5.334 m.

Glass-backed courts may negate the need for a gallery. If a timber strip floor is used, the boards run parallel to side walls.

Squash Court Dimensions

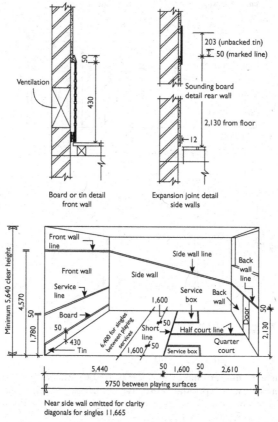

Dimensions of Singles Squash Courts

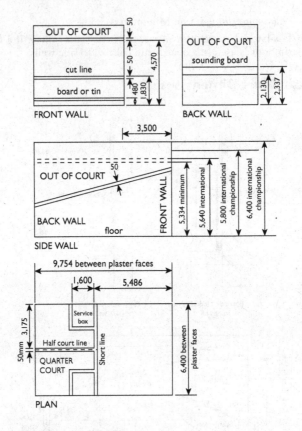

6.38 TENNIS (PADDLE)

Recommended area. 3,200 SF.

Surface and drainage. Surface is to be concrete or bituminous material with optional protective color coating. Drainage is to be end to end, side to side, or corner to corner diagonally at a minimum slope of 1" in 10'.

Special considerations. 10' high fencing is recommended on all sides of the court.

General Specifications

Court orientation. Year-round play—Long axis northwest to southeast 22 degrees off true north. April to October play—Long axis north to south.

Slope. A slope of 0.05 to 0.15% is acceptable depending on the type of surface. Courts should be sloped side to side or end to end.

Dimensions. Singles (65' 7⁷⁄₁₆" long × 32' 9¾" wide), with a tolerance of 1%.

Enclosure. Paddle court is completely enclosed with blackwalls and fencing. At 13' 1½" from the back of the wall, the side walls decrease in height from 9' 10⅛" to 5' at a 38 degree angle.

Surfaces. Hard court (concrete or asphalt) or artificial grass.

Paddle Tennis Court

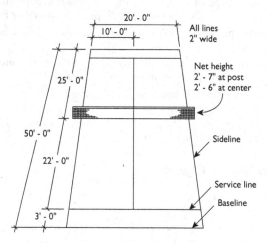

NOTES

1. Space behind each baseline to back fence is 15 ft. minimum.
2. Space from each sideline to side fence is 10 ft. minimum.
3. Fence shall be 8 ft. high unless otherwise noted.
4. Ball is a punctured tennis ball.

6.39 TENNIS (PLATFORM)

Recommended area. 1,800 SF.

Surface and drainage. Raised level platform is normally constructed of treated wood or aluminum superstructure with carriage set on concrete piers to permit construction slopes. Drainage is provided by ¼" space between 6" deck planks or channels. Snow removal is facilitated by hinged panels (snow gates) between posts around the bottom of perimeter fence.

Special considerations. Tension fencing that is 12' high, 16-gauge, hexagonal, galvanized, 1" flat wire-mesh fabric must be on all sides of the court. Lights should be provided, since the game is played throughout the year. Prefabricated courts are available from several manufacturers.

General Specifications

Court orientation. For year-round play, long axis is northwest to southeast 22 degrees off true north; for April to October play, long axis is north to south.

Slope. A slope of 0.05 to 0.15% is acceptable depending on the type of surface. Courts should be sloped side to side or end to end.

Dimensions. Singles (44' long × 20' wide), total surface is 68' × 38' or 2,584 SF.

Fence. 16-gauge hexagonal, galvanized, 1" flat wire-mesh fabric, 10' to 12' high, posts no more than 10' apart.

Preferred surface. Douglas fir 2' × 6' planks.

Surrounding space. All space around the platform is covered with wire, except for a 12' opening in the center of each side, which is covered with netting.

Platform Tennis Court

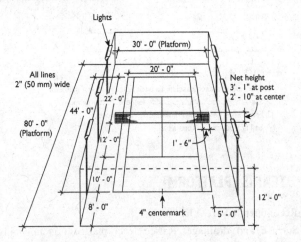

6.40   TENNIS

International

Tennis is a racket game in which individuals (singles) or pairs (doubles) compete against each other. The game is played on a court divided by a net. The object of the game is to strike the ball over the net with the racket in such a way that it bounces in the court and beats any attempt by an opponent to return it.

Court Dimensions

Singles. 23.77 m long × 10.97 m wide. All dimensions are to the outside edges of the lines.

Space about the Court

Davis Cup standard: 6.4 m backrun and 3.66 m from side fence. Playing area is 36.6 m × 18.3 m.

Club standard. 5.5 m backrun and 3.05 m from side fence. Playing area is 34.8 m × 17.1 m.

For two courts side by side, a clearance space of 3.66 m is acceptable.

Indoor Tennis

Height above the net minimum 10.67 m and at end of backrun minimum 4.27 m.

North America

General Specifications

Court orientation. For year-round play, long axis is northwest to southeast 22 degrees off true north. For April to October play, long axis is north to south.

Slope. A slope of 0.05 to 0.15% is acceptable depending on the type of surface. Courts should be sloped side to side or end to end.

Dimensions. Singles (78' long × 36' wide), doubles (minimum space 122' × 66').

Fence. 6 or 9 gauge, 10' or 12' high, posts no more than 10' apart.

Minimum size for court. 114' × 56'

Standard size court. 120' × 60'

Stadium size court. 132' × 66'

Types of Courts

Porous. Allows water to drain through the court surface.

Clay

Grass

Soft composition

Porous concrete

Synthetics

Nonporous

Concrete

Asphalt (cushioned and noncushioned)

Hard composition (liquid applied synthetic)

Synthetics

Selection Considerations

- Initial cost
- Maintenance cost
- Amount of use
- Geographic location
- Maintenance personnel needed
- Level of competition

The following concerns the three most popular types of tennis courts:

1. Hard court (asphalt or concrete base with possible acrylic coating)
2. Clay
3. Grass

Hard Court Construction (Acrylic)

A hard court is one made of asphalt or concrete, usually covered with an acrylic coating. The coating protects the court from the elements, enhances its appearance, and affects the playing characteristics of the court. Generally, a hard court yields what is known as a "fast" game, meaning that a tennis ball bounces off the court surface at a low angle. The speed and the angle of the tennis ball coming off a bounce are determined by the power and the spin of the hit and are relatively unaffected by the surface of the court. This speed, however, can be adjusted depending on the amount, type, and size of the sand used in the color coating. "Slow" playing, textured surfaces are available. Properly installed, hard courts are generally considered to be durable and require relatively low maintenance. Installation costs range from $18,000 to $40,000 depending on the specific construction.

When a resilient layer (or layers) of cushioning material is applied over an asphalt or concrete court, a cushioned court results. "Cushioned courts" usually have excellent playing characteristics and an all-weather surface for year-round play. These attributes make them popular with players, but such courts are considerably more expensive than hard courts; cushioning adds $5,000 to $25,000 to the cost of the court, over and above the cost of the asphalt or concrete base.

Clay Court Construction

Site Preparation: Grading shall be done to a tolerance of plus or minus one inch (1") of the final subgrade elevation. Rate and direction of slope shall be one inch (1") in forty feet (40') all in one plane. A compaction of 95% (Modified Proctor) is required, and the soil shall be free of all roots and vegetation. The material should be free of rocks.

Grass Court Construction

A grass court should be constructed with turf maintenance and wear tolerance in mind. The key to achieving this result is to provide the following:

- Adequate under-court drainage (see the diagram on page 267)
- An adequate irrigation system (see the diagram on page 267)
- Selection of the best turfgrass for your location, level of competition, and expected amount of play
- An adequate maintenance program (taking into account all of the above)

The sub-base of the tennis court area should be "cored out" similar to USGA golf green construction with a subgrade that slopes at least 1% from outer edge to the center (see side-view diagram). At this point, herringbone drainage is to be installed and covered with ⅜" pea gravel. Then the rest of the cavity should be filled with sand (greensmix) and graded perfectly level. Install the irrigation system (see the diagram on page 267), and water the sand "in" to settle the area (nothing settles sand or anything else like water). After a period of time watering in the playing surface, regrade the finished playing surface level.

"Sprigg" or seed the playing surface and maintain. Good maintenance practices include watering, fertilization, topdressing, applications of insecticides and chemicals, and frequent mowing. You should contact a turf maintenance professional for your particular needs, as well as your local tennis association for their requirements for mowing height, particularly for competitions.

Tennis Court

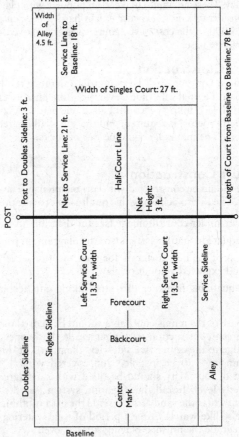

The Irrigation System for a Tennis Court

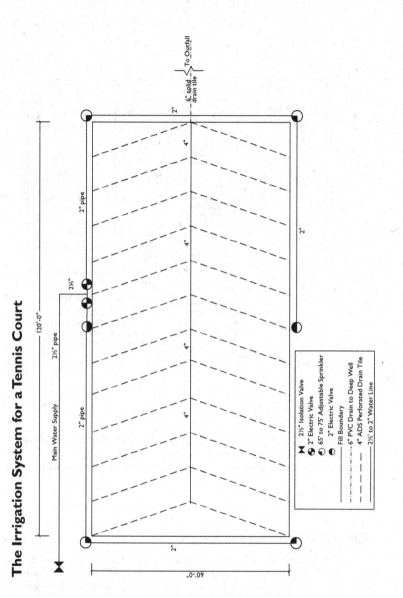

Legend:

- ⊠ 2½" Isolation Valve
- ◑ 2" Electric Valve
- ◐ 65' to 75' Adjustable Sprinkler
- ◑ 2" Electric Valve
- —— Fill Boundary
- ·—·—· 6" PVC Drain to Deep Well
- —··— 4" ADS Perforated Drain Tile
- —— 2½" to 2" Water Line

Main Water Supply

120'-0"

2½" pipe

2" pipe

2½"

90'-0"

2"

4"

6" solid drain tile

To Outfall

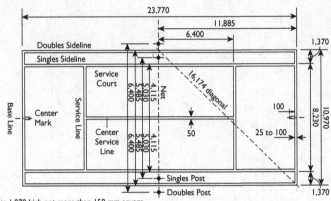

Posts 1,070 high not more than 150 mm square or 150 mm dig.

Net held down taut at center by 50 mm strap (white)

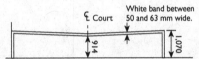

Center Service Line and Center Mark must be 50 mm wide.

Baseline to be between 50 and 100 mm wide.

All other lines to be between 25 and 50 mm wide (commonly all lines are marked 50 mm wide).

6.41 VOLLEYBALL

International volleyball is a team game played by six players on each side. The team aims to deliver the ball over the net and to ground it on their opponent's court while preventing it from touching the floor on their own.

Dimensions of the Court

The court is 18 m long × 9 m wide for all levels of play.

Clear Space above the Court

Olympic/World Championships. Backline clear space is 8 m. Sideline clear space is 3 m. Official's space additional to one side is 2 m. Overall area is 24 m × 17 m. Clear unobstructed height: National—9 m; Club—7 m.

Recreational. Backline clear space 2 m. Sideline clear space 2 m. Overall area 22 m × 13 m. Clear unobstructed height 7 m.

Net.

Height of net for men: 2.43 m.

Height of net for women: 2.24 m.

Height of net for schoolchildren: Boys ages 13 to 15 (2.35 m); Girls ages 13 to 15 (2.15 m).

Boys ages 16 to 17 (2.43 m); Girls ages 16 to 17 (2.24 m).

Mini Volleyball

Court size. 13.4 m × 6.1 m.

Net height. 2 m.

North America

Recommended area. Ground space is 4000 SF.

Surface and drainage. Recommended surface for intensive use is to be bituminous material or concrete, but sand-clay or turf may be used for informal play. Drainage is to be end to end, side to side, or corner to corner at a minimum slope of 1" in 10'.

General Specifications

- **Court orientation.** For year-round play, long axis is northwest to southeast 22 degrees off true north. For April to October play, long axis is north to south.

- **Slope.** A slope of 0.05 to 0.15% is acceptable depending on the type of surface. Courts should be sloped side to side or end to end.

- **Dimensions.** Traditional in the United States (60' long × 30' wide) or U.S. Volleyball rules (59' × 29.5'). Total area 80' × 50'.

- **Boundary lines.** Should be brightly colored ¼" rope or 1½" webbing. No centerline is required (177' needed for court lines).

- **Net height.** 7' 11⅝" for men's, 7' 4⅛" for women's, measured at the center of the playing court. 6'6" for girls 10 years and under. 7' for boys 10 years and under.. For both boys and girls 12 years and under, net is 7'.

- **Minimum clearance.** If possible, there should be a minimum clearance of 6' (preferably 10') completely around the court, and 23' above.
- The height of the net varies greatly, dependent upon level of play, and men's or women's competition. Consult league officials for legal dimensions.

Volleyball Court

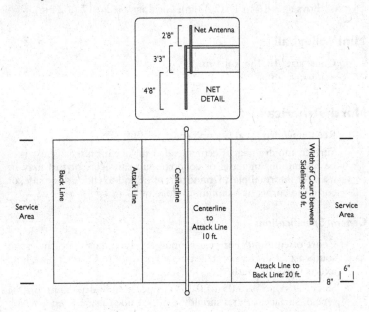

Volleyball Dimensions

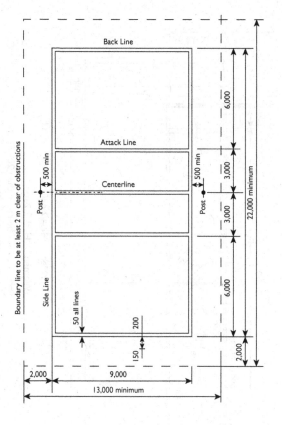

Back Line

Attack Line

Centerline

Side Line

Post

Post

Boundary line to be at least 2 m clear of obstructions

500 min

500 min

6,000

3,000

3,000

6,000

2,000

22,000 minimum

50 all lines

200

150

2,000

9,000

13,000 minimum

Volleyball Net

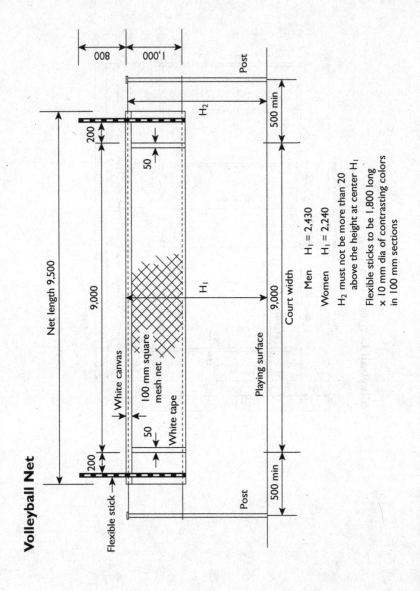

Net length 9,500

9,000

200

200

800

1,000

White canvas

100 mm square mesh net

White tape

50

50

Flexible stick

Post

Post

H₂

H₁

500 min

500 min

9,000

Court width

Playing surface

Men H₁ = 2,430

Women H₁ = 2,240

H₂ must not be more than 20 above the height at center H₁

Flexible sticks to be 1,800 long × 10 mm dia of contrasting colors in 100 mm sections

Section Seven

GENERAL INFORMATION

7.1 TEMPERATURE CONVERSION

Fahrenheit to Centigrade (°F to °C.)	Centigrade to Fahrenheit (°F to °C.)
−40 = −40	−40 = −40
−30 = −34.4	−30 = −22
−20 = −28.9	−20 = −4
−10 = −23.3	−10 = +14
0 = −17.8	0 = +32
+10 = −12.2	+10 = +50
+20 = −6.67	+20 = +68
+32 = 0.00	+30 = +86
+50 = +10.0	+50 = +122
+60 = +15.6	+60 = +140
+80 = +26.7	+80 = +176
+100 = +37.8	+100 = +212
+212 = +100.0	

$$\text{Centigrade} = (F - 32 \text{ deg.}) \times \tfrac{5}{9}$$
$$= F - 32 \times 0.5666$$
$$\text{Fahrenheit} = (C \times \tfrac{9}{5} + 32 \text{ deg.})$$
$$= C \times 1.80 + 32 \text{ F}$$

7.2 ANGULAR AND CIRCULAR MEASURE

60 seconds = 1 minute
60 minutes = 1 degree
90 degrees = 1 right angle
180 degrees = 1 straight angle
360 degrees = Full circle

7.3 MEASUREMENT OF TIME

60 seconds = 1 minute
60 minutes = 1 hour
3,600 seconds = 1 hour
24 hours = 1 day

1,440 minutes = 1 day
86,400 seconds = 1 day
7 days = 1 week
168 hours = 1 week
604,800 seconds = 1 week
52 weeks = 1 year
8,760 hours = 1 year
31,536,000 seconds = 1 year
365 days = 1 normal year
366 days = 1 leap year
360 days = 1 commercial year
10 years = 1 decade
100 years = 1 century

7.4 MEASUREMENT OF TIME: CONVERTING MINUTES TO DECIMAL HOURS

Minutes = Hour	Minutes = Hour	Minutes = Hour	Minutes = Hour
1 = 0.017	16 = 0.267	31 = 0.517	46 = 0.767
2 = 0.034	17 = 0.284	32 = 0.534	47 = 0.784
3 = 0.050	18 = 0.300	33 = 0.550	48 = 0.800
4 = 0.067	19 = 0.317	34 = 0.567	49 = 0.817
5 = 0.084	20 = 0.334	35 = 0.584	50 = 0.834
6 = 0.100	21 = 0.350	36 = 0.600	51 = 0.850
7 = 0.117	22 = 0.368	37 = 0.617	52 = 0.867
8 = 0.135	23 = 0.384	38 = 0.634	53 = 0.884
9 = 0.150	24 = 0.400	39 = 0.650	54 = 0.900
10 = 0.167	25 = 0.417	40 = 0.667	55 = 0.917
11 = 0.184	26 = 0.434	41 = 0.684	56 = 0.934
12 = 0.200	27 = 0.450	42 = 0.700	57 = 0.950
13 = 0.217	28 = 0.467	43 = 0.717	58 = 0.967
14 = 0.232	29 = 0.484	44 = 0.734	59 = 0.984
15 = 0.250	30 = 0.500	45 = 0.750	60 = 1.000

7.5 CONVERTING INCHES AND FRACTIONS TO DECIMAL PARTS OF A FOOT

	0"	⅛"	¼"	⅜"	½"	⅝"	¾"	⅞"
0"	0.000	0.010	0.021	0.031	0.042	0.052	0.063	0.073
1"	0.083	0.094	0.104	0.115	0.125	0.135	0.146	0.156
2"	0.167	0.177	0.188	0.198	0.208	0.219	0.229	0.240
3"	0.250	0.260	0.271	0.281	0.292	0.302	0.313	0.323
4"	0.333	0.344	0.354	0.365	0.375	0.385	0.396	0.406
5"	0.417	0.427	0.438	0.448	0.458	0.469	0.479	0.490
6"	0.500	0.510	0.521	0.531	0.542	0.552	0.563	0.573
7"	0.583	0.594	0.604	0.615	0.625	0.635	0.646	0.656
8"	0.667	0.677	0.688	0.698	0.708	0.719	0.729	0.740
9"	0.750	0.760	0.771	0.781	0.792	0.802	0.813	0.823
10"	0.833	0.844	0.854	0.865	0.875	0.885	0.896	0.906
11"	0.918	0.927	0.938	0.948	0.958	0.969	0.979	0.990

To use the chart: Find the point where the full inch wanted in the left column intersects with the fraction wanted in its column.

Example

6½" = 0.542'

7.6 DECIMAL EQUIVALENTS TABLE

Decimal	Fractional Equivalent	Decimal	Fractional Equivalent
0.01562	¹⁄₆₄	0.51562	³³⁄₆₄
0.03125	¹⁄₃₂	0.53125	¹⁷⁄₃₂
0.04687	³⁄₆₄	0.54687	³⁵⁄₆₄
0.0625	¹⁄₁₆	0.5625	⁹⁄₁₆
0.07812	⁵⁄₆₄	0.57812	³⁷⁄₆₄
0.09375	³⁄₃₂	0.59375	¹⁹⁄₃₂
0.10937	⁷⁄₆₄	0.60937	³⁹⁄₆₄
0.1250	⅛	0.6250	⅝
0.14062	⁹⁄₆₄	0.64062	⁴¹⁄₆₄
0.15625	⁵⁄₃₂	0.65625	²¹⁄₃₂
0.17187	¹¹⁄₆₄	0.67187	⁴³⁄₆₄
0.1785	³⁄₁₆	0.6875	¹¹⁄₁₆

(continued)

Decimal	Fractional Equivalent	Decimal	Fractional Equivalent
0.20312	$^{13}/_{64}$	0.70312	$^{45}/_{64}$
0.21875	$^{7}/_{32}$	0.71875	$^{23}/_{32}$
0.23437	$^{15}/_{64}$	0.73437	$^{47}/_{64}$
0.2500	$^{1}/_{4}$	0.7500	$^{3}/_{4}$
0.26562	$^{17}/_{64}$	0.76562	$^{49}/_{64}$
0.28125	$^{9}/_{32}$	0.78125	$^{25}/_{32}$
0.29687	$^{19}/_{64}$	0.79687	$^{51}/_{64}$
0.3125	$^{5}/_{16}$	0.8125	$^{13}/_{16}$
0.32812	$^{21}/_{64}$	0.82812	$^{53}/_{64}$
0.34375	$^{11}/_{32}$	0.84375	$^{27}/_{32}$
0.35937	$^{22}/_{64}$	0.85937	$^{55}/_{64}$
0.3750	$^{3}/_{8}$	0.8750	$^{7}/_{8}$
0.39062	$^{25}/_{64}$	0.89062	$^{57}/_{64}$
0.40625	$^{13}/_{32}$	0.90625	$^{29}/_{32}$
0.42187	$^{27}/_{64}$	0.92187	$^{59}/_{64}$
0.4375	$^{7}/_{16}$	0.9375	$^{15}/_{16}$
0.45312	$^{29}/_{64}$	0.95312	$^{61}/_{64}$
0.46875	$^{15}/_{32}$	0.96875	$^{31}/_{32}$
0.48437	$^{31}/_{64}$	0.98437	$^{63}/_{64}$
0.5000	$^{1}/_{2}$	1.00000	1

7.7 CIRCUMFERENCES AND AREAS OF CIRCLES

Diameters 1 ft. to 100 ft.

Diameter (ft.)	Circumference (ft.)	Area (ft.)
1	3.1416	0.7854
2	6.2832	3.1416
3	9.4248	7.0686
4	12.5664	12.5664
5	15.708	19.635
6	18.850	28.274
7	21.991	38.485
8	25.133	50.266
9	28.274	63.617

(continued)

7.7 CIRCUMFERENCES AND AREAS OF CIRCLES *(continued)*

Diameter (ft.)	Circumference (ft.)	Area (ft.)
10	31.416	78.540
11	34.558	95.033
12	37.699	133.10
13	40.841	132.73
14	43.982	153.94
15	47.124	176.71
16	50.265	201.06
17	53.407	226.98
18	56.549	254.47
19	59.690	283.53
20	62.832	314.16
21	65.973	346.36
22	69.115	380.13
23	72.257	415.48
24	75.398	452.39
25	78.540	490.87
26	81.681	530.93
27	84.823	572.56
28	87.965	615.75
29	91.106	660.52
30	94.248	760.86
31	97.389	754.77
32	100.53	804.25
33	103.67	855.30
34	106.81	907.92
35	109.96	962.11
36	113.10	1,017.88
37	116.24	1,075.21
38	119.38	1,134.11
39	122.52	1,194.59
40	125.66	1,256.61
41	128.81	1,320.25
42	131.95	1,385.44
43	135.09	1,452.20
44	138.23	1,520.53
45	141.37	1,590.43
46	144.51	1,661.90

(continued)

Diameter (ft.)	Circumference (ft.)	Area (ft.)
47	147.65	1,734.94
48	150.80	1,809.56
49	153.94	1,885.74
50	157.08	1,963.50
51	160.22	2,042.82
52	163.36	2,123.72
53	166.50	2,206.18
54	169.65	2,290.22
55	172.79	2,375.83
56	175.93	2,463.01
57	179.07	2,551.76
58	182.21	2,642.08
59	185.35	2,733.97
60	188.50	2,827.43
61	191.64	2,922.47
62	194.78	3,019.07
63	197.92	3,117.25
64	201.06	3,216.99
65	204.20	3,318.31
66	207.34	3,421.19
67	210.49	3,525.65
68	213.63	3,631.68
69	216.77	3,739.28
70	219.91	3,848.45
71	223.05	3,959.19
72	226.19	4,071.50
73	229.34	4,185.39
74	232.48	4,300.84
75	235.62	4,417.86
76	238.76	4,536.46
77	241.90	4,656.63
78	245.04	4,778.36
79	248.19	4,901.67
80	251.33	5,026.55
81	254.47	5,153.00
82	257.61	5,281.02

(continued)

7.7 CIRCUMFERENCES AND AREAS OF CIRCLES *(continued)*

Diameter (ft.)	Circumference (ft.)	Area (ft.)
83	260.75	5,410.61
84	263.89	5,541.77
85	267.04	5,674.50
86	270.18	5,808.80
87	273.32	5,944.68
88	276.46	6,082.12
89	279.60	6,221.14
90	282.74	6,361.73
91	285.88	6,503.88
92	289.03	6,647.61
93	292.17	6,792.91
94	295.31	6,939.78
95	298.45	7,088.22
96	301.59	7,238.23
97	304.73	7,389.81
98	307.88	7,542.96
99	311.02	7,697.69
100	314.16	7,853.98

7.8 UNITS OF LENGTH

12 inches = 1 foot
3 feet = 1 yard
4 rods = 1 chain
80 chains = 1 statute mile
320 rods = 1 statute mile
1,760 yards = 1 statute mile
5,280 feet = 1 statute mile

Units of Length Conversions

Unit	Multiplied By	Equal
Inches	0.0833	Feet
Inches	0.02778	Yards
Inches	0.0050	Rods
Inches	0.0000158	Miles

(continued)

Unit	Multiplied By	Equal
Feet	0.333	Yards
Feet	0.0606	Rods
Feet	0.0001894	Miles
Yards	36.00	Inches
Yards	3.00	Feet
Yards	0.0005681	Miles
Miles	63,360.00	Inches
Miles	5,280.00	Feet
Miles	1,760.00	Yards
Miles	320.00	Rods
Miles	80.00	Chains

7.9   UNITS OF AREA

144 square inches = 1 square foot
1,296 square inches = 1 square yard
9 square feet = 1 square yard
43,560 square feet = 1 acre
30.25 square yards = 1 square rod
4,840 square yards = 1 acre
160 square rods = 1 acre
640 acres = 1 square mile

Units of Area Conversions

Units	Multiplied By	Equal
Square inches	0.007	Square feet
Square inches	0.00077	Square yards
Square feet	144.00	Square inches
Square feet	0.11111	Square yards
Square yards	1,296.00	Square inches
Square yards	9.0	Square feet
Square yards	0.033	Square rods
Square rods	0.00625	Acres
Square miles	640.00	Acres

7.10 UNITS OF LENGTH (METRIC)

10 millimeters = 1 centimeter
10 centimeters = 1 decimeter
10 decimeters = 1 meter
10 meters = 1 decameter
10 dekameters = 1 hectometer
10 hectometers = 1 kilometer

7.11 UNITS OF AREA (METRIC)

100 square millimeters = 1 square centimeter
10,000 square centimeters = 1 square meter
100 square meters = 1 acre
100 acres = 1 hectare
100 hectares = 1 square kilometer

7.12 SQUARE TRACTS OF LAND (ENGLISH)

In Square Feet and in Acres

Length Each Side in Feet[a]	Area (SF)	Area (acres)	
66.0	4,356	0.100	(1/10)
73.8	5,445	0.125	(1/8)
85.2	7,260	0.166	(1/6)
104.4	10,890	0.250	(1/14)
120.5	14,520	0.333	(1/3)
147.6	21,780	0.500	(1/2)
180.8	32,670	0.750	(3/4)
208.7	43,560	1.000	(1)
255.6	65,340	1.500	(1 1/2)
295.2	87,120	2.000	(2)
330.0	108,900	2.500	(2 1/2)
361.5	130,680	3.000	(3)
466.7	217,800	5.000	(5)

[a]All sides equal in length.

7.13 AREA CONVERSION: SQUARE FEET TO EQUIVALENT ACREAGE

Square Feet	Acres	Square Feet	Acres
1,000	0.0230	41,000	0.9412
2,000	0.0459	42,000	0.9642
3,000	0.0689	43,000	0.9871
4,000	0.0918	43,560	1.0000
5,000	0.1148	65,340	1.5000
6,000	0.1377	87,120	2.0000
7,000	0.1607	108,900	2.5000
8,000	0.1837	130,680	3.0000
9,000	0.2066	152,460	3.5000
10,000	0.2296	174,240	4.0000
11,000	0.2525	196,020	4.5000
12,000	0.2755	217,800	5.0000
13,000	0.2984	239,580	5.5000
14,000	0.3214	261,360	6.0000
15,000	0.3444	283,140	6.5000
16,000	0.3673	304,920	7.0000
17,000	0.3903	326,700	7.5000
18,000	0.4132	348,480	8.0000
19,000	0.4362	370,260	8.5000
20,000	0.4591	392,040	9.0000
21,000	0.4821	413,820	9.5000
22,000	0.5051	435,600	10.0000
23,000	0.5280	871,200	20.0000
24,000	0.5510	1,089,000	25.0000
25,000	0.5739	1,742,400	40.0000
26,000	0.5969	2,178,000	50.0000
27,000	0.6198	2,613,600	60.0000
28,000	0.6428	3,049,200	70.0000
29,000	0.6657	3,484,800	80.0000
30,000	0.6887	3,920,400	90.0000
31,000	0.7117	4,356,000	100.0000

(continued)

7.13 AREA CONVERSION: SQUARE FEET TO EQUIVALENT ACREAGE *(continued)*

Square Feet	Acres	Square Feet	Acres
32,000	0.7346	8,712,000	200.0000
33,000	0.7576	10,890,000	250.0000
34,000	0.7805	17,424,000	400.0000
35,000	0.8035	21,780,000	500.0000
36,000	0.8264	26,136,000	600.0000
37,000	0.8494	30,492,000	700.0000
38,000	0.8724	34,848,000	800.0000
39,000	0.8953	39,204,000	900.0000
40,000	0.9183	43,560,000	1,000.0000

7.14 AREA CONVERSION: ACRES TO EQUIVALENT SQUARE FOOTAGE

Acre as Fraction	Acre as Decimal	Equivalent Square Footage
1/1000	0.001	43.56
1/100	0.01	435.6
1/10	0.100	4,356.00
1/8	0.125	5,445.00
1/4	0.250	10,890.00
1/3	0.333	14,520.00
1/2	0.500	21,780.00
3/4	0.750	32,670.00
1	1.000	43,560.00
1 1/8	1.125	49,005.00
1 1/4	1.250	54,450.00
1 1/3	1.333	58,080.00
1 1/2	1.500	65,340.00
1 3/4	1.750	76,230.00
2	2.000	87,120.00
2 1/2	2.500	108,900.00
3	3.000	130,680.00
3 1/2	3.500	152,460.00
4	4.000	174,240.00
4 1/2	4.500	196,020.00
5	5.000	217,800.00

7.15 CONVERTING MILES TO KILOMETERS AND KILOMETERS TO MILES

English Statute Miles	to	Metric Kilometers	Metric Kilometers	to	English Statute Miles
1	=	1.6093	1	=	0.62137
2	=	3.2186	2	=	1.24274
3	=	4.8279	3	=	1.86411
4	=	6.4372	4	=	2.48548
5	=	8.0465	5	=	3.10685
6	=	9.6558	6	=	3.72822
7	=	11.2651	7	=	4.34959
8	=	12.8744	8	=	4.97096
9	=	14.4837	9	=	5.59233
10	=	16.0930	10	=	6.21370

7.16 CONVERTING FEET TO METERS AND METERS TO FEET

English Feet	to	Metric Meters	Metric Meters	to	English Feet
1	=	0.3048	1	=	3.2808
2	=	0.6096	2	=	6.5616
3	=	0.9144	3	=	9.8425
4	=	1.2192	4	=	13.1233
5	=	1.5240	5	=	16.4041
6	=	1.8288	6	=	19.6850
7	=	2.1336	7	=	22.9658
8	=	2.4384	8	=	26.2466
9	=	2.7432	9	=	29.5275
10	=	3.0480	10	=	32.8083

7.17 CONVERTING INCHES TO CENTIMETERS AND CENTIMETERS TO INCHES

English Inches	to	Metric Centimeters	Metric Centimeters	to	English Inches
1	=	2.54	1	=	0.3937
2	=	5.08	2	=	0.7874
3	=	7.62	3	=	1.1811
4	=	10.16	4	=	1.5748
5	=	12.70	5	=	1.9685
6	=	15.24	6	=	2.3622
7	=	17.78	7	=	2.7559
8	=	20.32	8	=	3.1496
9	=	22.86	9	=	3.5433
10	=	25.40	10	=	3.9370

7.18 MEASUREMENT OF AREA CONVERSIONS: ENGLISH SYSTEM TO METRIC SYSTEM

Unit		Multiplied By		Equal
Square inches	×	6.452	=	Square centimeters
Square feet	×	0.0929	=	Square meters
Square yards	×	0.8361	=	Square meters
Acres	×	0.4047	=	Hectares
Square statute miles	×	2.59	=	Square kilometers

7.19 MEASUREMENT OF AREA CONVERSIONS: METRIC SYSTEM TO ENGLISH SYSTEM

Unit		Multiplied By		Equal
Square centimeters	×	0.1550	=	Square inches
Square meters	×	10.764	=	Square feet
Square meters	×	1.196	=	Square yards
Hectares	×	2.417	=	Acres
Square kilometers	×	0.3861	=	Square statute miles

7.20 MEASUREMENT OF VOLUME CONVERSIONS: ENGLISH SYSTEM TO METRIC SYSTEM

Unit		Multiplied By		Equal
Cubic inches	×	16.4	=	Cubic centimeters
Cubic feet	×	0.0283	=	Cubic meters
Cubic yards	×	0.765	=	Cubic meters

7.21 UNITS OF WEIGHT (METRIC)

10 milligrams = 1 centigram
10 centigrams = 1 decigram
10 decigrams = 1 gram
10 grams = 1 decagram
10 decagrams = 1 hectogram
10 hectograms = 1 kilogram
1,000 kilograms = 1 metric ton

7.22 UNITS OF VOLUME (ENGLISH)

1,728 cubic inches = 1 cubic foot
46,656 cubic inches = 1 cubic yard
27 cubic feet = 1 cubic yard

7.23 UNITS OF VOLUME (ENGLISH): CONVERSIONS

Unit	Multiplied By	Equal
Cubic inches	0.0005787	Cubic feet
Cubic inches	0.0000214	Cubic yards
Cubic inches	0.0043	U.S. gallons
Cubic feet	1,728.00	Cubic inches
Cubic feet	0.03704	Cubic yards
Cubic feet	7.48	U.S. gallons
Cubic yards	46,656.00	Cubic inches
Cubic yards	27.00	Cubic feet
Cubic yards	202.00	U.S. gallons

7.24 UNITS OF CUBIC MEASURE (METRIC)

1,000 cubic millimeters = 1 cubic centimeter
1,000 cubic centimeters = 1 cubic decimeter
1,000 cubic decimeters = 1 cubic meter

7.25 UNITS OF WEIGHT (ENGLISH)

7,000 grains = 1 pound
437.5 grains = 1 ounce
16 ounces = 1 pound
100 pounds = 1 hundredweight
20 hundredweight = 1 ton
2,000 pounds = 1 ton

7.26 UNITS OF WEIGHT CONVERSIONS: ENGLISH SYSTEM TO METRIC SYSTEM

1 kilogram = 2.2 pounds
1,000 kilogram = 1 metric ton
1 metric ton = 1.1 U.S. ton
1 U.S. ton = 0.90 metric ton
1 U.S. ton = 909 kilograms

7.27 UNITS OF VOLUME (ENGLISH): CONVERSIONS

Unit	Multiplied By	Equal
Grains	0.002286	Ounces
Ounces	0.0625	Pounds
Ounces	0.00003125	Hundredweight
Pounds	16.00	Ounces
Pounds	0.01	Hundredweight
Pounds	0.0005	Tons
Tons	32,000.00	Ounces
Tons	2,000.00	Pounds

7.28 UNITS OF DRY MEASURE

1 pint = 0.5 quarts
1 pint = 33.6 cubic inches
2 pints = 1.0 quart
2 pints = 67.2 cubic inches
8 quarts = 1.0 peck
8 quarts = 537.6 cubic inches
4 pecks = 1.0 bushel
1 bushel = 32.0 quarts
1 bushel = 64.0 pints
1 bushel = 2,150.4 cubic inches

7.29 UNITS OF DRY MEASURE: CONVERSIONS

Units	Multiplied By	Equal
Pints	0.5	Quarts
Pints	33.6	Cubic inches
Pints	0.0625	Pecks
Pints	0.01562	Bushels
Quarts	2.0	Pints
Quarts	0.1250	Pecks
Quarts	0.03125	Bushels
Pecks	16.0	Pints
Pecks	8.0	Quarts
Pecks	0.2500	Bushels
Bushels	4.0	Pecks
Bushels	32.0	Quarts
Bushels	64.0	Pints
Bushels	2,150.4	Cubic inches

7.30 DRY MATERIALS: CONVERSION USE FOR SMALL AREAS

Rate/Acre (lb.)	Rate/1,000 SF (ounces)	Rate/100 SF
1	0.35	0.25 teaspoon
2	0.7	0.50 teaspoon
3	1.1	0.75 teaspoon
4	1.4	1.00 teaspoon
5	1.8	1.25 teaspoon
6	2.1	1.50 teaspoon
8	2.8	1.75 teaspoon
10	3.7	2.00 teaspoon
20	7.3	0.73 ounces
40	14.0	1.40 ounces
50	18.0	1.80 ounces
100	37.0	3.70 ounces
200	73.0	7.30 ounces
300	110.0	11.00 ounces
400	147.0	14.70 ounces
500	184.0	18.40 ounces

7.31 UNITS OF LIQUID MEASURE

1 gill = 0.5 cup
1 gill = 7.23 cubic inches
2 gills = 1.0 cup
2 gills = 14.46 cubic inches
2 cups = 1.0 pint
4 gills = 1.0 pint
1 pint = 28.875 cubic inches
2 pints = 1.0 quart
4 quarts = 1.0 gallon
1 gallon = 8.0 pints
1 gallon = 231.0 cubic inches
1 gallon = 32.0 gills

7.32 UNITS OF LIQUID MEASURE: CONVERSIONS

Units	Multiplied By	Equal
Gills	0.5	Cups
Gills	7.23	Cubic inches
Gills	0.2500	Pints
Gills	0.1250	Quarts
Gills	0.03125	Gallons
Cups	2.00	Gills
Cups	0.5	Pints
Pints	4.0	Gills
Pints	2.0	Cups
Pints	0.5	Quarts
Pints	0.1250	Gallons
Quarts	2.0	Pints
Quarts	0.250	Gallons
Gallons	4.0	Quarts
Gallons	8.0	Pints
Gallons	231.0	Cubic inches

7.33 UNITS OF LIQUID MEASURE (METRIC)

10 milliliters = 1 centiliter
10 centiliters = 1 deciliter
10 deciliters = 1 liter
10 liters = 1 decaliter
10 decaliters = 1 hectoliter
10 hectoliters = 1 kiloliter

7.34 UNITS OF LIQUID MEASURE: WEIGHT EQUIVALENTS

1 pint = 16 ounces
1 quart = 32 ounces
1 gallon = 128 ounces
1 teaspoon = 0.167 ounces
1 tablespoon = 0.5 ounces

7.35 UNITS OF WATER MEASUREMENT AND EQUIVALENCIES

1 U.S. gallon	=	231 cubic inches	= 1.34 cubic feet	= 8.33 pounds
1 cubic foot	=	7.48 U.S. gallons	= 62.4 pounds	
1 acre foot	=	43,560 cubic feet	= 325,850 U.S. gallons	= 12 acre inches
1 acre inch	=	27,154 U.S. gallons		
1 U.S. gallon per minute	=	0.00223 cubic feet per second		
1 miner's inch	=	11.25 U.S. gallons per minute @ 40 inches per second feet		
1 miner's inch	=	9.0 U.S. gallons per minute @ 50 inches per second feet		
1 cubic foot per second	=	7.48 gallons/second		
	=	448.8 gallons/minute		
	=	646,272 gallons/24 hours		
	=	1.983 acre feet/24 hours		
	=	40 miner's inches (in California)		
1 cubic meter	=	1,000 liters	= 264.2 U.S. gallons	= 2,200.78 pounds
1 gallon	=	3.787 liters		
1 liter	=	1.0567 quarts		
1 liter	=	0.2642 gallon		

7.36 UNITS OF LIQUID MEASURE EQUIVALENCIES: GALLONS/POUNDS/CUBIC FEET

Gallons	Pounds	Cubic Feet
0.12	1.0	0.016
1.00	8.33	0.134
7.48	62.3	1.0
26.0	216.67	3.48
26.5	220.83	3.54
27.0	225.00	3.61
27.5	229.17	3.68
28.0	233.33	3.75
28.5	237.50	3.81
29.0	241.67	3.88
29.5	245.83	3.95
30.0	250.00	4.01
30.5	254.17	4.08
31.0	258.33	4.15
31.5	262.50	4.21
32.0	266.67	4.28
32.5	270.83	4.35
33.0	275.00	4.41
33.5	279.17	4.48
34.0	283.33	4.55
34.5	287.50	4.61
35.0	291.67	4.68
35.5	295.83	4.75
36.0	300.00	4.82
36.5	304.17	4.88
37.0	308.33	4.95
37.5	312.50	5.02
38.0	316.67	5.08
38.5	320.83	5.15
39.0	325.00	5.22
39.5	329.17	5.28
40.0	333.33	5.35

7.37 CAPACITY OF SQUARE POOLS PER FOOT OF DEPTH

Size (ft.)	Gallons	Size (ft.)	Gallons
4 × 4	119.68	12 × 12	1,007.12
5 × 5	187.00	14 × 14	1,466.08
6 × 6	269.28	16 × 16	1,914.88
7 × 7	366.52	20 × 20	2,992.00
8 × 8	478.72	30 × 30	6,732.00
9 × 9	605.88	40 × 40	11,968.00
10 × 10	748.00	50 × 50	18,700.00

7.38 CAPACITY OF RECTANGULAR POOLS PER FOOT OF DEPTH

Size (ft.)	Gallons	Size (ft.)	Gallons
4 × 8	239.36	8 × 16	957.44
4 × 12	359.04	8 × 24	1,436.16
4 × 16	478.72	10 × 20	1,496.00
5 × 10	374.00	15 × 30	3,366.00
5 × 15	561.00	20 × 40	5,984.00
5 × 20	748.00	30 × 60	13,464.00
6 × 12	538.56	40 × 80	23,936.00
6 × 18	807.84	50 × 100	37,400.00
6 × 24	1,077.02		

7.39 CAPACITY OF ROUND POOLS PER FOOT OF DEPTH

Diameter	Gallons	Diameter	Gallons
3'0"	52.88	15'0"	1,321.90
3'6"	71.97	16'0"	1,504.10
4'0"	94.00	19'0"	2,120.90
4'6"	118.97	20'0"	2,350.10
5'0"	155.00	21'0"	2,591.00

(continued)

Diameter	Gallons	Diameter	Gallons
6'0"	211.51	22'0"	2,843.60
7'0"	287.88	23'0"	3,108.00
8'0"	376.01	24'0"	3,384.10
9'0"	475.89	25'0"	3,672.00
10'0"	587.52	26'0"	3,971.60
11'0"	710.90	27'0"	4,283.00
12'0"	846.03	28'0"	4,606.20
13'0"	992.91	29'0"	4,941.00
14'0"	1,151.50	30'0"	5,287.70

To find the capacity of round pools greater than shown above: Find a tank of one-half the size desired and multiply its capacity by 4, or find one one-third the size desired and multiply its capacity by 9.

To find capacity in gallons per foot depth on oval, kidney, elliptical, or other shapes of pools: Multiply the total area of square feet by 7.48.

7.40 LIQUID MATERIALS: CONVERSIONS FOR USE FOR SMALL AREAS

Rate/Acre	Rate/1,000 SF	Rate/100 SF
1 pint	0.36 fl. oz.	0.25 teaspoon
1 quart	0.72 fl. oz.	0.50 teaspoon
1 gallon	2.90 fl. oz.	2.00 teaspoons
2 gallons	5.90 fl. oz.	4.00 teaspoons
3 gallons	8.80 fl. oz.	2.00 tablespoons
5 gallons	14.70 fl. oz.	3.00 tablespoons
10 gallons	29.40 fl. oz.	6.00 tablespoons
15 gallons	2.80 pints	9.00 tablespoons
20 gallons	3.70 pints	12.00 tablespoons
25 gallons	2.30 quarts	15.00 tablespoons
50 gallons	4.60 quarts	1.84 cups
75 gallons	6.90 quarts	2.76 cups
100 gallons	9.19 quarts	3.67 cups
200 gallons	18.37 quarts	7.35 cups

7.41 VOLUMES BASED ON AREAS OF WATER BY DEPTH (CU./YDS.)

	Area in 1,000 SF									
Depth	**1**	**1.5**	**2.0**	**2.5**	**3.0**	**3.5**	**4.0**	**4.5**	**5.0**	**5.5**
1 foot	37	56	74	93	111	130	148	167	185	204
1 yard	111	167	222	278	333	389	444	500	556	611
1 meter	121	182	243	304	364	425	486	547	607	668

	Area in 1,000 SF								
Depth	**6.0**	**6.5**	**7.0**	**7.5**	**8.0**	**8.5**	**9.0**	**9.5**	**10.0**
1 foot	222	241	259	278	296	315	333	352	370
1 yard	667	722	778	833	889	944	1,000	1,056	1,111
1 meter	729	790	850	911	972	1,032	1,093	1,154	1,215

Formula

$$\frac{\text{Depth}(\text{Decimal of foot}) \times \text{Area} = \text{Cubic yards}}{27} = \text{Cubic yards}$$

$$\text{Meters} \times 1.094 = \text{Yards}$$

$$\text{Meter} = 3.28 \text{ Feet}$$

$$\text{Cubic yards} = 229 \text{ Gallons}$$

$$\text{Cubic yards} \times 0.765 = \text{Cubic meter}$$

$$\text{Cubic meter} = 264.2 \text{ Gallons}$$

$$\text{Cubic foot} = 7.48 \text{ Gallons}$$

7.42 CALCULATION OF VOLUME BASED ON AREA OF WATER BY DEPTH

Example
Determine area by

1. Cross section.
2. Measure perimeter and refer to Section 6.7.

55,000 SF
Lake 6,111 sq. yd.
5,109 m²

If you have a lake area of 55,000 SF, with a depth of 2 yd., what would be the volume of water present?

Chart Approach

Find (10,000 SF × 5)	+ 5,000 SF volumes
(111) (cu. yd. × 5)	+ (556 cu. yd.)
5,555 + 556	= 6,111 cu. yd. per 1 yd. depth
6,111 cu. yd. × 2	= 12,222 cu. yd.

Formula Approach

Depth (decimal of yd.) × Area in square yards

= Cubic yards

2.0 × 6,111 = 12,222 cu. yd.

Metric

Depth (decimal of meter) × Area (square meters)

= Cubic meters

7.43 CONVERSION TABLE FOR U.S. AND METRIC SYSTEMS

Metric to U.S.				U.S. to Metric			
Multiply			To Obtain	Multiply			To Obtain
Millimeters (mm)	×	0.03937	= Inches	Inches (in.)	×	25.4	= Millimeters
Centimeters (cm)	×	0.3937	= Inches	Inches (in.)	×	2.54	= Centimeters
Meters (m)	×	39.37	= Inches	Inches (in.)	×	0.0254	= Meters
Meters (m)	×	3.281	= Feet	Feet (ft.)	×	0.3048	= Meters
Meters (m)	×	1.094	= Yards	Yards (yd.)	×	0.9144	= Meters
Kilometers (km)	×	0.62137	= Miles	Miles (mi.)	×	1.6093	= Kilometers
Kilometers (km)	×	1,093.62	= Yards	Yards (yd.)	×	0.00091	= Kilometers
Kilometers (km)	×	3,280.87	= Feet	Feet (ft.)	×	0.0003	= Kilometers
Liters (l)	×	1.0567	= Quarts (liq.)	Quarts (qt.)	×	0.945	= Liters
Liters (l)	×	0.2642	= Gallons (U.S.)	Gallons (gal.)	×	3.78	= Liters
Liters (l)	×	0.455	= Pounds	Pounds	×	2.2	= Liters

(Degrees Celsius × 1.80) + 32° = Degrees Fahrenheit
(Degrees Fahrenheit −32°) × 0.5666= Degrees Celsius
Kilograms per square centimeters (kg/cm²) × 14.223 = Pounds per square inch (psi)
Cubic foot (cu. ft.) × 28.316= Liters (l)

7.44 USGA SAND SPECIFICATION

Ideal Putting Green Soil Mixture Specification

In 1960, after years of research, the Green Section of the United States Golf Association (USGA) published its "Specification for a Method of Putting Green Construction." The method advocated the use of much sand in the topsoil mixture in order to resist compaction and ensure good drainage. It recommended that all soil mixtures be laboratory-tested to determine particle size distribution, infiltration rates, percent capillary (small) and noncapillary (large) pore space, and moisture retention.

Since 1960, the Green Section, through research, has refined the original specification. However, the method for determining these physical properties has remained the same. Current USGA specifications for putting green topsoil mixtures are as follows:

1. **Water permeability or infiltration rate.** 4 to 8 in./hr. for Bermuda greens; 8 to 12 in./hr. for bent greens.

2. **Pore space.** 15 to 25% large (noncapillary) pore space; 15 to 25% small (capillary) pore space.

3. **Bulk density.** 1.25 to 1.45 g/cm; minimum of 1.20; maximum of 1.60.

4. **Water retention.** 12 to 25% at 40 cm tension (field capacity).

5. **Topsoil particle size distribution.** See the following table.

Textural Name	Particle Size (mm)	Topsoil Mix Limits
Fine gravel	>2.00	<3% max.
Sand fractions		
Very coarse	1.00–2.00	<10% max.
Coarse	0.50–1.00	<30% max.
Medium	0.5–0.50	>50% min.
Fine	0.10–0.25	<25% max.
Very fine	0.05–0.10	
Silt	0.002–0.05	<5% max.
Clay	<0.002	

6. **Organic matter.** 10 to 20%.

The depth of the topsoil mixture should be a minimum of 12 in. The 12 in. top mix is above a 4 in. blanket of ⅜" pea gravel, which covers the subsurface drain tile that drains the water away from the green, preferably into a ditch, pond, or other hazard.

Where specifications meeting these requirements are followed, consistent and satisfactory golf greens can be more easily produced. When these specifications are not met, a higher level of maintenance will be needed to maintain consistently good golf greens. The golf club always suffers the consequences of poorly constructed greens; that is, increased maintenance cost and poor playing conditions. According to reports, greens built with these initial specifications, and properly maintained, can expect 20 or more years of satisfactory service.

7.45 EQUIPMENT AMORTIZATION TABLE

The Amount of Fee Increase Necessary to Compensate for New Equipment; Cost In Cents/Hr. Based on 250 (8-Hour) Days

Purchase Price	If the Equipment Will Last at Least:					
	1 yr.	2 yrs.	3 yrs.	4 yrs.	5 yrs.	6 yrs.
10.00	0.500	0.250	0.1667	0.1250	0.1000	0.0833
20.00	1.000	0.500	0.3333	0.2500	0.2000	0.1667
30.00	1.500	0.750	0.5000	0.3750	0.3000	0.2500
40.00	2.000	2.000	0.6667	0.5000	0.4000	0.3333
50.00	2.500	1.250	0.8333	0.6250	0.5000	0.4167
60.00	3.000	1.500	1.0000	0.7500	0.6000	0.5000
70.00	3.500	1.750	1.1670	0.8750	0.7000	0.5833
80.00	4.000	2.000	1.3330	1.0000	0.8000	0.6667
90.00	4.500	2.250	1.5000	1.1250	0.9000	0.7500
100.00	5.00	2.500	1.6670	1.2500	1.0000	0.8300
200.00	10.00	5.000	3.3330	2.5000	2.0000	1.6670
300.00	15.00	7.500	5.0000	3.7500	3.0000	2.5000
400.00	20.00	10.00	6.6670	5.000	4.0000	3.3330
500.00	25.00	12.50	8.3330	6.2500	5.0000	4.1670
750.00	37.50	18.75	12.50	9.3750	7.5000	6.2500
1,000.00	50.00	25.00	16.67	12.50	10.00	8.3330
5,000.00	250.00	125.00	83.33	62.50	50.00	41.67

7.46 SIMPLE INTEREST TABLE

Amount	Time	Interest Rate				
		4%	5%	6%	7%	8%
$1.00	1 month	$0.003	$0.004	$0.005	$0.005	$0.006
$1.00	2 months	0.007	0.008	0.010	0.011	0.013
$1.00	3 months	0.010	0.013	0.150	0.017	0.020
$1.00	6 months	0.020	0.025	0.030	0.035	0.040
$1.00	12 months	0.040	0.500	0.600	0.070	0.080
$100.00	1 day	0.011	0.013	0.016	0.019	0.022
$100.00	2 days	0.022	0.027	0.032	0.038	0.044
$100.00	3 days	0.034	0.041	0.050	0.058	0.067
$100.00	4 days	0.045	0.053	0.066	0.077	0.089
$100.00	5 days	0.056	0.069	0.082	0.097	0.111
$100.00	6 days	0.067	0.083	0.100	0.116	0.133
$100.00	1 month	0.334	0.416	0.500	0.583	0.667
$100.00	2 months	0.667	0.832	1.000	1.166	1.333
$100.00	3 months	1.000	1.250	1.500	1.750	2.000
$100.00	6 months	2.000	2.500	3.000	3.500	4.000
$100.00	12 months	4.000	5.000	6.000	7.000	8.000

7.47 30 DAYS INTEREST TABLE

Amount	Interest Rate										
	6%	6.5%	6.6%	7%	7.2%	7.5%	8%	8.5%	9%	9.5%	10%
$1,000.00	$5.00	$5.42	$5.50	$5.83	$6.00	$6.25	$6.67	$7.08	$7.50	$7.92	$8.33
$2,000.00	10.00	10.83	11.00	11.67	12.00	12.50	13.33	14.17	15.00	15.83	16.67
$3,000.00	15.00	16.25	16.50	17.50	18.00	18.75	20.00	21.25	22.50	23.75	25.00
$4,000.00	20.00	21.66	22.00	23.33	24.00	25.00	26.67	28.33	30.00	31.67	33.33
$5,000.00	25.00	27.08	27.50	29.17	30.00	31.25	33.33	35.42	37.50	35.58	41.67
$6,000.00	30.00	32.50	33.00	35.00	36.00	37.50	40.00	42.50	45.00	47.50	50.00
$7,000.00	35.00	37.91	38.50	40.83	42.00	43.75	46.67	49.58	52.50	55.42	58.33
$8,000.00	40.00	43.33	44.00	46.67	48.00	50.00	53.33	56.67	60.00	63.33	67.67
$9,000.00	45.00	48.74	49.50	52.50	54.00	56.25	60.00	63.75	67.50	71.25	75.00
$100.00	0.50	0.54	0.55	0.58	0.60	0.62	0.67	0.71	0.75	0.79	0.83
$200.00	1.00	1.08	1.10	1.17	1.20	1.25	1.33	1.42	1.50	1.58	1.67
$300.00	1.50	1.63	1.65	1.75	1.80	1.88	2.00	2.13	2.25	2.38	2.50
$400.00	2.00	2.17	2.20	2.33	2.40	2.50	2.67	2.83	3.00	3.17	3.33
$500.00	2.50	2.71	2.75	2.92	3.00	3.13	3.33	3.54	3.75	3.96	4.17
$600.00	3.00	3.25	3.30	3.50	3.60	3.75	4.00	4.25	4.50	4.75	5.00
$700.00	3.50	3.79	3.85	4.08	4.20	4.37	4.67	4.96	5.25	5.54	5.83
$800.00	4.00	4.33	4.40	4.67	4.80	5.00	5.33	5.67	6.00	6.33	6.67
$900.00	4.50	4.88	4.95	5.25	5.40	5.63	6.00	6.38	6.75	7.13	7.50
$10.00	0.05	0.05	0.06	0.06	0.06	0.06	0.07	0.07	0.08	0.08	0.08
$20.00	0.10	0.11	0.11	0.12	0.12	0.13	0.13	0.14	0.15	0.16	0.17
$30.00	0.15	0.16	0.17	0.18	0.18	0.19	0.20	0.21	0.23	0.24	0.25
$40.00	0.20	0.22	0.22	0.23	0.24	0.25	0.27	0.28	0.30	0.32	0.33

7.48 WORLD TIME ZONE MAP

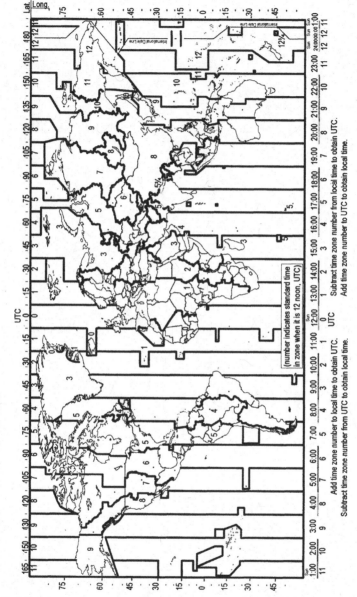

(number indicates standard time
in zone when it is 12 noon, UTC)

Add time zone number to local time to obtain UTC.
Subtract time zone number from UTC to obtain local time.

Subtract time zone number from local time to obtain UTC.
Add time zone number to UTC to obtain local time.

303

Section Eight

FORMULAS FOR AREAS AND VOLUMES OF VARIOUS GEOMETRIC FIGURES

8.1 RECTANGLE

Area = (Base)(Altitude) = ab

$$\text{Diagonal} = \sqrt{(\text{Altitude})^2 + (\text{Base})^2}$$

$$c = \sqrt{a^2 + b^2}$$

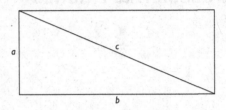

8.2 PARALLELOGRAM

Area = (Base) (Altitude)

Altitude h is perpendicular to base AB.

Angles $A + B + C + D = 360°$

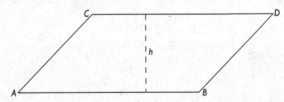

8.3   TRAPEZOID

Area = ½ (Altitude) (Sum of bases)

Altitude h is perpendicular to sides AB and CD.

Side AB is parallel to side CD.

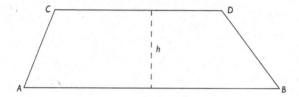

8.4   RECTANGULAR PRISM

Volume = Length × Width × Height

Volume = Area of base × Altitude

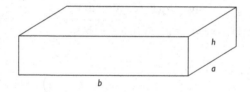

8.5   ANY PRISM

Axis either perpendicular or inclined to base

Volume = (Area of base) (Perpendicular height)

Volume = (Lateral length) (Area of perpendicular cross section)

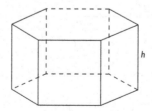

8.6 PYRAMID

Axis either perpendicular or inclined to base

Volume = ⅓ (Area of base) (Perpendicular height)

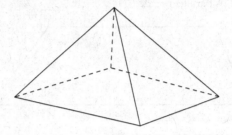

8.7 RIGHT TRIANGLE

Angle A + Angle B = Angle C = 90°

Area = ½(Base)(Altitude)

Hypotenuse = $\sqrt{(\text{Altitude})^2 + (\text{Base})^2}$

$C = \sqrt{a^2 + b^2}$

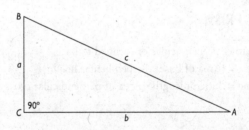

8.8 ANY TRIANGLE

Angle $A + B + C = 180°$
Altitude h is perpendicular to base c.
Area = ½ (Base) (Altitude)

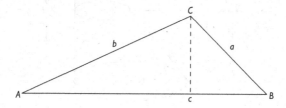

8.9 REGULAR POLYGON

$$\text{Area} = \tfrac{1}{2}\left[\begin{array}{c}\text{Length of}\\\text{one side}\end{array}\right]\left[\begin{array}{c}\text{Number}\\\text{of sides}\end{array}\right]\left[\begin{array}{c}\text{Distance}\\OA\\\text{to center}\end{array}\right]$$

A regular polygon has equal angles and equal sides and can be inscribed in or circumscribed about a circle.

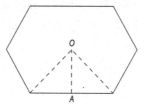

8.10 ANY CYLINDER

Axis either perpendicular or inclined to base.
Volume = (Area of base) (Perpendicular height)
Volume = (Length of axis) (Area of section perpendicular to axis)
Area of cylindrical surface = (Perimeter of base) (Perpendicular height)

8.11 ANY CONE AND FRUSTUM OF ANY CONE

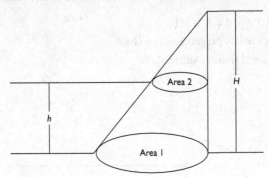

Volume of cone = ⅓ Area 1 × H

Volume of frustum = ⅓$\left[\text{Area 1} + \text{Area 2} + \sqrt{\text{Area 1} \times \text{Area 2}}\right] h$

Area 1 and Area 2 may be any shape but must be parallel.

8.12 ELLIPSE

Area = π (Long radius OA) (Short radius OC)

Area = $\frac{\pi}{4}$ (Long diameter AB) (Short diameter CD)

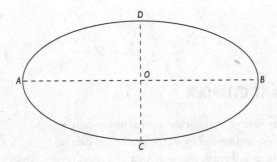

8.13 CIRCLE

AB = diameter, CD = radius

$$\text{Area} = \pi\left(\text{Radius}\right)^2 = \pi\,\frac{\left(\text{Diameter}\right)^2}{4}$$

$$\text{Circumference} = \pi\left(\text{Diameter}\right)$$

$$C = 2\pi\left(\text{Radius}\right)$$

$$\frac{\text{Arc } BC}{\text{Circumference}} = \frac{\text{Angle } BDC}{360°}$$

$$1\ \text{Radian} = \frac{180°}{\pi} = 57.2958$$

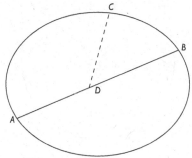

8.14 SECTOR OF A CIRCLE

$$\text{Area} = \frac{\left(\text{Arc } AB\right)\left(\text{Radius}\right)}{2}$$

$$= \pi\,\frac{\left(\text{Radius}\right)^2\left(\text{Angle } ACB\right)}{360}$$

$$= \frac{\left(\text{Radius}\right)^2\left(\text{Angle } ACB \text{ in radians}\right)}{2}$$

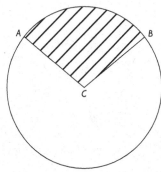

8.15 SEGMENT OF A CIRCLE

$$\text{Area} = \frac{(\text{Radius})^2}{2}\left[\frac{\pi > ACB}{180} - \sin\,ACB\right]$$

$$\text{Area} = \frac{(\text{Radius})^2}{2} > (ACB \text{ in radians} - \sin\,ACB)$$

Area = Area of sector ACB − Area of triangle ABC

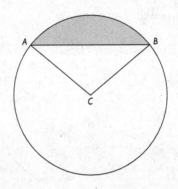

Section Nine

MAP AND SURVEYOR INFORMATION

9.1 SCALE EQUIVALENTS: SCALE 1/16" TO 1'

Scale	Feet/Inch	Square Feet/Inch
1/16" = 1'	16	256
1/8" = 1'	8	64
3/16" = 1'	5.33	28.4
1/4" = 1'	4	16
5/16" = 1'	3.2	10.24
3/8" = 1'	2.67	7.13
7/16" = 1'	2.3	5.29
1/2" = 1'	2	4
9/16" = 1'	1.77	3.13
5/8" = 1'	1.6	2.56
11/16" = 1'	1.45	2.1
3/4" = 1'	1.33	1.77
13/16" = 1'	1.23	1.51
7/8" = 1'	1.14	1.3
15/16" = 1'	1.06	1.12
1" = 1'	1	1

9.2 SCALE EQUIVALENTS: SCALES 1" TO 10'/1" TO 80'

Scale	Feet/Inch	Square Feet/Inch	Acres
1" = 10'	10'	100	0.0023
1" = 20'	20'	400	0.0092
1" = 30'	30'	900	0.0207
1" = 40'	40'	1,600	0.0367
1" = 50'	50'	2,500	0.0574
1" = 60'	60'	3,600	0.0826
1" = 66'	66'	4,356	0.1000
1" = 70'	70'	4,900	0.1124
1" = 80'	80'	6,400	0.1470

9.3   SCALE EQUIVALENTS: SCALES 1"/100' TO 1"/1,000'

Representative Fraction	Feet/Inch	Sq. Ft./Sq. Inch	Acres
1:1,200	100	10,000	0.2296
1:2,400	200	40,000	0.9184
1:3,600	300	90,000	2.0661
1:4,800	400	160,000	3.6736
1:6,000	500	250,000	5.7392
1:7,200	600	360,000	8.2644
1:12,000	1,000	1,000,000	22.9568

9.4   SCALE EQUIVALENTS: SCALES: 1" = 50' TO 1" = 10 MILES

Representative Fraction	Inch/Mile	Mile/Inch	Feet/Inch
1:600	105.60	0.0095	50
1:1,200	52.80	0.0189	100
1:2,400	26.40	0.0379	200
1:3,600	17.60	0.0568	300
1:4,800	13.20	0.0758	400
1:6,000	10.56	0.0947	500
1:7,200	8.80	0.1136	600
1:12,000	5.28	0.1894	1,000
1:63,360	1.00	1.0000	5,280
1:316,800	.2	5.0000	26,400
1:633,600	.1	10.0000	52,800

9.5 SLOPE STAKE

Distances from Side Stakes for Cross-Sectioning Roadway of Any Width, Side Slope 1½ to 1

Cut or Fill	Distances out from Side or Shoulder Stake										Cut or Fill
	0	0.1	0.2	0.3	0.4	0.5	0.6	0.7	0.8	0.9	
0	0.00	0.15	0.3	0.45	0.6	0.75	0.9	1.05	1.2	1.35	0
1	1.50	1.65	1.80	1.95	2.10	2.25	2.40	2.55	2.70	2.85	1
2	3.00	3.15	3.30	3.45	3.60	3.75	3.90	4.05	4.20	4.35	2
3	4.50	4.65	4.80	4.95	5.10	5.25	5.40	5.55	5.70	5.85	3
4	6.00	6.15	6.30	6.45	6.60	6.75	6.90	7.05	7.20	7.35	4
5	7.50	7.65	7.80	7.95	8.10	8.25	8.40	8.55	8.70	8.85	5
6	9.00	9.15	9.30	9.45	9.60	9.75	9.90	10.05	10.20	10.35	6
7	10.50	10.65	10.80	10.95	11.10	11.25	11.40	11.55	11.70	11.85	7
8	12.00	12.15	12.30	12.45	12.60	12.75	12.90	13.05	13.20	13.35	8
9	13.50	13.65	13.80	13.95	14.10	14.25	14.40	14.55	14.70	14.85	9
10	15.00	15.15	15.30	15.45	15.60	15.75	15.90	16.05	16.20	16.35	10
11	16.50	16.65	16.80	16.95	17.10	17.25	17.40	17.55	17.70	17.85	11
12	18.00	18.15	18.30	18.45	18.60	18.75	18.90	19.05	19.20	19.35	12
13	19.50	19.65	19.80	19.95	20.10	20.25	20.40	20.55	20.70	20.85	13
14	21.00	21.15	21.30	21.45	21.60	21.75	21.90	22.05	22.20	22.35	14
15	22.50	22.65	22.80	22.95	23.10	23.25	23.40	23.55	23.70	23.85	15
16	24.00	24.15	24.30	24.45	24.60	24.75	24.90	25.05	25.20	25.35	16
17	25.50	25.65	25.80	25.95	26.10	26.25	26.40	26.55	26.70	26.85	17

18	27.00	27.15	27.30	27.45	27.60	27.75	27.90	28.05	28.20	28.35	18
19	28.50	28.65	28.80	28.95	29.10	29.25	29.40	29.55	29.70	29.85	19
20	30.00	30.15	30.30	30.45	30.60	30.75	30.90	31.05	31.20	31.35	20
21	31.50	31.65	31.80	31.95	32.10	32.25	32.40	32.55	32.70	32.85	21
22	33.00	33.15	33.30	33.45	33.60	33.75	33.90	34.05	34.20	34.35	22
23	34.50	34.65	34.80	34.95	35.10	35.25	35.40	35.55	35.70	35.85	23
24	36.00	36.15	36.30	36.45	36.60	36.75	36.90	37.05	37.20	37.35	24
25	37.50	37.65	37.80	37.95	38.10	38.25	38.40	38.55	38.70	38.85	25
26	39.00	39.15	39.30	39.45	39.60	39.75	39.90	40.05	40.20	40.35	26
27	40.50	40.65	40.80	40.95	41.10	41.25	41.40	41.55	41.70	41.85	27
28	42.00	42.15	42.30	42.45	42.60	42.75	42.90	43.05	43.20	43.35	28
29	43.50	43.65	43.80	43.95	44.10	44.25	44.40	44.55	44.70	44.85	29
30	45.00	45.15	45.30	45.45	45.60	45.75	45.90	46.05	46.20	46.35	30
31	46.50	46.65	46.80	46.95	47.10	47.25	47.40	47.55	47.70	47.85	31
32	48.00	48.15	48.30	48.45	48.60	48.75	48.90	49.05	49.20	49.35	32
33	49.50	49.65	49.80	49.95	50.10	50.25	50.40	50.55	50.70	50.85	33
34	51.00	51.15	51.30	51.45	51.60	51.75	51.90	52.05	52.20	52.35	34
35	52.50	52.65	52.80	52.95	53.10	53.25	53.40	53.55	53.70	53.85	35
36	54.00	54.15	54.30	54.45	54.60	54.75	54.90	55.05	55.20	55.35	36
37	55.50	55.65	55.80	55.95	56.10	56.25	56.40	56.55	56.70	56.85	37
38	57.00	57.15	57.30	57.45	57.60	57.75	57.90	58.05	58.20	58.35	38
39	58.50	58.65	58.80	58.95	59.10	59.25	59.40	59.55	59.70	59.85	39
40	60.00	60.15	60.30	60.45	60.60	60.75	60.90	61.05	61.20	61.35	40

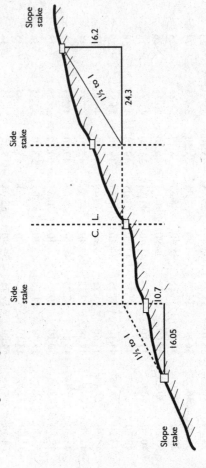

Distances from Side Stakes for Cross-Sectioning Roadway of Any Width, Side Slope 1½ to 1

318

9.6 STADIA CORRECTION AND HORIZONTAL DISTANCES

Stadia Reductions for Reading 100

Vertical Angle	Horizontal Correction	Difference Elevation	Vertical Angle	Horizontal Correction	Difference Elevation
2°00'	0.1	03.50	18°30'	10.1	30.1
3°00'	0.3	5.3	19°00'	10.6	30.8
4°00'	0.5	7.0	19°30'	11.2	31.5
5°00'	0.8	8.7	20°00'	11.7	32.1
6°00'	1.1	10.4	20°30'	12.3	32.8
7°00'	1.5	12.1	21°00'	12.8	33.5
8°00'	1.9	13.8	21°30'	13.4	34.1
9°00'	2.5	15.5	22°00'	14.0	34.7
10°00'	3.0	17.10	22°30'	14.7	35.4
10°30'	3.3	17.9	23°00'	15.3	36.0
11°00'	3.6	18.7	23°30'	15.3	36.0
11°30'	4.0	19.5	24°00'	15.9	36.6
12°00'	4.3	20.3	24°30'	16.5	37.2

(continued)

9.6 STADIA CORRECTION AND HORIZONTAL DISTANCES (continued)

Vertical Angle	Horizontal Correction	Difference Elevation	Vertical Angle	Horizontal Correction	Difference Elevation
12°30'	4.7	21.1	25°00'	17.2	37.7
13°00'	5.1	21.9	25°30'	17.9	38.3
13°30'	5.5	22.7	26°00'	19.2	39.4
14°00'	5.9	23.4	26°30'	19.9	39.9
14°30'	6.3	24.2	27°00'	20.6	40.5
15°00'	6.7	25.0	27°30'	21.3	41.0
15°30'	7.2	25.8	28°00'	22.0	42.0
16°00'	7.6	26.5	28°30'	22.8	41.9
16°30'	8.1	27.2	29°00'	23.5	42.4
17°00'	8.5	28.0	29°30'	24.3	42.9
17°30'	9.0	28.7	30°00'	25.0	43.3
18°00'	9.5	29.4			

9.7   CHAINS TO FEET AND FEET TO CHAINS CONVERSIONS

Chains to Feet		Feet to Chains	
Chains	Feet	Feet	Chains
1	66	100	1.515
2	132	200	3.030
3	198	300	4.545
4	264	400	6.060
5	330	500	7.575
6	396	600	9.090
7	462	700	10.606
8	528	800	12.121
9	594	900	13.636
10	660	1,000	15.151

9.8   TRIGONOMETRIC FORMULAS

For Angle A: $\sin = a/c$, $\cos = b/c$, $\tan = a/b$, $\cot = b/a$, $\sec = c/b$, $\operatorname{cosec} = c/a$.

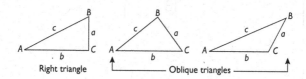

Right triangle   ————— Oblique triangles —————

Solution of Right Triangles

Given	Required	Formulas
a, b	A, B, c	$\tan A = \dfrac{a}{b}$, $\quad = \cot B$, $\quad c = \sqrt{a^2 + b^2} = a\sqrt{1 + \dfrac{b^2}{a^2}}$
a, c	A, B, b	$\sin A = \dfrac{a}{c}$, $\quad = \cos B$, $\quad b = \sqrt{(c+a)(c-a)} = c\sqrt{1 - \dfrac{a^2}{c^2}}$
A, a	B, b, c	$B = 90° - A$, $\quad b = a \cot A$, $\quad c = \dfrac{a}{\sin A}$
A, b	B, a, c	$B = 90° - A$, $\quad a = b \tan A$, $\quad c = \dfrac{b}{\cos A}$
A, c	B, a, b	$B = 90° - A$, $\quad a = c \sin A$, $\quad b = c \cos A$
A, B, a	b, c, C	$b = \dfrac{a \sin B}{\sin A}$, $\quad C = 180° - (A + B)$, $\quad C = \dfrac{a \sin C}{\sin A}$

Given	Required	Formulas
A, a, b	B, c, C	$\sin B = \dfrac{b\sin A}{a}$, $\qquad C = 180° - (A+B),$ $\qquad c = \dfrac{a\sin C}{\sin A}$
a, b, C	A, B, c	$A+B = 180° - C,\ \tan\tfrac{1}{2}(A-B) = \dfrac{(a-b)\tan\frac{1}{2}(A+B)}{(a+b)}\quad c = \dfrac{a\sin C}{\sin A}$
a, b, c	A, B, C	$S = \dfrac{a+b+c}{2},\ \sin\tfrac{1}{2}A = \sqrt{\dfrac{(S-b)(S-c)}{bc}}\ ,\quad \sin\tfrac{1}{2}B = \sqrt{\dfrac{(S-a)(S-c)}{ac}}\quad C = 180° - (A+B)$
a, b, c	Area	$S = \dfrac{a+b+c}{2},\ \text{Area} = \sqrt{S(S-a)(S-b)(S-c)}$
A, b, c	Area	$\text{Area} = \dfrac{bc\sin A}{2}$
A, B, C, a	Area	$\text{Area} = \dfrac{a^2\sin B\sin C}{2\sin A}$

9.9   REDUCTION TO HORIZONTAL

Horizontal distance is equal to slope distance multiplied by the cosine of the vertical angle. Given that slope distance = 319.4 ft., and vertical angle = 5°10'. From Section 9.8, cos 5°10' = 0.9959, so

$$319.4 \times .9959 = 318.09 \text{ ft.}$$

An alternate formula is this: Horizontal distance equals slope distance minus slope distance multiplied by (1 – cosine of vertical angle). With the same figures as in the preceding example, the following result is obtained. Cosine 5°10' = 0.9959. Therefore

$$1 - 0.9959 = 0.0041$$
$$319.4 \times 0.0041 = 1.31$$
$$319.4 - 1.31 = 318.09 \text{ ft.}$$

When the rise is known, the horizontal distance is approximately equal to slope distance less the square of the rise divided by twice the slope distance. Thus:

$$\text{Rise} = 14 \text{ ft.}$$
$$\text{Slope distance} = 302.6 \text{ ft.}$$

$$\text{Horizontal distance} = 302.6 - \frac{14 \times 14}{2 \times 302.6} = 302.6 - 0.32 = 302.28 \text{ ft.}$$

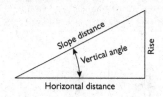

9.10 NATURAL TRIGONOMETRICAL FUNCTIONS

Angle ° '	Sine	Tangent	Secant	Cosecant	Cotangent	Cosine	Angle ° '
0	0	0	1.			1.	90
10	0.0029	0.0029	1.	343.8	343.8	1.	50
20	0.0058	0.0058	1.	171.9	171.9	0.99998	40
30	0.0087	0.0087	1.	114.6	114.6	0.99996	30
40	0.0116	0.0116	1.0001	85.94	85.94	0.99993	20
50	0.0145	0.0145	1.0001	68.75	68.76	0.99989	10
1	0.0175	0.0175	1.0002	57.30	57.29	0.99985	89
10	0.0204	0.0204	1.0002	49.11	49.10	0.99979	50
20	0.0233	0.0233	1.0003	42.98	42.96	0.99973	40
30	0.0262	0.0262	1.0003	38.20	38.19	0.99966	30
40	0.0291	0.0291	1.0004	34.38	34.37	0.99958	20
50	0.0320	0.0320	1.0005	31.26	31.24	0.99949	10
2	0.0349	0.0349	1.0006	28.65	28.64	0.99939	88
10	0.0378	0.0378	1.0007	26.45	26.43	0.99929	50
20	0.0407	0.0407	1.0008	24.56	24.54	0.99917	40
30	0.0436	0.0437	1.0010	22.93	22.90	0.99905	30
40	0.0465	0.0466	1.0011	21.49	21.47	0.99892	20
50	0.0494	0.0495	1.0012	20.23	20.21	0.99878	10

(continued)

9.10 NATURAL TRIGONOMETRICAL FUNCTIONS (continued)

Angle ° '	Sine	Tangent	Secant	Cosecant	Cotangent	Cosine	Angle ° '
3	0.0523	0.0524	1.0014	19.11	19.08	0.99863	87
10	0.0552	0.0553	1.0015	18.10	18.07	0.99847	50
20	0.0581	0.0582	1.0017	17.20	17.17	0.99831	40
30	0.0610	0.0612	1.0019	16.38	16.35	0.99813	30
40	0.6400	0.0641	1.0020	15.64	15.60	0.99795	20
50	0.0669	0.0670	1.0022	14.96	14.92	0.99776	10
4	0.0698	0.0699	1.0024	14.34	14.30	0.99756	86
10	0.0727	0.0729	1.0027	13.76	13.73	0.99736	50
20	0.0756	0.0758	1.0029	13.23	13.20	0.99714	40
30	0.0785	0.0787	1.0031	12.75	12.71	0.99692	30
40	0.0814	0.0816	1.0033	12.29	12.25	0.99668	20
50	0.0843	0.0846	1.0036	11.87	11.83	0.99644	10
5	0.0872	0.0875	1.0038	11.47	11.43	0.99619	85
10	0.0901	0.0904	1.0041	11.10	11.06	0.99594	50
20	0.0929	0.0934	1.0043	10.76	10.71	0.99567	40
30	0.0958	0.0963	1.0046	10.43	10.39	0.99540	30
40	0.0987	0.0992	1.0049	10.13	10.08	0.99511	20
50	0.1016	0.1022	1.0052	9.839	9.788	0.99482	10
6	0.1045	0.1051	1.0055	9.567	9.514	0.99452	84
10	0.1074	0.1080	1.0058	9.309	9.255	0.99421	50

20	0.1103	0.1110	1.0061	9.065	9.010	0.99390	40
30	0.1132	0.1139	1.0065	8.834	8.777	0.99357	30
40	0.1161	0.1169	1.0068	8.614	8.556	0.99324	20
50	0.1190	0.1198	1.0072	8.405	8.345	0.99290	10
7	0.1219	0.1228	1.0075	8.206	8.144	0.99255	83
10	0.1248	0.1257	1.0079	8.016	7.953	0.99219	50
20	0.1276	0.1287	1.0082	7.834	7.770	0.99182	40
30	0.1305	0.1317	1.0086	7.661	7.596	0.99144	30
40	0.1334	0.1346	1.0090	7.496	7.429	0.99106	20
50	0.1363	0.1376	1.0094	7.337	7.269	0.99067	10
8	0.1392	0.1405	1.0098	7.185	7.115	0.99027	82
10	0.1421	0.1435	1.0102	7.040	6.968	0.98986	50
20	0.1449	0.1465	1.0107	6.900	6.827	0.98944	40
30	0.1478	0.1495	1.0111	6.766	6.691	0.98902	30
40	0.1507	0.1524	1.0115	6.636	6.561	0.98858	20
50	0.1536	0.1554	1.0120	6.512	6.435	0.98814	10
9	0.1564	0.1584	1.0125	6.394	6.314	0.98769	81
10	0.1593	0.1614	1.0129	6.277	6.197	0.98723	50
20	0.1622	0.1644	1.0134	6.166	6.084	0.98676	40
30	0.1650	0.1673	1.0139	6.059	5.976	0.98629	30
40	0.1679	0.1703	1.0144	5.955	5.871	0.98580	20
50	0.1708	0.1733	1.0149	5.855	5.769	0.98531	10
10	0.1736	0.1763	1.0154	5.759	5.671	0.98481	80

(continued)

9.10 NATURAL TRIGONOMETRICAL FUNCTIONS (continued)

Angle ° '	Sine	Tangent	Secant	Cosecant	Cotangent	Cosine	Angle ° '
10	0.1765	0.1793	1.0160	5.665	5.576	0.98430	50
20	0.1794	0.1823	1.0165	5.575	5.485	0.98378	40
30	0.1822	0.1853	1.0170	5.488	5.396	0.98325	30
40	0.1851	0.1883	1.0176	5.403	5.309	0.98272	20
50	0.1880	0.1914	1.0181	5.320	5.226	0.98218	10
11	0.1908	0.1944	1.0187	5.241	5.145	0.98163	79
10	0.1937	0.1974	1.0193	5.164	5.066	0.98107	50
20	0.1965	0.2004	1.0199	5.089	4.989	0.98050	40
30	0.1994	0.2035	1.0205	5.016	4.915	0.97992	30
40	0.2022	0.2065	1.0211	4.945	4.843	0.97934	20
50	0.2051	0.2095	1.0217	4.877	4.773	0.97875	10
12	0.2079	0.2126	1.0223	4.810	4.705	0.97815	78
10	0.2108	0.2156	1.0230	4.745	4.638	0.97754	50
20	0.2136	0.2186	1.0236	4.682	4.574	0.97692	40
30	0.2164	0.2217	1.0243	4.620	4.511	0.97630	30
40	0.2193	0.2247	1.0249	4.560	4.449	0.97566	20
50	0.2221	0.2278	1.0256	4.502	4.390	0.97502	10
13	0.2250	0.2309	1.0263	4.445	4.331	0.97437	77
10	0.2278	0.2339	1.0270	4.39	4.275	0.97371	50
20	0.2306	0.2370	1.0277	4.336	4.219	0.97304	40

30	0.2334	0.2401	1.0284	4.284	4.165	0.97237	30
40	0.2363	0.2432	1.0291	4.232	4.113	0.97169	20
50	0.2391	0.2462	1.0299	4.182	4.061	0.97100	10
14	0.2419	0.2493	1.0306	4.133	4.011	0.97030	76
10	0.2447	0.2524	1.0314	4.086	3.962	0.96959	50
20	0.2476	0.2555	1.0321	4.039	3.914	0.96887	40
30	0.2504	0.2586	1.0329	3.994	3.867	0.96815	30
40	0.2532	0.2617	1.0337	3.949	3.821	0.96742	20
50	0.2560	0.2648	1.0345	3.906	3.776	0.96667	10
15	0.2588	0.2679	1.0353	3.864	3.732	0.96593	75
10	0.2616	0.2711	1.0361	3.822	3.689	0.96517	50
20	0.2644	0.2742	1.0369	3.782	3.647	0.96440	40
30	0.2672	0.2773	1.0377	3.742	3.606	0.96363	30
40	0.2700	0.2805	1.0386	3.703	3.566	0.96285	20
50	0.2728	0.2836	1.0394	3.665	3.526	0.96206	10
16	0.2756	0.2867	1.0403	3.628	3.487	0.96126	74
10	0.2784	0.2899	1.0412	3.592	3.450	0.96046	50
20	0.2812	0.2931	1.0423	3.556	3.412	0.95964	40
30	0.2840	0.2962	1.0429	3.521	3.376	0.95882	30
40	0.2868	0.2994	1.0438	3.487	3.340	0.95799	20
50	0.2896	03026	1.0448	3.453	3.305	0.95715	10
17	0.2924	0.3057	1.0457	3.420	3.271	0.95630	73
10	0.2952	0.3089	1.0466	3.388	3.237	0.95545	50

(continued)

9.10 NATURAL TRIGONOMETRICAL FUNCTIONS (continued)

Angle ° '	Sine	Tangent	Secant	Cosecant	Cotangent	Cosine	Angle ° '
20	0.2979	0.3121	1.0476	3.357	3.204	0.95459	40
30	0.3007	0.3153	1.0485	3.326	3.172	0.95372	30
40	0.3035	0.3185	1.0495	3.295	3.140	0.95284	20
50	0.3062	0.3217	1.0505	3.265	3.108	0.95195	10
18	0.3090	0.3249	1.0515	3.236	3.078	0.95106	72
10	0.3118	0.3281	1.0525	3.207	3.048	0.95015	50
20	0.3145	0.3314	1.0535	3.179	3.018	0.94924	40
30	0.3173	0.3346	1.0545	3152	2.989	0.94832	30
40	0.3201	0.3378	1.0555	3.124	2.960	0.94740	20
50°	0.3228	0.3411	1.0566	3.098	2.932	0.94646	10
19	0.3256	0.3443	1.0576	3.072	2.904	0.94552	71
10	0.3283	0.3476	1.0587	3.046	2.877	0.94457	50
20	0.3311	0.3508	1.0598	3.020	2.850	0.94361	40
30	0.3338	0.3541	1.0608	2.996	2.824	0.94264	30
40	0.3365	0.3574	1.0619	2.971	2.798	0.94167	20
50	0.3393	0.3607	1.0631	2.947	2.773	0.94068	10
20	0.3420	0.3640	1.0642	2.924	2.747	0.93969	70
10	0.3448	0.3673	1.0653	2.900	2.723	0.93869	50
20	0.3475	0.3706	1.0665	2.878	2.699	0.93769	40
30	0.3502	0.3739	1.0676	2.856	2.675	0.93667	30

40	0.3529	0.3772	1.0688	2.833	2.651	0.93565	20
50	0.3557	0.3805	1.0700	2.811	2.628	0.93462	10
21	0.3584	0.3839	1.0711	2.790	2.605	0.93358	69
10	0.3611	0.3872	1.0723	2.769	2.583	0.93253	50
20	0.3638	0.3906	1.0736	2.749	2.560	0.93148	40
30	0.3665	0.3939	1.0748	2.729	2.539	0.93042	30
40	0.3692	0.3973	1.0760	2.709	2.517	0.92935	20
50	0.3719	0.4006	1.0773	2.689	2.496	0.92827	10
22	0.3746	0.4040	1.0785	2.670	2.475	0.92718	68
10	0.3773	0.4074	1.0798	2.650	2.455	0.92609	50
20	0.3800	0.4108	1.0811	2.632	2.434	0.92499	40
30	0.3827	0.4142	1.0824	2.613	2.414	0.92388	30
40	0.3854	0.4176	1.0837	2.595	2.394	0.92276	20
50	0.3881	0.4210	1.0850	2.577	2.375	0.92164	10
23	0.3907	0.4245	1.0864	2.559	2.356	0.92050	67
10	0.3934	0.4279	1.0877	2.542	2.337	0.91936	50
20	0.3961	0.4314	1.0891	2.525	2.318	0.91822	40
30	0.3987	0.4348	1.0904	2.508	2.300	0.91706	30
40	0.4014	0.4383	1.0918	2.491	2.282	0.91590	20
50	0.4041	0.4417	1.0932	2.475	2.264	0.91472	10
24	0.4067	0.4452	1.0946	2.459	2.246	0.91355	66

(continued)

331

9.10 NATURAL TRIGONOMETRICAL FUNCTIONS (continued)

Angle ° '	Sine	Tangent	Secant	Cosecant	Cotangent	Cosine	Angle ° '
10	0.4094	0.4487	1.0961	2.443	2.229	0.91236	50
20	0.4120	0.4522	1.0975	2.427	2.211	0.91116	40
30	0.4147	0.4557	1.0989	2.411	2.194	0.90996	30
40	0.4173	0.4592	1.1004	2.396	2.177	0.90875	20
50	0.4200	0.4628	1.1019	2.381	2.161	0.90753	10
25	0.4226	0.4663	1.1034	2.366	2.145	0.90631	65
10	0.4253	0.4699	1.1049	2.351	2.128	0.90507	50
20	0.4279	0.4734	1.1064	2.337	2.112	0.90383	40
30	0.4305	0.4770	1.1079	2.323	2.097	0.90259	30
40	0.4331	0.4806	1.1095	2.309	2.081	0.90133	20
50	0.4358	0.4841	1.1110	2.295	2.066	0.90007	10
26	0.4384	0.4877	1.1126	2.281	2.050	0.89879	64
10	0.4410	0.4913	1.1142	2.268	2.035	0.89752	50
20	0.4436	0.4950	1.1158	2.254	2.020	0.89623	40
30	0.4462	0.4986	1.1174	2.241	2.006	0.89493	30
40	0.4488	0.5022	1.1190	2.228	1.991	0.89363	20
50	0.4514	0.5052	1.1207	2.215	1.977	0.89232	10
27	0.4540	0.5095	1.1223	2.203	1.963	0.89101	63
10	0.4566	0.5132	1.1240	2.190	1.949	0.88968	50
20	0.4592	0.5169	1.1257	2.178	1.935	0.88835	40

30	0.4617	0.5206	1.1274	2.166	1.921	0.88701	30
40	0.4643	0.5243	1.1291	2.154	1.907	0.88566	20
50	0.4669	0.5280	1.1308	2.142	1.894	0.88431	10
28	0.4695	0.5317	1.1326	2.130	1.881	0.88295	62
10	0.4720	0.5354	1.1343	2.119	1.868	0.88158	50
20	0.4746	0.5392	1.1361	2.107	1.855	0.88020	40
30	0.4772	0.5430	1.1379	2.096	1.842	0.87882	30
40	0.4797	0.5467	1.1397	2.085	1.829	0.87743	20
50	0.4823	0.5505	1.1415	2.073	1.816	0.87603	10
29	0.4848	0.5543	1.1434	2.063	1.804	0.87462	61
10	0.4874	0.5581	1.1452	2.052	1.792	0.87321	50
20	0.4899	0.5619	1.1471	2.041	1.780	0.87178	40
30	0.4924	0.5658	1.1490	2.031	1.767	0.87036	30
40	0.4950	0.5696	1.1509	2.020	1.756	0.86892	20
50	0.4975	0.5735	1.1528	2.010	1.744	0.86748	10
30	0.5000	0.5774	1.1547	2.000	1.732	0.86603	60
10	0.5025	0.5812	1.1566	1.990	1.720	0.86457	50
20	0.5050	0.5851	1.1586	1.980	1.709	0.86310	40
30	0.5075	0.5890	1.1606	1.970	1.698	0.86163	30
40	0.5100	0.5930	1.1626	1.961	1.686	0.86015	20
50	0.5125	0.5969	1.1646	1.951	1.675	0.85866	10
31	0.5150	0.6009	1.1666	1.924	1.664	0.85717	59

(continued)

9.10 NATURAL TRIGONOMETRICAL FUNCTIONS (continued)

Angle ° '	Sine	Tangent	Secant	Cosecant	Cotangent	Cosine	Angle ° '
20	0.6202	0.7907	1.2748	1.612	1.265	0.78442	40
30	0.6225	0.7954	1.2778	1.606	1.257	0.78261	30
40	0.6248	0.8002	1.2808	1.601	1.250	0.78079	20
50	0.6271	0.8050	1.2838	1.595	1.242	0.77897	10
39	0.6293	0.8098	1.2868	1.589	1.235	0.77715	51
10	0.6316	0.8146	1.2898	1.583	1.228	0.77531	50
20	0.6338	0.8195	1.2929	1.578	1.220	0.77347	40
30	0.6361	0.8243	1.2959	1.572	1.213	0.77162	30
40	0.6383	0.8292	1.2991	1.567	1.206	1.76977	20
50	0.6406	0.8342	1.3022	1.561	1.199	0.76791	10
40	0.6428	0.8391	1.3054	1.556	1.192	0.76604	40
10	0.6450	0.8441	1.3086	1.550	1.185	0.76417	50
20	0.6472	0.8491	1.3118	1.545	1.178	0.76229	40
30	0.6494	0.8541	1.3151	1.540	1.171	0.76041	30
40	0.6517	0.8591	1.3184	1.535	1.164	0.75851	20
50	0.6539	0.8642	1.3217	1.529	1.157	0.75661	10
41	0.6561	0.8693	1.3251	1.524	1.150	0.75471	41
10	0.6583	0.8744	1.3284	1.519	1.144	0.75280	50
20	0.6604	0.8796	1.3318	1.514	1.137	0.75088	40

30	0.6626	0.8847	1.3352	1.509	1.130	0.74896	30
40	0.6648	0.8899	1.3386	1.504	1.124	0.74703	20
50	0.6670	0.8952	1.3421	1.499	1.117	0.74509	10
42	0.6691	0.9004	1.3456	1.494	1.111	0.74314	42
10	0.6713	0.9057	1.3492	1.490	1.104	0.74120	50
20	0.6734	0.9110	1.3527	1.485	1.098	0.73924	40
30	0.6756	0.9163	1.3563	1.480	1.091	0.73728	30
40	0.6777	0.9217	1.3600	1.476	1.085	0.73531	20
50	0.6799	0.9271	1.3636	1.471	1.079	0.73333	10
43	0.6820	0.9325	1.3673	1.466	1.072	0.73135	43
10	0.6841	0.9380	1.3711	1.462	1.066	0.72937	50
20	0.6862	0.9435	1.3748	1.457	1.060	0.72737	40
30	0.6884	0.9490	1.3786	1.453	1.054	0.72537	30
40	0.6905	0.9545	1.3824	1.448	1.048	0.72337	20
50	0.6926	0.9601	1.3863	1.444	1.042	0.72136	10
44	0.6947	0.9657	1.3902	1.440	1.036	0.71934	44
10	0.6967	0.9713	1.3941	1.435	1.030	0.71732	50
20	0.6988	0.9770	1.3980	1.431	1.024	0.71529	40
30	0.7009	0.9827	1.4020	1.427	1.018	0.71325	30
40	0.7030	0.9884	1.4061	1.422	1.012	0.71121	20
50	0.7050	0.9942	1.4101	1.418	1.006	0.70916	10
	0.7071	1.	1.4140	1.414	1.	0.70711	

9.11 CURVE TABLE

Table of Tangent and External to a 1-Degree Curve

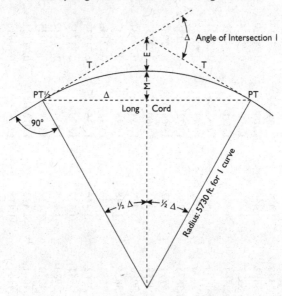

To find tangent and external for curve of any other degree, divide by degree of curve and add correction found in column of correction.

Degree of curve with a given *I* may be found by dividing tangent (or external) opposite *I* by given tangent (or external).

The distance from a point on the tangent to the curve is very nearly the square of the tangent length divided by twice the radius.

Curve Formulas

Radius

$$R = \frac{50}{\sin \frac{1}{2} D}$$

Length of Curve

$$L = 100 \frac{A}{D}$$

also

$$L = 0.0174533 \times \Delta \times R$$

Degree of Curve

$$D = 100 \frac{\Delta}{L}$$

Tangent

$$T = R \tan \frac{1}{2} \Delta$$

Long Cord

$$LC = 2R \sin \frac{1}{2} \Delta$$

Middle Ordinate

$$M = R \left(1 - \cos \frac{1}{2} \Delta \right)$$

External

$$E = T \tan \frac{1}{4} \Delta$$

9.12 TANGENTS AND EXTERNALS TO A 1-DEGREE CURVE

I	T	E
1°	50.00	0.218
10'	58.34	0.297
20'	66.67	0.388
30'	75.01	0.491
40'	83.34	0.606
50'	91.68	0.733
2°	100.01	0.873
10'	108.35	1.024
20'	116.68	1.188
30'	125.02	1.364
40'	133.36	1.552
50'	141.70	1.752
3°	150.04	1.964
10'	158.38	2.188
20'	166.72	2.425
30'	175.06	2.674
40'	183.40	2.934
50'	191.74	3.207
4°	200.08	3.492
10'	208.43	3.790
20'	216.77	4.099
30'	225.12	4.421
40'	233.47	4.755
50'	241.81	5.100
5°	250.16	5.549
10'	258.51	5.829
20'	266.86	6.211
30'	275.21	6.606
40'	283.57	7.013
50'	291.92	7.432

I = 10°
	+ 5°C	10°C	15°C
T	0.03	0.06	0.09
E	0.001	0.003	0.004

I	T	E
11°	551.70	26.500
10'	560.11	27.313
20'	568.53	28.137
30'	576.95	28.974
40'	585.36	29.824
50'	593.79	30.686
12°	602.21	31.561
10'	610.44	32.447
20'	619.07	33.347
30'	627.50	34.259
40'	635.93	35.183
50'	644.37	36.120
13°	652.81	37.070
10'	661.25	38.031
20'	669.70	39.006
30'	678.15	39.993
40'	686.60	40.992
50'	695.06	42.004
14°	703.51	43.029
10'	711.97	44.066
20'	720.44	45.116
30'	728.90	46.178
40'	737.37	47.253
50'	745.85	48.341
15°	754.32	49.441
10'	762.80	50.554
20'	771.29	51.679
30'	779.77	52.818
40'	788.26	53.969
50'	796.75	55.132

I = 20°
	+ 5°C	10°C	15°C
T	0.06	0.13	0.019
E	0.006	0.011	0.0017T

I	T	E
21°	1061.9	97.577
10'	1070.6	99.155
20'	1079.2	100.75
30'	1087.8	102.35
40'	1096.4	103.97
50'	1105.1	105.60
22°	1123.7	107.24
10'	1122.4	108.90
20'	1131.0	110.57
30'	1139.7	112.25
40'	1148.4	113.95
50'	1157.0	115.66
23°	1165.7	11738
10'	1174.4	119.12
20'	1183.1	120.87
30'	1191.8	122.63
40'	1200.5	124.41
50'	1209.2	126.20
24°	1217.9	128.00
10'	1226.6	129.82
20'	1235.3	131.65
30'	1244.0	133.50
40'	1252.8	135.35
50'	1261.5	137.23
25°	1270.2	139.11
10'	1279.0	141.01
20'	1287.7	142.93
30'	1296.5	144.85
40'	1305.3	246.79
50'	1314.0	148.75

I = 34°
	+ 5°C	10°C	15°C
T	0.01	0.19	0.29
E	0.0013	0.025	0.38
			0.29

6°	300.28	7.863		**16°**	805.25	56.309		**26°**	1322.8	150.71	0.038
10'	308.64	8.307		10'	813.75	57.498		10'	1331.6	152.69	
20'	316.99	8.762		20'	822.25	58.699		20'	1340.4	154.69	
30'	325.35	9.230	20° C	30'	830.76	59.914	20° C	30'	1349.2	156.70	20° C
40'	333.71	9.710	T	40'	839.27	61.141	T	40'	1358.0	158.72	T
50'	342.08	10.202	0.13	50'	847.78	62.381	0.26	50'	1366.8	160.76	0.39
7°	350.44	10.707	E	**17°**	856.30	63.634	E	**27°**	1375.6	162.81	E
10'	358.81	11.224	0.006	10'	864.82	64.900	0.022	10'	1384.4	164.86	0.051
20'	367.17	11.753		20'	873.35	66.178		20'	1393.2	166.95	
30'	375.54	12.294		30'	881.88	67.470		30'	1402.0	169.04	
40'	383.91	12.847		40'	890.41	68.774		40'	1410.9	171.15	
50'	392.28	13.413		50'	898.95	70.091		50'	1419.7	173.27	
8°	400.66	13.991	25° C	**18°**	907.49	71.421	25° C	**28°**	1428.6	175.41	25° C
10'	409.23	14.582	T	10'	916.03	72.764	T	10'	1437.4	177.55	T
20'	417.41	15.184	0.16	20'	924.58	74.119	0.32	20'	1446.3	179.72	0.49
30'	425.79	15.799	E	30'	933.13	75.488	E	30'	1455.1	181.89	E
40'	437.17	16.426	0.007	40'	941.69	76.869	0.028	40'	1464.0	184.08	0.065
50'	442.55	17.065		50'	950.25	78.264		50'	1472.9	186.29	
9°	450.93	17.717		**19°**	958.91	79.671		**29°**	1481.8	188.51	
10'	459.32	18.381		10'	967.38	81.092		10'	1490.7	190.74	
20'	467.71	19.058		20'	975.96	82.525		20'	1499.6	192.99	
30'	476.10	19.746		30'	984.53	83.972		30'	1508.5	195.25	
40'	484.49	20.447		40'	993.12	85.431		40'	1517.4	197.53	
50'	492.88	21.161		50'	1001.70	86.904		50'	1526.3	199.82	
10°	501.28	21.287	30° C	**20°**	1010.30	88.389	30° C	**30°**	1535.3	202.12	30° C
10'	509.68	22.624	T	10'	1018.90	88.888		10'	1544.2	204.44	T
20'	518.08	23.375	0.19	20'	1027.50	91.399	0.39	20'	1553.1	206.77	0.59
30'	526.48	24.138	E	30'	1036.10	92.924		30'	1562.1	209.12	E
40'	534.89	24.913	0.008	40'	1044.70	94.642		40'	1571.0	211.48	0.078
50'	543.29	25.700		50'	1053.30	96.013		50'	1580.0	213.86	

T = R tan I/2I *E = R exsec I/2I*

(continued)

9.12 TANGENTS AND EXTERNALS TO A 1-DEGREE CURVE (continued)

I	E	T	I=10°	I	T	E	I=20°	I	T	E	I=34°
31°	216.3	1589.0	+	41°	2142.2	387.4	+	51°	2732.9	618.4	+
10'	218.7	1598.0	5°C	10'	2151.7	390.7	5°C	10'	2743.1	622.8	5°C
20'	221.1	1606.9	T	20'	2161.2	394.1	T	20'	2753.4	627.2	T
30'	223.5	1615.9	0.13	30'	2170.8	397.4	0.17	30'	2763.7	631.7	0.21
40'	226.0	1624.9	E	40'	2180.3	400.8	E	40'	2773.9	636.2	E
50'	228.4	1633.9	0.023	50'	2189.9	404.2	0.037	50'	2784.2	640.7	0.056
32°	230.9	1643.0		42°	2199.4	407.6		52°	2794.5	645.2	
10'	233.4	1652.0		10'	2209.0	411.1		10'	2804.9	649.7	
20'	235.9	1661.0		20'	2218.6	414.5		20'	2815.2	654.3	
30'	238.4	1670.0		30'	2228.1	418.0		30'	2825.6	658.8	
40'	241.0	1679.1		40'	2237.7	421.4		40'	2835.9	663.4	
50'	243.5	1688.1		50'	2247.3	425.0		50'	2846.3	668.0	
33°	246.1	1697.2	10°C	43°	2257.0	428.5	10°C	53°	2856.7	672.7	10°C
10'	248.7	1706.3	T	10'	2266.6	432.0	T	10'	2867.1	677.3	T
20'	251.3	1715.3	0.26	20'	2276.2	435.6	0.34	20'	2877.5	682.0	0.42
30'	253.9	1724.4	E	30'	2285.9	439.2	E	30'	2888.0	686.7	E
40'	256.5	1733.5	0.046	40'	2295.6	442.8	0.075	40'	2898.4	691.4	0.112
50'	259.1	1742.6		50'	2305.2	446.4		50'	2908.9	696.1	
34°	261.8	1751.7		44°	2314.9	450.0		54°	2919.4	700.9	
10'	264.5	1760.8		10'	2324.6	453.6		10'	2929.9	705.7	
20'	267.2	1770.0		20'	2334.3	457.3		20'	2940.4	710.5	
30'	269.9	1779.1		30'	2344.1	461.0		30'	2951.0	715.3	
40'	272.6	1788.2		40'	2353.8	464.6		40'	2961.5	720.1	
50'	275.3	1797.4		50'	2363.5	468.4		50'	2972.1	725.0	
35°	278.1	1806.6	15°C	45°	2373.3	472.1	15°C	55°	2982.7	729.9	15°C
10'	280.8	1815.7	T	10'	2383.1	475.8	T	10'	2993.3	734.8	T
20'	283.6	1824.9	0.4	20'	2392.8	479.6	0.51	20'	3003.9	739.7	0.63
30'	286.4	1834.1	E	30'	2402.6	483.4	E	30'	3014.5	744.6	E
40'	289.2	1843.3	0.07	40'	2412.4	487.2	0.116	40'	3025.2	749.6	0.168
50'	292.0	1852.5		50'	2422.3	491.0		50'	3035.8	754.6	

(continued)

	36°	1861.7	294.9	46°	2432.1	494.8	56°	3046.5	759.6
	10'	1870.9	297.7	10'	2441.9	498.7	10'	3057.9	764.6
	20'	1880.1	300.6	20'	2451.8	502.5	20'	3067.9	769.7
	30'	1889.4	303.5	30'	2461.7	506.4	30'	3078.7	774.7
	40'	1898.6	306.4	40'	2471.5	510.3	40'	3089.4	779.8
	50'	1907.9	309.3	50'	2481.4	514.3	50'	3100.2	784.9
	37°	1917.1	312.2	47°	2491.3	518.2	57°	3110.9	790.1
	10'	1926.4	315.2	10'	2501.2	522.2	10'	3121.7	795.2
	20'	1935.7	318.1	20'	2511.2	526.1	20'	3132.6	800.4
	30'	1945.0	321.1	30'	2521.1	530.1	30'	3143.4	805.6
	40'	1954.3	324.1	40'	2531.1	534.2	40'	3154.2	810.9
	50'	1963.6	327.1	50'	2541.0	538.2	50'	3165.1	816.1
	38°	1972.9	330.2	48°	2551.0	542.2	58°	3176.0	821.4
	10'	1982.2	333.2	10'	2561.0	546.3	10'	3186.9	826.7
	20'	1991.5	336.3	20'	2571.0	550.4	20'	3197.8	832.0
	30'	2000.9	339.3	30'	2581.0	554.5	30'	3208.8	837.3
	40'	2010.2	342.4	40'	2591.0	558.6	40'	3219.7	842.7
	50'	2019.6	345.5	50'	2601.1	562.8	50'	3230.7	848.1
	39°	2029.0	348.6	49°	2611.2	566.9	59°	3241.7	853.5
	10'	2038.4	351.8	10'	2621.2	571.1	10'	3252.7	858.9
	20'	2047.8	354.9	20'	2631.3	575.3	20'	3263.7	864.3
	30'	2057.2	358.1	30'	2641.4	579.5	30'	3274.8	869.8
	40'	2066.6	361.3	40'	2651.5	583.8	40'	3285.8	875.3
	50'	2076.0	364.5	50'	2661.6	588.0	50'	3296.9	880.8
	40°	2085.4	367.7	50°	2671.8	592.3	60°	3308.0	886.4
	10'	2094.9	371.0	10'	2681.9	596.6	10'	3319.1	892.0
	20'	2104.3	374.2	20'	2692.1	600.9	20'	3330.3	897.5
	30'	2113.8	377.5	30'	2702.3	605.3	30'	3341.4	903.2
	40'	2123.3	380.8	40'	2712.5	609.6	40'	3352.6	908.8
	50'	2132.7	384.1	50'	2722.7	614.0	50'	3363.8	914.5

Header blocks (left section): 20° C, T 0.53, E 0.093; 25° C, T 0.67, E 0.117; 30° C, T 0.8, E 0.141

Header blocks (middle section): 20° C, T 0.68, E 0.151; 25° C, T 0.85, E 0.189; 30° C, T 1.02, E 0.227

Header blocks (right section): 20° C, T 0.84, E 0.225; 25° C, T 1.05, E 0.283; 30° C, T 1.27, E 0.34

$T = R \tan \tfrac{1}{2} I$ $E = R \operatorname{exsec} \tfrac{1}{2} I$

9.12 TANGENTS AND EXTERNALS TO A 1-DEGREE CURVE (continued)

I	T	E	I = 10°
61°	3375.0	920.2	
10'	3386.3	925.9	
20'	3397.5	931.3	
30'	3408.8	937.3	5°C
40'	3420.1	943.1	T 0.25
50'	3431.4	948.9	E 0.08
62°	3442.7	954.8	
10'	3454.1	960.6	
20'	3465.4	966.5	
30'	3476.8	972.4	
40'	3488.3	978.3	
50'	3499.7	984.3	
63°	3511.1	990.2	10°C
10'	3522.6	996.2	
20'	3534.1	1002.3	T 0.51
30'	3545.6	1008.3	
40'	3557.2	1014.4	E 0.159
50'	3568.7	1020.5	
64°	3580.3	1026.6	
10'	3591.9	1032.8	
20'	3603.5	1039.0	
30'	3615.1	1045.2	
40'	3626.8	1051.4	
50'	3638.5	1057.7	
65°	3650.2	1063.9	15°C
10'	3661.9	1070.2	
20'	3673.7	1076.6	T 0.76
30'	3685.4	1082.9	
40'	3697.2	1089.3	E 0.24
50'	3709.0	1095.7	

I	T	E	I = 20°
71°	4086.9	1308.2	
10'	4099.5	1315.6	
20'	4112.1	1322.9	
30'	4124.8	1330.3	5°C
40'	4137.4	1337.7	T 0.3
50'	4150.1	1345.1	E 0.11
72°	4162.8	1352.6	
10'	4175.6	1360.1	
20'	4188.5	1367.6	
30'	4201.2	1375.2	
40'	4214.0	1382.8	
50'	4226.8	1390.4	
73°	4239.7	1398.0	10°C
10'	4252.6	1405.7	
20'	4265.6	1413.5	T 0.61
30'	4278.5	1421.2	
40'	4291.5	1429.0	E 0.22
50'	4304.6	1436.8	
74°	4317.6	1444.6	
10'	4330.7	1452.5	
20'	4343.8	1460.4	
30'	4356.9	1468.4	
40'	4370.1	1476.4	
50'	4383.3	1484.4	
75°	4396.5	1492.2	15°C
10'	4409.8	1500.5	T 0.91
20'	4423.1	1508.6	
30'	4436.4	1516.7	E 0.332
40'	4449.7	1524.9	
50'	4463.1	1533.1	

I	T	E	I = 34°
81°	4893.6	1805.3	
10'	4908.0	1814.7	
20'	4922.5	1824.1	
30'	4937.0	1833.6	5°C
40'	4951.5	1843.1	T 0.36
50'	4966.1	1852.6	E 0.149
82°	4980.7	1862.2	
10'	4995.4	1871.8	
20'	5010.0	1881.5	
30'	5024.8	1891.2	
40'	5039.5	1900.9	
50'	5054.3	1910.7	
83°	5069.2	1920.5	10°C
10'	5084.0	1930.4	
20'	5099.0	1940.3	T 0.72
30'	5113.9	1950.3	
40'	5128.9	1960.2	E 0.299
50'	5143.9	1970.3	
84°	5159.0	1980.4	
10'	5174.1	1990.5	
20'	5189.3	2000.6	
30'	5204.4	2010.8	
40'	5219.7	2021.1	
50'	5234.9	2031.4	
85°	5250.3	2041.7	15°C
10'	5265.6	2052.1	T 1.09
20'	5281.0	2062.5	
30'	5296.4	2073.0	E 0.45
40'	5311.9	2083.5	
50'	5327.4	2094.1	

Angle			°C	T	E
66°	3720.9	1102.2	20°C	1.02	0.321
10'	3732.7	1108.6			
20'	3744.6	1115.1			
30'	3756.5	1121.7			
40'	3768.5	1128.2			
50'	3780.4	1134.8			
67°	3792.4	1141.4			
10'	3804.4	1148.0			
20'	3816.4	1154.7			
30'	3828.4	1161.3			
40'	3840.5	1168.1			
50'	3852.6	1174.8			
68°	3864.7	1181.6	25°C	1.28	0.403
10'	3876.8	1188.4			
20'	3889.0	1195.2			
30'	3901.2	1202.0			
40'	3913.4	1208.9			
50'	3925.6	1215.8			
69°	3937.9	1222.7			
10'	3950.2	1229.7			
20'	3962.5	1236.7			
30'	3974.8	1243.7			
40'	3987.2	1250.8			
50'	3999.5	1257.9			
70°	4011.9	1265.0	30°C	1.54	0.485
10'	4024.4	1272.1			
20'	4036.8	1279.3			
30'	4049.3	1286.5			
40'	4061.8	1293.6			
50'	4074.4	1300.9			

Angle			°C	T	E
76°	4476.5	1541.4	20°C	1.22	0.445
10'	4489.9	1549.7			
20'	4503.4	1558.0			
30'	4516.9	1566.3			
40'	4530.4	1574.7			
50'	4544.0	1583.1			
77°	4557.6	1591.6			
10'	4571.2	1600.1			
20'	4584.8	1608.6			
30'	4598.5	1617.1			
40'	4612.2	1625.7			
50'	4626.0	1634.4			
78°	4639.8	1643.0	25°C	1.53	0.558
10'	4653.6	1651.7			
20'	4667.4	1660.5			
30'	4681.3	1669.2			
40'	4695.2	1678.1			
50'	4709.2	1686.9			
79°	4723.2	1695.8			
10'	4737.2	1704.7			
20'	4751.2	1713.7			
30'	4765.3	1722.7			
40'	4779.4	1731.7			
50'	4793.6	1740.8			
80°	4807.7	1749.9	30°C	1.54	0.485
10'	4822.0	1759.0			
20'	4836.2	1768.2			
30'	4850.5	1777.4			
40'	4864.8	1786.7			
50'	4879.2	1796.0			

Angle			°C	T	E
86°	5343.0	2104.7	20°C	1.45	0.603
10'	5358.6	2115.3			
20'	5374.2	2126.0			
30'	5389.9	2136.7			
40'	5405.6	2147.5			
50'	5421.4	2158.4			
87°	5437.2	2169.2			
10'	5453.1	2180.2			
20'	5469.0	2191.1			
30'	5484.9	2202.2			
40'	5500.9	2213.2			
50'	5517.0	2224.3			
88°	5533.1	2235.5	25°C	1.83	0.756
10'	5549.2	2246.7			
20'	5565.4	2258.0			
30'	5581.6	2269.3			
40'	5597.8	2280.6			
50'	5614.2	2292.0			
89°	5630.5	2303.5			
10'	5646.9	2315.0			
20'	5663.4	2326.6			
30'	5679.9	2338.2			
40'	5696.4	2349.8			
50'	5713.0	2361.5			
90°	5729.7	2373.3	30°C	2.2	0.91
10'	5746.3	2385.1			
20'	5763.1	2397.0			
30'	5779.9	2408.9			
40'	5796.7	2420.9			
50'	5813.6	2432.9			

$T = R \tan \frac{1}{2} I \qquad E = R \operatorname{exsec} \frac{1}{2} I$

(continued)

343

9.12 TANGENTS AND EXTERNALS TO A 1-DEGREE CURVE (continued)

I = 10°

I	T	E
91°	5830.5	2444.9
10'	5847.5	2457.1
20'	5864.6	2469.3
30'	5881.7	2481.5
40'	5898.8	2493.8
50'	5916.0	2506.1
92°	5933.2	2518.5
10'	5950.5	2531.0
20'	5967.9	2543.5
30'	5985.3	2556.0
40'	6002.7	2568.6
50'	6020.2	2581.3
93°	6037.8	2594.0
10'	6055.4	2606.8
20'	6073.1	2619.7
30'	6090.8	2632.6
40'	6108.6	2645.5
50'	6126.4	2658.5
94°	6144.3	2671.6
10'	6162.2	2684.7
20'	6180.2	2697.9
30'	6198.3	2711.2
40'	6216.4	2724.5
50'	6234.6	2737.9
95°	6252.8	2751.3
10'	6271.1	2764.8
20'	6289.4	2778.3
30'	6307.9	2792.0
40'	6326.3	2805.6
50'	6344.8	2819.4

Corrections:
+ 5°C T 0.43 E 200
10°C T 0.86 E 0.401
15°C T 1.3 E 0.604

I = 20°

I	T	E
101°	6950.6	3278.1
10'	6971.3	3294.1
20'	6992.0	3310.1
30'	7012.7	3326.1
40'	7033.6	3342.3
50'	7054.5	3358.5
102°	7075.5	3374.9
10'	7096.6	3391.2
20'	7117.8	3407.7
30'	7139.0	3424.3
40'	7160.3	3440.9
50'	7181.7	3457.6
103°	7203.2	3474.4
10'	7224.7	3491.3
20'	7246.3	3508.2
30'	7268.0	3525.2
40'	7289.8	3542.4
50'	7311.7	3559.6
104°	7333.6	3576.8
10'	7355.6	3594.2
20'	7377.8	3611.7
30'	7399.9	3629.2
40'	7422.2	3646.8
50'	7444.6	3664.5
105°	7467.0	3682.3
10'	7489.6	3700.2
20'	7512.2	3718.2
30'	7354.9	3736.2
40'	7557.7	3754.4
50'	7580.5	3772.6

Corrections:
+ 5°C T 0.51 E 0.268
10°C T 0.103 E 0.536
15°C T 1.56 E 0.806

I = 34°

I	T	E
111°	8336.7	4386.1
10'	8362.7	4407.6
20'	8388.9	4429.2
30'	8415.1	4450.9
40'	8441.5	4472.7
50'	8468.0	4494.6
112°	8494.6	4516.6
10'	8521.3	4538.8
20'	8548.1	4561.1
30'	8575.0	4583.4
40'	8602.1	4606.0
50'	8629.3	4628.6
113°	8656.6	4651.3
10'	8684.0	4674.2
20'	8711.5	4697.2
30'	8739.2	4720.3
40'	8767.0	4743.6
50'	8794.9	4766.9
114°	8822.9	4790.4
10'	8851.0	4814.1
20'	8879.3	4837.8
30'	8907.7	4861.7
40'	8936.3	4885.7
50'	8965.0	4909.9
115°	8993.8	4934.1
10'	9022.7	4958.6
20'	9051.7	4983.1
30'	9080.9	5007.8
40'	9110.3	5032.6
50'	9139.8	5057.6

Corrections:
+ 5°C T 0.62 E 360
10°C T 1.25 E 0.721
15°C T 1.93 E 1.09

96° – 100°

Temperature correction:

	20° C	25° C	30° C
T	1.74	2.18	2.62
E	0.809	1.02	1.22

Angle	T	E
96°	6363.4	2833.2
10'	6382.1	2847.0
20'	6400.8	2861.0
30'	6419.5	2875.0
40'	6438.4	2889.0
50'	6457.3	2903.1
97°	6476.2	2917.3
10'	6495.2	2931.6
20'	6514.3	2945.9
30'	6533.4	2960.3
40'	6552.6	2974.7
50'	6571.9	2989.2
98°	6591.2	3003.8
10'	6610.6	3018.4
20'	6630.1	3033.1
30'	6649.6	3047.9
40'	6669.2	3062.8
50'	6688.8	3077.7
99°	6708.6	3092.7
10'	6728.4	3107.7
20'	6748.2	3122.9
30'	6768.1	3138.1
40'	6788.1	3153.3
50'	6808.2	3168.7
100°	6828.3	3184.1
10'	6848.5	3199.6
20'	6868.8	3215.1
30'	6889.2	3230.8
40'	6909.6	3246.5
50'	6930.1	3262.3

106° – 110°

Temperature correction:

	20° C	25° C	30° C
T	2.08	2.61	3.14
E	1.08	1.36	1.63

Angle	T	E
106°	7603.5	3791.0
10'	7626.6	3809.4
20'	7649.7	3827.9
30'	7672.9	3846.5
40'	7696.3	3865.2
50'	7719.7	3884.0
107°	7743.2	3902.9
10'	7766.8	3921.9
20'	7790.5	3940.9
30'	7814.3	3960.1
40'	7838.1	3979.4
50'	7862.1	3998.7
108°	7886.2	4018.2
10'	7910.4	4037.8
20'	7934.6	4057.4
30'	7959.0	4077.2
40'	7983.5	4097.1
50'	8008.0	4117.0
109°	8032.7	4137.1
10'	8057.4	4157.3
20'	8082.3	4177.5
30'	8107.3	4197.9
40'	8132.3	4218.4
50'	8157.5	4239.0
110°	8182.8	4259.7
10'	8208.2	4280.5
20'	8233.7	4301.4
30'	8259.3	4322.4
40'	8285.0	4343.6
50'	8310.8	4364.8

116° – 120°

Temperature correction:

	20° C	25° C	30° C
T	1.4	3.16	3.81
E	1.46	1.83	2.2

Angle	T	E
116°	9169.4	5082.7
10'	9199.1	5107.9
20'	9229.0	5133.3
30'	9259.0	5158.8
40'	9289.2	5184.5
50'	9319.5	5210.3
117°	9349.9	5236.2
10'	9380.5	5262.3
20'	9411.3	5288.6
30'	9442.2	5315.0
40'	9473.2	5341.5
50'	9504.4	5368.2
118°	9535.7	5395.1
10'	9567.2	5422.1
20'	9598.9	5449.2
30'	9630.7	5476.5
40'	9662.6	5504.0
50'	9694.7	5531.7
119°	9727.0	5559.4
10'	9759.4	5587.4
20'	9792.0	5615.5
30'	9824.8	5643.8
40'	9857.7	5672.3
50'	9890.8	5700.9
120°	9924.0	5729.7
10'	9957.5	5758.6
20'	9991.0	5787.7
30'	10025.0	5817.0
40'	10059.0	5846.5
50'	10093.0	5876.1

$T = R \tan \tfrac{1}{2} I \qquad E = R \operatorname{exsec} \tfrac{1}{2} I$

9.13 USEFUL RELATIONS

Lineal feet	×	0.00019	=	Miles
Lineal yards	×	0.0006	=	Miles
Square inches	×	0.007	=	Square feet
Square feet	×	0.111	=	Square yards
Square yards	×	0.0002067	=	Acres
Acres	×	4,840	=	Square yards
Cubic inches	×	0.00058	=	Cubic feet
Cubic feet	×	0.03704	=	Cubic yards
Links	×	0.22	=	Yards
Links	×	0.66	=	Feet
Feet	×	1.5	=	Links

$360° = 21,600' = 12,960,000"$

Radius = Arc of 57.2957790

Arc of 1° (Radius = 1) = 0.017453292

Arc of 1' (Radius = 1) = 0.000290888

Arc of 1" (Radius = 1) = 0.000004848

Curvative of Earth's surface = About 0.7 ft. in 1 mile.

Curvative in feet = 0.667 (Distance in miles).^{2}

Difference between arc and chord length, 0.05 ft. in 11½ miles.

Probable error of a single observation $= 0.6754 \sqrt{\dfrac{\sum \nabla 2}{n-1}}$

Error in chaining of 0.01 ft. in 100' due to

1. Length of tape error of 0.01 ft.
2. Alignment: One end 1.4 ft. out of line
3. Sag of tape at center of 0.61 ft.
4. Temperature differences of 15°
5. Difference of pull of 15 lb.

9.14 SQUARE MEASURE AND SURVEYOR'S MEASURE

Square Measure

> 144 sq. in. = 1 SF
> 9 SF = 1 sq. yd.
> 30¼ sq. yd. = 1 sq. rd.
> 40 sq. rds. = 1 rood
> 4 roods = 1 acre
> 640 acres = 1 sq. mile

Surveyor's Measure

> 7.92 in. = 1 link
> 25 links = 1 rod
> 4 rds. = 1 chain
> 10 sq. chains or 160 sq. rds. = 1 acre
> 640 acres = 1 sq. mile
> 36 sq. miles (6 miles) = 1 township

9.15 INCHES TO DECIMALS OF A FOOT

Inches	0	1	2	3	4	5	6	7	8	9	10	11	Inches
a	Foot	0.0833	0.1667	0.2500	0.3333	0.4167	0.5000	0.5833	0.6667	0.7500	0.8333	0.9167	0
1/32	0.0026	0.0859	0.1693	0.2526	0.3359	0.4193	0.5026	0.5859	0.6693	0.7526	0.8359	0.9193	1/32
1/16	0.0052	0.0885	0.1719	0.2552	0.3385	0.4219	0.5052	0.5885	0.6719	0.7552	0.8385	0.9219	1/16
3/32	0.0078	0.0911	0.1745	0.2578	0.3411	0.4245	0.5078	0.5911	0.6745	0.7578	0.8411	0.9245	3/32
1/8	0.0104	0.0938	0.1771	0.2604	0.3438	0.4271	0.5104	0.5938	0.6771	0.7604	0.8438	0.9271	1/8
5/32	0.0130	0.0964	0.1797	0.2630	0.3464	0.4297	0.5130	0.5964	0.6797	0.7630	0.8464	0.9297	5/32
3/16	0.0156	0.0990	0.1823	0.2656	0.3490	0.4323	0.5156	0.5990	0.6823	0.7656	0.8490	0.9323	3/16
7/32	0.0182	0.1016	0.1849	0.2682	0.3516	0.4349	0.5182	0.6016	0.6849	0.7682	0.8516	0.9349	7/32
1/4	0.0208	0.1042	0.1875	0.2708	0.3542	0.4375	0.5208	0.6042	0.6875	0.7708	0.8542	0.9375	1/4
9/32	0.0234	0.1068	0.1901	0.2734	0.3568	0.4401	0.5234	0.6068	0.6901	0.7734	0.8568	0.9401	9/32
5/16	0.0260	0.1094	0.1927	0.2760	0.3594	0.4427	0.5260	0.6094	0.6927	0.7760	0.8594	0.9427	5/16
11/32	0.0286	0.1120	0.1953	0.2786	0.3620	0.4453	0.5286	0.6120	0.6953	0.7786	0.8620	0.9453	11/32
3/8	0.0313	0.1146	0.1979	0.2813	0.3646	0.4479	0.5313	0.6146	0.6979	0.7813	0.8646	0.9479	3/8
13/32	0.0339	0.1172	0.2005	0.2839	0.3672	0.4505	0.5339	0.6172	0.7005	0.7839	0.8672	0.9505	13/32
7/16	0.0365	0.1198	0.2031	0.2865	0.3698	0.4531	0.5365	0.6198	0.7031	0.7865	0.8698	0.9531	7/16

Inches	0	1	2	3	4	5	6	7	8	9	10	11	Inches
15/32	0.0391	0.1224	0.2057	0.2891	0.3724	0.4557	0.5391	0.6224	0.7057	0.7891	0.8724	0.9557	15/32
1/2	0.0417	0.1250	0.2083	0.2917	0.3750	0.4583	0.5417	0.6250	0.7083	0.7917	0.8750	0.9583	1/2
17/32	0.0443	0.1276	0.2109	0.2943	0.3776	0.4609	0.5443	0.6276	0.7109	0.7943	0.8776	0.9609	17/32
9/16	0.0469	0.1302	0.2135	0.2969	0.3802	0.4635	0.5469	0.6302	0.7135	0.7969	0.8802	0.9635	9/16
19/32	0.0495	0.1328	0.2161	0.2995	0.3828	0.4661	0.5495	0.6328	0.7161	0.7995	0.8828	0.9661	19/32
5/8	0.0521	0.1354	0.2188	0.3021	0.3854	0.4688	0.5521	0.6354	0.7188	0.8021	0.8854	0.9668	5/8
21/32	0.0547	0.1380	0.2214	0.3047	0.3880	0.4714	0.5547	0.6380	0.7214	0.8047	0.8880	0.9714	21/32
11/16	0.0573	0.1406	0.2240	0.3073	0.3906	0.4740	0.5573	0.6406	0.7240	0.8073	0.8906	0.9740	11/16
23/32	0.0599	0.1432	0.2266	0.3099	0.3932	0.4766	0.5599	0.6432	0.7266	0.8099	0.8932	0.9766	23/32
3/4	0.0625	0.1458	0.2292	0.3125	0.3958	0.4792	0.5625	0.6458	0.7292	0.8125	0.8958	0.9792	3/4
25/32	0.0651	0.1484	0.2318	0.3151	0.3984	0.4818	0.5651	0.6484	0.7318	0.8151	0.8984	0.9818	25/32
13/16	0.0677	0.1510	0.2344	0.3177	0.4010	0.4844	0.5677	0.6510	0.7344	0.8177	0.9010	0.9844	13/16
27/32	0.0703	0.1536	0.2370	0.3203	0.4036	0.4870	0.5703	0.6536	0.7370	0.8203	0.9036	0.9870	27/32
7/8	0.0729	0.1563	0.2396	0.3229	0.4063	0.4896	0.5729	0.6563	0.7396	0.8229	0.9063	0.9896	7/8
29/32	0.0755	0.1589	0.2422	0.3255	0.4089	0.4922	0.5755	0.6589	0.7422	0.8255	0.9089	0.9922	29/32
15/16	0.0781	0.1615	0.2448	0.3281	0.4115	0.4948	0.5781	0.6615	0.7448	0.8281	0.9115	0.9948	15/16
31/32	0.0807	0.1614	0.2474	0.3307	0.4141	0.4974	0.5087	0.6641	0.7474	0.0831	0.9141	0.9974	31/32

9.16 MINUTES IN DECIMALS OF A DEGREE

Minutes	Seconds	Decimals	Minutes	Seconds	Decimals	Minutes	Seconds	Decimals
0'	30"	0.00833	20'	30"	0.34167	40'	30"	0.67500
1	00	0.01667	21	00	0.35000	41	00	0.68333
	30	0.02500		30	0.35833		30	0.69167
2	00	0.03333	22	00	0.36667	42	00	0.70000
	30	0.04167		30	0.37500		30	0.70833
3	00	0.05000	23	00	0.38333	43	00	0.71667
	30	0.05833		30	0.39167		30	0.72500
4	00	0.06667	24	00	0.40000	44	00	0.73333
	30	0.07500		30	0.40833		30	0.74167
5	00	0.08333	25	00	0.41667	45	00	0.75000
	30	0.09167		30	0.42500		30	0.75833
6	00	0.01000	26	00	0.43333	46	00	0.76667
	30	0.10833		30	0.44167		30	0.77500
7	00	0.11667	27	00	0.45000	47	00	0.78333
	30	0.12500		30	0.45833		30	0.79167
8	00	0.13333	28	00	0.46667	48	00	0.80000
	30	0.14167		30	0.47500		30	0.80833
9	00	0.15000	29	00	0.48333	49	00	0.81667
	30	0.15833		30	0.49167		30	0.82500

Minutes	Seconds	Decimals	Minutes	Seconds	Decimals	Minutes	Seconds	Decimals
10	00	0.16667	30	00	0.50000	50	00	0.83333
	30	0.17500		30	0.50833		30	0.84167
11	00	0.18333	31	00	0.51667	51	00	0.85000
	30	0.19167		30	0.52500		30	0.85833
12	00	0.20000	32	00	0.53333	52	00	0.86667
	30	0.20833		30	0.54167		30	0.87500
13	00	0.21667	33	00	0.55000	53	00	0.88333
	30	0.22500		30	0.55833		30	0.89167
14	00	0.24167	34	00	0.56667	54	00	0.90000
	30	0.25000		30	0.57500		30	0.90833
15	00	0.25833	35	00	0.58333	55	00	0.91667
	30	0.26670		30	0.59167		30	0.92500
16	00	0.26667	36	00	0.60000	56	00	0.93333
	30	0.27500		30	0.60833		30	0.94167
17	00	0.28333	37	00	0.61667	57	00	0.95000
	30	0.29167		30	0.62500		30	0.95833
18	00	0.30000	38	00	0.63333	58	00	0.96667
	30	0.30833		30	0.64167		30	0.97500
19	00	0.31667	39	00	0.65000	59	00	0.98333
	30	0.32500		30	0.65833		30	0.99167
20	00	0.33333	40	00	0.66667	60	00	1.00000

9.17 MIDDLE ORDINATES OF LENGTH OF RAIL (FEET)

C 0'	R (ft.)	30 In.	28 In.	26 In.	24 In.	22 In.	20 In.	C 0'	R (ft.)	30 In.	28 In.	26 In.	24 In.	22 In.	20 In.
0-20	17,189	0.08	0.07	0.06	0.05	0.04	0.03	8	716.8	1.88	1.64	1.42	1.20	1.01	0.84
0-40	8,594	0.16	0.14	0.12	0.10	0.08	0.07	9	637.3	2.12	1.84	1.60	1.35	1.14	0.94
1-0	5,730	0.24	0.20	0.18	0.15	0.13	0.10	10	573.7	2.36	2.05	1.78	1.50	1.27	1.04
1-20	4,297	0.31	0.27	0.23	0.20	0.17	0.13	11	521.7	2.59	2.26	1.95	1.65	1.39	1.15
1-40	3,438	0.39	0.34	0.29	0.25	0.21	0.17	12	478.3	3.83	2.47	2.15	1.81	1.54	1.26
2-0	2,865	0.47	0.41	0.35	0.30	0.25	0.20	13	441.7	3.05	2.66	2.30	1.96	1.66	1.36
2-20	2,456	0.55	0.48	0.41	0.35	0.29	0.23	14	410.3	3.30	2.87	2.48	2.10	1.78	1.46
2-40	2,149	0.63	0.55	0.47	0.40	0.33	0.27	15	383.1	3.54	3.08	2.68	2.26	1.91	1.57
3-0	1,910	0.71	0.62	0.53	0.45	0.38	0.31	16	359.3	3.76	3.28	2.83	2.40	2.04	1.67
3-20	1,719	0.78	0.68	0.59	0.50	0.42	0.35	17	338.3	4.00	3.48	3.02	2.57	2.16	1.78
3-40	1,563	0.86	0.75	0.65	0.55	0.46	0.38	18	319.6	4.21	3.67	3.18	2.70	2.28	1.87
4-0	1,433	0.94	0.82	0.71	0.60	0.50	0.42	19	302.9	4.45	3.89	3.36	2.86	2.41	1.98
4-20	1,323	1.02	0.89	0.77	0.65	0.55	0.45	20	287.9	4.70	4.09	3.55	3.00	2.54	2.09
4-40	1,228	1.10	0.96	0.83	0.70	0.59	0.48	22	262.0	5.16	4.44	3.84	3.30	2.80	2.29
5	1,146	1.18	1.03	0.89	0.75	0.63	0.52	24	240.5	5.64	4.92	4.20	3.59	3.04	2.50
6	955.3	1.41	1.23	1.06	0.90	0.76	0.62	26	222.3	6.07	5.29	4.58	3.88	3.29	2.70
7	819.0	1.65	1.44	1.24	1.05	0.89	0.73								

9.18   SHORT RADIUS CURVES

Radius (ft.)	Chord (ft.)	Central Angle	Deflection Angle	Deflection for 1 ft.
35	10	16-26	8-13	49.3
45	10	12-46	6-23	38.3
50	15	17-16	8-38	34.5
60	15	14-22	7-11	28.8
75	15	11-30	5-45	23.0
100	20	11-30	5-45	17.3
120	20	9-34	4-47	14.3
150	20	7-39	3-49	11.5
190	25	7-32	3-46	9.15
200	25	7-10	3-35	8.6
225	25	6-25	3-12	7.7
240	25	5-58	2-59	7.2
250	25	5-44	2-52	6.9
275	25	5-12	2-36	6.2
288	50	9-58	4-59	6.0
300	50	9-32	4-46	5.7
350	50	8-12	4-06	4.9
376	50	7-40	3-50	4.6
400	50	7-10	3-35	4.3
410	50	7-00	3-30	4.2

To find length of curve, divide angle from *P.C.* to *P.T.* by central angle of chord and multiply by length of chord.

9.19 RODS IN FEET, TENTHS, AND HUNDREDTHS OF FEET

Rods	Feet	Rods	Feet	Rods	Feet	Rods	Feet	Rods	Feet
1	16.50	21	346.50	41	676.50	61	1,006.50	81	1,336.50
2	33.00	22	363.00	42	693.00	62	1,023.00	82	1,353.00
3	49.50	23	379.50	43	709.50	63	1,039.50	83	1,369.50
4	66.00	24	396.00	44	726.00	64	1,056.00	84	1,386.00
5	82.50	25	412.50	45	742.50	65	1,072.50	85	1,402.50
6	99.00	26	429.00	46	759.00	66	1,089.00	86	1,419.00
7	115.50	27	445.50	47	775.50	67	1,105.50	87	1,435.50
8	132.00	28	462.00	48	792.00	68	1,122.00	88	1,452.00
9	148.50	29	478.50	49	808.50	69	1,138.50	89	1,468.50
10	165.00	30	495.00	50	825.00	70	1,155.00	90	1,485.00
11	181.50	31	511.50	51	841.50	71	1,171.50	91	1,501.50
12	198.00	32	528.00	52	858.00	72	1,188.00	92	1,518.00
13	214.50	33	544.50	53	874.50	73	1,204.50	93	1,534.50
14	231.00	34	561.00	54	891.00	74	1,221.00	94	1,551.00
15	247.50	35	577.50	55	907.50	75	1,237.50	95	1,567.50
16	264.00	36	594.00	56	924.00	76	1,254.00	96	1,584.00
17	280.50	37	610.50	57	940.50	77	1,270.50	97	1,600.50
18	297.00	38	627.00	58	957.00	78	1,287.00	98	1,617.00
19	313.50	39	643.50	59	973.50	79	1,303.50	99	1,633.50
20	330.00	40	660.00	60	990.00	80	1,320.00	100	1,650.00

9.20 LINKS IN FEET, TENTHS, AND HUNDREDTHS OF FEET

Links	Feet	Links	Feet	Links	Feet	Links	Feet	Links	Feet
1	0.66	18	11.88	35	23.10	52	34.32	69	45.54	86	56.76
2	1.32	19	12.54	36	23.76	53	34.98	70	46.20	87	57.42
3	1.98	20	13.20	37	24.42	54	35.64	71	46.86	88	58.08
4	2.64	21	13.86	38	25.08	55	36.30	72	47.52	89	58.74
5	3.30	22	14.52	39	25.74	56	36.96	73	48.18	90	59.40
6	3.96	23	15.18	40	26.40	57	37.62	74	48.84	91	60.06
7	4.62	24	15.84	41	27.06	58	38.28	75	49.30	92	60.72
8	5.28	25	16.50	42	27.72	59	38.94	76	50.16	93	61.38
9	5.94	26	17.16	43	28.38	60	39.60	77	50.82	94	62.04
10	6.60	27	17.82	44	29.04	61	40.26	78	51.48	95	62.70
11	7.26	28	18.48	45	29.70	62	40.92	79	52.14	96	63.36
12	7.92	29	19.14	46	30.36	63	41.58	80	52.80	97	64.02
13	8.58	30	19.80	47	31.02	64	42.24	81	53.46	98	64.68
14	9.24	31	20.46	48	31.68	65	42.90	82	54.12	99	65.34
15	9.90	32	21.12	49	32.34	66	43.56	83	54.78	100	66.00
16	10.56	33	21.78	50	33.00	67	44.22	84	55.44	101	66.66
17	11.22	34	22.44	51	33.66	68	44.88	85	56.10	102	67.32

9.21   INCHES IN DECIMALS OF A FOOT

$\frac{1}{16}$	$\frac{3}{32}$	$\frac{1}{8}$	$\frac{3}{16}$	$\frac{1}{4}$	$\frac{5}{16}$	$\frac{3}{8}$	$\frac{1}{2}$	$\frac{5}{8}$	$\frac{3}{4}$	$\frac{7}{8}$
0.0052	0.0078	0.0104	0.0156	0.0208	0.0260	0.0313	0.0417	0.0521	0.0625	0.0729
1	2	3	4	5	6	7	8	9	10	11
0.0833	0.1667	0.2500	.3333	0.4167	0.5000	0.5833	0.6667	0.7500	0.8333	0.9167

9.22   CURVE FORMULAS

D	=	Degree of curve
$1°$	=	1 degree of curve
$2°$	=	2 degree of curve
P.C.	=	Point of curve
P.T.	=	Point of tangent
P.I.	=	Point of intersection
I	=	Intersection of angle, angle between two tangents
L	=	Length of curve from P.C. to P.T.
T	=	Tangent distance
E	=	External distance
R	=	Radius
L.C.	=	Length of chord
M	=	Length of middle ordinate
c	=	Length of subchord
d	=	Angle of subchord

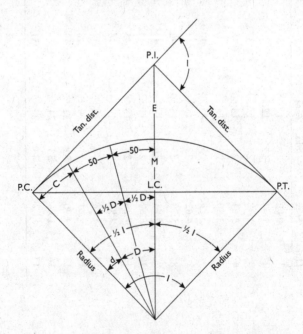

$$R = \frac{L.C.}{2 \sin \frac{1}{2} I} \quad T = R \tan \frac{1}{2} I = \frac{L.C.}{2 \cos \frac{1}{2} I}$$

$$\frac{L.C.}{2} = R \sin \frac{I}{2}, \quad D1^\circ = R5{,}730,$$

$$D2^\circ = \frac{5{,}730}{2}, \quad D = \frac{5{,}730}{R}$$

$$M = R\left(1 - \cos \frac{1}{2} I\right) = R - R \cos \frac{I}{2}$$

$$\frac{E + R}{R} = \sec \frac{I}{2}, \quad \frac{R - M}{R} = \cos \frac{I}{2}$$

$$c = 2R \sin \frac{1}{2} d, \quad d = \frac{c}{2R}$$

$$L.C. = 2R \sin \frac{1}{2} I, \quad E = R\left(\sec \frac{1}{2} I - 1\right) = R \sec \frac{I}{2} - R$$

9.23 TABLE OF POWERS AND ROOTS (1–100)

Number	Square	Cube	Square Root	Cube Root
1	1	1	1.000	1.000
2	4	8	1.414	1.260
3	9	27	1.732	1.442
4	16	64	2.000	1.587
5	25	125	2.236	1.710
6	36	216	2.449	1.817
7	49	343	2.616	1.913
8	64	512	2.828	2.000
9	81	729	3.000	2.080
10	100	1,000	3.162	2.154
11	121	1,331	3.317	2.224
12	144	1,728	3.464	2.289
13	169	2,197	3.606	2.351
14	196	2,744	3.742	2.410
15	225	3,375	3.873	2.466
16	256	4,096	4.000	2.520
17	289	4,913	4.123	2.571
18	324	5,832	4.243	2.621
19	361	6,859	4.359	2.668
20	400	8,000	4.472	2.714
21	441	9,261	4.583	2.759
22	484	10,648	4.690	2.802
23	529	12,167	4.796	2.844
24	576	13,824	4.899	2.884

(continued)

9.23 TABLE OF POWERS AND ROOTS (1–100) *(continued)*

Number	Square	Cube	Square Root	Cube Root
25	625	15,625	5.000	2.924
26	676	17,576	5.099	2.962
27	729	19,683	5.196	3.000
28	784	21,952	5.292	3.037
29	841	24,389	5.385	3.072
30	900	27,000	5.477	3.107
31	961	29,791	5.568	3.141
32	1,024	32,768	5.657	3.175
33	1,089	35,937	5.745	3.208
34	1,156	39,304	5.831	3.210
35	1,225	42,875	5.916	3.271
36	1,296	46,656	6.000	3.302
37	1,369	50,653	6.083	3.332
38	1,444	54,872	6.164	3.362
39	1,521	59,319	6.245	3.391
40	1,600	64,000	6.325	3.420
41	1,681	68,921	6.403	3.448
42	1,764	74,088	6.481	3.476
43	1,849	79,507	6.557	3.503
44	1,936	85,184	6.633	3.530
45	2,025	91,125	6.708	3.557
46	2,116	97,336	6.782	3.583
47	2,209	103,823	6.856	3.609
48	2,304	110,592	6.928	3.634
49	2,401	117,649	7.000	3.659
50	2,500	125,000	7.071	3.684
51	2,601	132,651	7.141	3.708
52	2,704	140,608	7.211	3.732
53	2,809	148,877	7.280	3.756
54	2,916	157,464	7.348	3.780
55	3,025	166,375	7.416	3.803
56	3,136	175,616	7.483	3.826
57	3,249	185,193	7.550	3.848
58	3,364	195,112	7.616	3.871
59	3,481	205,379	7.681	3.893
60	3,600	216,000	7.746	3.915
61	3,721	226,981	7.810	3.936
62	3,844	238,328	7.874	3.958

(continued)

Number	Square	Cube	Square Root	Cube Root
63	3,969	250,047	7.937	3.979
64	4,096	262,144	8.000	4.000
65	4,225	274,625	8.062	4.021
66	4,356	287,496	8.124	4.041
67	4,489	300,763	8.185	4.062
68	4,624	314,432	8.246	4.082
69	4,761	328,509	8.307	4.102
70	4,900	343,000	8.367	4.121
71	5,041	357,911	8.426	4.141
72	5,184	373,248	8.485	4.160
73	5,329	389,017	8.544	4.179
74	5,476	405,224	8.602	4.198
75	5,625	421,875	8.660	4.217
76	5,776	438,976	8.718	4.236
77	5,929	456,533	8.775	4.254
78	6,084	474,552	8.832	4.273
79	6,241	493,039	8.888	4.291
80	6,400	512,000	8.911	4.309
81	6,561	531,441	9.000	4.327
82	6,724	551,368	9.055	4.344
83	6,889	571,787	9.110	4.362
84	7,056	592,704	9.165	4.380
85	7,225	614,125	9.220	4.397
86	7,396	636,056	9.271	4.414
87	7,569	658,503	9.327	4.431
88	7,744	681,472	9.381	4.448
89	7,921	704,969	9.134	4.465
90	8,100	729,000	9.487	4.481
91	8,281	753,571	9.539	4.498
92	8,464	778,688	9.592	4.514
93	8,649	804,357	9.613	4.531
94	8,836	830,584	9.695	4.547
95	9,025	857,375	9.747	4.563
96	9,216	881,736	9.798	4.579
97	9,409	912,673	9.819	4.595
98	9,604	941,192	9.899	4.610
99	9,801	970,299	9.950	4.626
100	10,000	1,000,000	10.000	4.642

9.24 SQUARE ROOTS AND CUBE ROOTS (1,000–2,000)

Number	Square Root	Cube Root	Number	Square Root	Cube Root
1,000	31.62	10.00	1,210	34.79	10.66
1,005	31.70	10.02	1,215	34.86	10.67
1,010	31.78	10.03	1,220	34.93	10.69
1,020	31.94	10.07	1,225	35.00	10.70
1,025	32.02	10.08	1,235	35.14	10.73
1,030	32.09	10.10	1,245	35.28	10.76
1,035	32.17	10.12	1,255	35.43	10.79
1,045	32.33	10.15	1,260	35.50	10.80
1,050	32.40	10.16	1,265	35.57	10.82
1,060	32.56	10.20	1,275	35.71	10.84
1,065	32.63	10.21	1,280	35.78	10.86
1,075	32.79	10.24	1,285	35.85	10.87
1,080	32.86	10.26	1,290	35.92	10.89
1,085	32.94	10.28	1,300	36.06	10.91
1,090	33.02	10.29	1,305	36.12	10.93
1,095	33.09	10.31	1,315	36.26	10.96
1,100	33.17	10.32	1,320	36.33	10.97
1,105	33.24	10.34	1,330	36.47	11.00
1,110	33.32	10.35	1,335	36.54	11.01
1,115	33.39	10.37	1,340	36.61	11.02
1,120	33.47	10.38	1,345	36.67	11.04
1,125	33.54	10.40	1,350	36.74	11.05
1,130	33.62	10.42	1,355	36.81	11.07
1,135	33.69	10.43	1,360	36.88	11.08
1,140	33.76	10.45	1,365	36.95	11.09
1,145	33.84	10.46	1,370	37.01	11.11
1,150	33.91	10.48	1,375	37.08	11.12
1,155	33.99	10.49	1,380	37.15	11.13
1,160	34.06	10.51	1,385	37.22	11.15
1,165	34.13	10.52	1,390	37.28	11.16
1,170	34.21	10.54	1,395	37.35	11.17
1,175	34.28	10.55	1,400	37.42	11.19
1,180	34.35	10.57	1,405	37.48	11.20
1,185	34.42	10.58	1,410	37.55	11.21
1,190	34.50	10.60	1,415	37.62	11.23
1,195	34.57	10.61	1,420	37.68	11.24

(continued)

Number	Square Root	Cube Root	Number	Square Root	Cube Root
1,200	34.64	10.63	1,425	37.75	11.25
1,205	34.71	10.64	1,430	37.82	11.27
1,435	37.88	11.28	1,655	40.68	11.83
1,440	37.95	11.29	1,660	40.74	11.84
1,445	38.01	11.31	1,665	40.80	11.85
1,450	38.08	11.32	1,670	40.87	11.86
1,455	38.14	11.33	1,675	40.93	11.88
1,460	38.21	11.34	1,680	40.99	11.89
1,465	38.28	11.36	1,685	41.05	11.90
1,470	38.34	11.37	1,690	41.11	11.91
1,475	38.41	11.38	1,695	41.17	11.92
1,480	38.47	11.40	1,700	41.23	11.93
1,490	38.60	11.42	1,705	41.29	11.95
1,500	38.73	11.45	1,710	41.35	11.96
1,510	38.86	11.47	1,715	41.41	11.97
1,515	38.92	11.49	1,720	41.47	11.98
1,520	38.99	11.50	1,725	41.53	11.99
1,530	39.12	11.52	1,730	41.59	12.00
1,535	39.18	11.54	1,735	41.65	12.02
1,540	39.24	11.55	1,745	41.77	12.04
1,545	39.31	11.56	1,755	41.89	12.06
1,555	39.43	11.59	1,765	42.01	12.09
1,560	39.50	11.60	1,770	42.07	12.10
1,570	39.62	11.62	1,775	42.13	12.11
1,575	39.69	11.63	1,785	42.25	12.13
1,585	39.81	11.66	1,790	42.31	12.14
1,590	39.87	11.67	1,795	42.37	12.15
1,595	39.94	11.68	1,800	42.43	12.16
1,600	40.00	11.70	1,810	42.54	12.19
1,605	40.06	11.71	1,815	42.60	12.20
1,610	40.12	11.72	1,825	42.72	12.22
1,615	40.19	11.73	1,830	42.78	12.23
1,620	40.25	11.74	1,840	42.90	12.25
1,625	40.31	11.76	1,845	42.95	12.26
1,630	40.37	11.77	1,850	43.01	12.28
1,635	40.44	11.78	1,855	43.07	12.29

(continued)

9.24 SQUARE ROOTS AND CUBE ROOTS (1,000–2,000) (continued)

Number	Square Root	Cube Root	Number	Square Root	Cube Root
1,640	40.50	11.79	1,860	43.13	12.30
1,645	40.56	11.80	1,865	43.19	12.31
1,650	40.62	11.82	1,870	43.24	12.32
1,875	43.30	12.33	1,940	44.05	12.47
1,880	43.36	12.34	1,945	44.10	12.48
1,885	43.42	12.35	1,950	44.16	12.49
1,890	43.47	12.36	1,955	44.22	12.50
1,895	43.53	12.37	1,960	44.27	12.51
1,900	43.59	12.39	1,965	44.33	12.53
1,905	43.65	12.41	1,970	44.38	12.54
1,910	43.70	12.42	1,975	44.44	12.55
1,915	43.76	12.43	1,980	44.50	12.56
1,920	43.82	12.44	1,985	44.55	12.57
1,925	43.87	12.45	1,990	44.61	12.58
1,930	43.93	12.46	1,995	44.67	12.59
1,935	43.99	12.47	2,000	44.72	12.60

Section Ten

SANITARY SEWERS

10.1 SEPARATION DISTANCES

Treatment plants should be located as far as possible from human habitation. The table below contains a list of minimum separation distances that must be maintained between treatment facilities and dwellings, public roads, or other areas of substantial use by the public unless special designs or considerations warrant a reduced distance.

Minimum Separation Distance from Treatment Facility

Treatment	Distance (ft.)
Aeration	500
Aerated lagoon	1,000
RBCs	400
Open sand filter	500
Buried or covered sand filter	200

10.2 MINIMUM SEPARATION BETWEEN SANITARY FACILITIES AND OTHER FEATURES

	Minimum Separation Distances (ft.) from:			
	Septic Tank	Absorption Fields	Seepage Pits	Sewer Line
Drilled well—Public	100	200	200	50
Drilled well—Private	50	100	150	50
Dug well	75	150	150	50
Water line (pressure)	10	10	10	10
Water line (suction)	50	100	150	50
Foundation	10	20	20	—
Surface water	50	100	100	25
Open drainage	25	35	35	25
Culvert (tight pipe)	25	35	35	10
Culvert opening	25	50	50	25
Catch basin	25	50	50	—
Interceptor drain	25	35	35	25
Swimming pool—In-ground	20	35	50	10
Reservoir	50	100	100	50
Property line	10	10	10	10
Top of embankment or steep slope	25	25	25	25

10.3 EXPECTED HYDRAULIC LOADING RATES

Type of Facility	Flow Rate per Person (gal./day)	Flow Rate per Unit (gal./day)
Airports		
Per passenger	3	
Per employee	15	
Apartments	75	
1 bedroom		150
2 bedroom		300
3 bedroom		400
Bathhouse, per swimmer	10	
Boarding house	75	
Bowling alley		75
Per lane—no food		
With food—add food-service value		
Campgrounds (Recreational vehicle—per site)		
Sewered sites		100
Central facilities		
Served sites, 300' radius		100
Peripheral sites, 500' radius		75
Subtractions from above		
No showers		25
Dual service (central facilities and sewered facilities overlapping the central)		25
Campground (summer camp)		
Central facilities	50	
Separate facilities		
Toilet	10	
Shower	25	
Kitchen	10	
Campground dumping stations		
Per unsewered site		10
Per sewered site		5
Camp, day	13	
Add for lunch	3	
Add for showers	5	

(continued)

10.3 EXPECTED HYDRAULIC LOADING RATES *(continued)*

Type of Facility	Flow Rate per Person (gal./day)	Flow Rate per Unit (gal./day)
Carwashes, assuming no recycle		
Tunnel, per car		80
Rollover, per car		40
Wandwash, per 5-minute cycle		20
Churches—per seat		
with catering—add food-service value		3
Clubs		
Country		
Per resident member		75
Per nonresident member		25
Racquet (per court per hour)		80
Factories		
Per person/shift	25	
Add for showers	10	
Food-service operations (per seat)		
Ordinary restaurant		35
24-hour restaurant		50
Restaurant along freeway		70
Tavern (little food service)		20
Curb service (drive-in, per car space)		50
Catering, or banquet facilities	20	
Hairdresser (per station)		170
Hospitals (per bed)		175
Hotels (per room)		120
Add for banquet facilities, theater, night club, as applicable		
Homes		
1 bedroom		150
2 bedroom		300
3 bedroom		400
4 bedroom		475
5 bedroom		550
Institutions (other than hospitals)	125	
Laundromats (per machine)		580

(continued)

Type of Facility	Flow Rate per Person (gal./day)	Flow Rate per Unit (gal./day)
Mobile home parks		
Fewer than five units: use flow rates for homes		
Twenty or more units		
Per trailer		200
Double wide		300
Five to twenty units—use prorated scale		
Motels		
Per living unit		100
With kitchen		150
Office buildings		
Per employee	15	
Per square foot		0.1
Dentist—per chair/day		750
Parks (per picnicker)		
Restroom only	5	
Showers and restroom	10	
Schools (per students)		
Boarding	75	
Day	10	
Cafeteria—add	5	
Showers—add	5	
Service stations		
Per toilet (not including carwash)		400
Shopping center (per sq. ft.—food extra)		0.1
Per employee	15	
Per toilet		400
Swimming pools (per swimmer)	10	
Sports stadium	5	
Theater		
Drive-in (per space)		3
Movie (per seat)		3
Dinner theater		
Individual (per seat)	20	
With hotel	10	

10.4 INTERMITTENT STREAM EFFLUENT LIMITS

Discharge to an intermittent stream will be allowed only when all other methods of disposal have been considered and judged unacceptable. Data should be supplied to show that the discharge from any sewage treatment facility will not contravene stream standards. As a minimum, discharge to intermittent streams must meet the following limits:

BOD	5.0 mg/l
Suspended solids	10.0 mg/l
Dissolved oxygen	≥7.0 mg/l
Ammonia	Limit to be set by review agency
pH	Limit to be set by review agency
Chlorine residual	Limit to be set by review agency

Conventional Gravity Sewers

Design Factors

Systems conveying raw sewage from institutional/commercial facilities may use a minimum size 6 in. diameter collector sewer with a minimum slope of 1/8 in./ft. (1.0%). Trunk sewers shall be a minimum of 8 in. diameter with a minimum 0.4% slope. In very small installations, 4 in. diameter sewers may be used for raw sewage if a minimum slope of 1/4 in./ft. (2%) is maintained and a velocity of at least 2 ft./sec. is achieved when the sewer is flowing full. The use of smooth interior pipe is recommended.

Where velocities greater than 15 ft./sec. are expected, special provision shall be made to protect against displacement by erosion or shock.

10.5 MINIMUM SIZE OF SEWER APPURTENANCES

Fixture	Diameter (in.)
Standard manholes	42
Nonstandard manholes	24
Inspection pipes	24
Cleanouts	8

10.6 RECOMMENDED HAZEN-WILLIAMS COEFFICIENTS FOR SEWER PIPE

Head loss determination should be based on total dynamic head under the maximum flow to occur infrequently. Frictional head loss should be determined using the Hazen-Williams coefficients given in the following table.

Pipe Description	Hazen-Williams Coefficient C
Plastic	150
Concrete/cement	120
Cast iron	100
Welded steel	100
Riveted steel	90

10.7 SYSTEM PRESSURES AND PUMP TYPES

System Pressures
Operating pressures in general should be in the range of 40 to 60 psi and shall not exceed 60 psi for any appreciable amount of time. Provision shall be made in both the system and grinder pumps to protect against the creation of any long-term high-pressure situations.

Pump Types
Both stable-curve centrifugal and progressing cavity semipositive displacement pumps may be used in pressure sewer systems.

The stable-curve centrifugal, a pump having maximum head at no flow, may be considered for its ability to compensate with reduced or zero delivery against excessive high pressures, and the ability to deliver at a high rate during low-flow situations in the system, thus enhancing scouring during low-flow periods.

10.8 SEPTIC TANKS

Design and Installations
Septic tanks must be followed by subsurface disposal or by polishing treatment prior to surface discharge.

The table below shows the calculation that should be used to determine minimum effective tank capacity for commercial/institutional and multihome wastewaters. Tanks larger than this minimum will show enhanced performance. No tanks shall have a capacity less than 1,000 gallons. For multihome purposes, calculated flows must be based upon the maximum occupancy of the homes. For commercial/institutional purposes, the tank must be able to treat continuous wastewater flows for 8 to 16 hours per day, as well as expected peak loading. It may be necessary to increase tank size when a commercial or institutional facility has a short significant delivery period to prevent excessive rate of flow through the system.

Septic Tank Sizing for Multihome and
Commercial/Institutional Applications

Daily Flow, Q (gpd)	Tank Size (gal.)
Under 5,000	1.5 Q
5,000–15,000	3,750 + 0.075 Q
15,000+	Q

10.9 ESTIMATING THE SIZE GREASE TRAP FOR RESTAURANTS AND HOSPITALS

The following two equations should be used for estimating the size grease trap necessary for restaurants and other types of commercial kitchens. The minimum-size grease trap should be 750 gallons.

Restaurants

$(D) (GL) (ST) (HR/2) (LE)$ = Size of grease trap (gal.)

where D = Number of seats in dining area

GL = Gallons of wastewater per meal (normally 5 gal.)

ST = Storage capacity factor (minimum = 1.7, on-site disposal = 2.5)

HR = Number of hours open

LF = Loading factor (interstate freeways = 1.25,

other freeways and recreational areas = 1.0,

main highways = 0.8, other highways = 0.5)

Hospitals, Nursing Homes, and So On

$(M) (GL) (ST (2.5) (LF))$ = Size of grease trap (gal.)

where M = Meals per day

GL = Gallons of wastewater per meal (normally 4.5 gal.)

ST = Storage capacity factor (minimum = 1.7, on-site disposal = 2.5)

10.10   BURIED FILTERS

Media

Recommended effective media size is 0.25 to 1.0 mm, and the uniformity coefficient shall be less than 4. If nitrification is not required, media size should be at least 0.5 mm.

Loading

The application rate for buried filters shall be no more than 1.0 gpd/SF for filters in continuous operation. Loading rates up to 2.0 gpd/SF may be allowed if a bed is operated such that it will rest for an amount of time equal to or greater than that for which it is in use on a yearly basis (i.e., seasonal operation). When nitrification is required, the application rate shall not exceed 1.0 gpd/SF. Efficient nitrification cannot be expected with filters that operate on a seasonal basis.

10.11   OPEN FILTERS—SINGLE PASS

Single-pass filters can be used as a secondary treatment method or for effluent polishing following package plants.

Media

The effective media size shall range from 0.25 to 1.0 mm with a uniformity coefficient less than 4. If nitrification is not required, media size should be at least 0.5 mm.

Loading

The loading rate shall not exceed 5 gpd/SF from septic tanks and other primary settling tanks. From trickling filters and package activated sludge plants, application rates up to 10 gpd/SF are acceptable. Decreased loading rates are recommended when nitrification is required.

10.12   FILTER MEDIA

The table below provides a listing of the normally acceptable range of media sizes and minimum media depths. The uniformity coefficient of the media shall be 1.7 or less. The designer has the responsibility for selection of media to meet specific conditions and treatment requirements relative to the project under consideration.

Media Sizes and Minimum Depths

	Single Media		Dual Media		Multimedia	
Material	Size (mm)	Depth (in.)	Size (mm)	Depth (in.)	Size (mm)	Depth (in.)
Anthracite	—	—	1.0–2.0	20	1.0–2.0	20
Sand	1.0–4.0	48	0.5–1.0	12	0.6–0.8	10
Garnet or similar material	—	—	—	—	0.3–0.6	2

10.13   RECOMMENDED SEWAGE APPLICATION RATES

Percolation Rate (min./in.)	Soil Type	Application Rate (gal./day/SF)
<1	Gravel, coarse sand	Not suitable
1–5	Coarse-medium sand	1.2
6–7	Fine sand, loamy sand	1
8–10	Fine sand, loamy sand	0.9
11–15	Fine sand, loamy sand	0.8
16–20	Sandy loam, loam	0.7
21–30	Sandy loam, loam	0.6
31–45	Loam, porous silt loam	0.5
46–60	Loam, porous silt loam	0.45
61–120	Silty clay loam, clay loam	0.2
>120	Clay	Not suitable

10.14   SITE CRITERIA FOR MOUNDS

Criteria	Nonsevere Sites	Severe Sites
Landscape position	Well-drained level or sloping areas; crests of slope or convex slopes are preferred; avoid depressions, bases of slopes, and concave slopes.	Depressions, bases of slopes, concave slopes may be considered if suitable drainage is provided.
Percolation rate	0–60 min./in.	60–120 min./in.

(continued)

Criteria	Nonsevere Sites	Severe Sites
Slope	0–6% if percolation rate slower than 30 min./in.	Up to 20%
Depth to water table (minimum)	12 in.	>0 in.
Depth to pervious or creviced Bedrock (minimum)	24 in.	24 in.
Depth to impermeable barrier (minimum)	3 ft.	3 ft.

10.15 ALLOWABLE LATERAL LENGTHS (FT.)

Hole Spacing (in.)	Hole Diameter (in.)	Pipe Diameter		
		1 in.	1 ¼ in.	1 ½ in.
30	³⁄₁₆	34	52	70
	⁷⁄₃₂	30	45	57
	¼	25	38	50
36	³⁄₁₆	36	60	75
	⁷⁄₃₂	33	51	63
	¼	27	42	54

10.16 SIDEWALL AREAS OF CIRCULAR SEEPAGE PITS

It is recommended that pits have an effective diameter at least equal to the depth of the pit. The effective diameter shall not be less than 6 ft. Only the sidewall area of the pit structure may be used for sizing the absorption area. Application rates for sizing the necessary sidewall area are given below. Soil layers with percolation rates slower than 30 min./in. must be excluded from the effective depth. The table on the following page enumerates the effective absorption area of pits' various dimensions.

Sidewall Areas of Circular Seepage Pits (SF)

Diameter of Seepage Pit (ft.)	Effective Strata Depth Below Inlet (Feet)									
	1	2	3	4	5	6	7	8	9	10
3	9.4	19	28	38	47	57	66	75	85	94
4	12.6	25	38	50	63	75	88	101	113	126
5	15.8	31	47	63	79	94	110	126	141	157
6	18.8	38	57	75	94	113	132	151	170	188
7	22.0	44	66	88	110	132	154	176	198	220
8	25.1	50	75	101	126	151	176	201	226	251
9	28.3	57	85	113	141	170	198	226	254	283
10	31.4	63	94	126	157	188	220	251	283	314
11	34.6	68	104	138	173	207	242	276	311	346
12	37.7	75	113	151	188	226	264	302	339	377

For depths greater than 10 ft., find the area by adding sections.

Example

Area of 15 ft. deep pit = Area of 10 ft. pit + Area of 5 ft. pit

10.17 CASCADE AERATION

Step or cascade aerators are especially useful when the needed dissolved oxygen increment is small or moderate. Cascades consist of a series of weirs or concrete or metal steps over which the effluent flows in a thin sheet. The edges of metal steps may have low weirs. The objective of a cascade is to maximize turbulence, thus increasing oxygen transfer. As the number of descents increases for a given head loss, the quality of exposure may decrease because of the tendency of droplets to break away from jets of falling water as soon as the jets strike air decreases.

Head requirements generally vary from 3 to 10 ft., and effluent pumping may be necessary if the required head is not available. Normally, less than 100 SF of surface area is needed per mgd (million gallons per day) capacity.

The following equation may be used to obtain an estimate of aeration potential:

$$h = \frac{r-1}{0.11ab\,(1 + 0.46T)}$$

where h = Height through which water falls (ft.)

r = Deficit ratio = $(C_s - C_o)/(C_s - C)$

C_s = DO saturation concentration of wastewater at temperature T (mg/l)

C_s = DO concentration of influent to cascade (mg/l)

C = Required DO level after aeration (mg/l)

T = Water temperature (C)

a = Water quality parameter equal to 0.8 for wastewater treatment plant effluent

b = Weir geometry parameters

The parameter b should be set equal to 1.0 for free weirs, 1.1 for concrete steps, and 1.3 for step weirs unless analysis of a particular design shows that individual characteristics justify the use of a different value.

10.18 CONVERSION FACTORS

Multiply	By	To Obtain
Acres (ac)	0.405	Hectares (ha.)
Feet (ft.)	0.3048	Meters (m)
Inches (in.)	2.54	Centimeters (cm)
Gallons (gal.)	0.003785	Cubic meters (m^3)
Gallons (gal.)	3.785	Liters (l)
Square feet (SF)	0.0929	Square meters (m^2)
Square inches (sq. in.)	6.45	Square centimeters (cm^2)
Million gallons/day (mgd)	0.0438	Cubic meter/second (m^3/s)
Million gallons/day (mgd)	3,785	Cubic meters/day (m^3/d)
Gallons/minute (gpm)	0.06308	Liters/second (l/s)
Cubic feet/second (cfs)	0.02832	Cubic meters/second (m^3/s)
Cubic feet/minute (cfm)	0.000472	Cubic meters/second (m^3/s)
Gallons/day/square foot (gpd/SF)	0.04074	Cubic meters/day/square meter (m^3/m^2 • d)
Gallons/minute/ square foot (gpm/SF)	0.67902	Liters/second/square meter (l/m^2 • s)
Feet/second (fps)	0.3048	Meters/second (m/s)
Gallons/day/lineal foot (gpd/lin ft.)	0.01242	Cubic meters/day/lineal meter (m^3/m • d)
Pounds of BOD/1,000 cubic feet (lb BOD/1,000 cu. ft.)	10,020	Kilograms of BOD/cubic meter (kg BOD/m^3)
Pounds/square inch (psi)	0.0703	Kilograms/square centimeters (kg/cm^2)
Gallons/minute/1,000 square feet (gpm/1,000 SF)	679.02	Liters/second/square meter (lm^2 • s)
Cubic foot of air/pound BOD (cu. ft./lb BOD)	0.06243	Cubic meters/kilogram BOD (m^3/kg BOD)

10.19 MOUND DESIGN EXAMPLE

Design a mound for a site with a percolation rate of 60 min./in. The slope of the land is 15%, so the site may be classified as severe. The design flow rate is 5,000 gpd of aerobically treated wastewater.

Step 1: Evaluate the Site

The soil horizon consists of an 18 in. layer of loam (the A horizon) and a 12 in. layer of fine silt loam (the B horizon) overlaying a slowly permeable clay layer (the C horizon). The vertical and horizontal unsaturated hydraulic conductivities are listed in the table below. Evidence of seasonally high groundwater exists at 15 in. below the ground surface.

	Unsaturated Hydraulic Conductivity (gpd/SF)	
Horizon	Vertical	Horizontal
A	0.7	60
B	0.6	40
C	0.25	1

Step 2: Boundary Conditions

(a) If we assume that the clogging mat will be insignificant because of the extent of pretreatment, the vertical acceptance rates (VARs) can be considered equal to the vertical unsaturated hydraulic conductivities. The vertical boundary is at the C horizon with a VAR of 0.25 gpd/SF.

(b) The horizontal boundary will be at or downslope from the base of the mound. Use

$$HAR = KA \frac{dH}{dx}$$

where HAR = Horizontal acceptance rate

K = Horizontal hydraulic conductivity = 60 gpd/SF in the A horizon and 40 gpd/SF in the B horizon

A = Area of horizontal flow = 1 SF

dH/dx = Hydraulic gradient = 0.15

Thus, HAR_A = 9.0 gpd/ft. and HAR_B = 6.0 gpd/ft. Since horizontal flow can be expected, the area downslope must be evaluated for any restrictions, such as a stream, change in slope, and so on, that will act as a horizontal boundary.

Step 3: Determine the Vertical Wastewater Application Width (AW)

The horizontal and vertical loading rates must not exceed HAR and VAR. The application width is width over which the vertical flow is applied in the C horizon upslope from the horizontal boundary. To keep the toe of the mound from getting too wet, there should be no horizontal flow in the A horizon. The basal loading for the A horizon must not exceed the basal loading for the B horizon; thus HAR_B is used in these calculations.

$$AW_C = \frac{HAR}{(VAR_B - VAR_C)}$$
$$= \frac{6 \text{ gpd/ft.}}{(0.60 - 0.25) \text{ gpd/SF}}$$
$$AW_C = 17.1 \text{ ft.}$$

Thus, unless a restriction exists upslope from this point, the horizontal boundary will be at the edge of the application width.

Step 4: Determine the Linear Loading Rate (LLR) of the Wastewater

$$LLR = VAR_C(AW_C) + HAR_B$$
$$= (0.25 \text{ gpd/SF})(17.1 \text{ ft.}) + 6 \text{ gpd/ft.}$$
$$LLR = 10.3 \text{ gpd/ft.}$$

Step 5: Determine the Basal Width (BW) of Each Horizon

(a) Basal width of A

$$BW_A = \frac{LLR}{VAR_A} = \frac{10.3 \text{ gpd ft.}}{0.7 \text{ gpd/SF}}$$
$$BW_A = 14.7 \text{ ft.}$$

Thus, the toe of the mound will end 2.4 ft. upslope from the horizontal boundary. To allow a 3:1 slope, this can be extended to 24.5 ft.

(b) Basal width of B:

$$BW_B = \frac{LLR}{VAR_B} = \frac{10.3 \text{ gpd/ft.}}{0.6 \text{ gpd/SF}}$$
$$BW_B = 17.1 \text{ ft.}$$

(c) Basal width of C:

$$BW_C = \frac{LLR}{VAR_C} = \frac{10.3 \text{ gpd/ft.}}{0.25 \text{ gpd/SF}}$$
$$BW_C = 41.2 \text{ ft.}$$

Therefore, water will move an additional 24.1 ft. downslope from the horizontal boundary, at which point all movement will be vertically downward. This area must be protected to prevent loss of infiltrative capacity.

Step 6. Determine Width of Absorption

$$A = \frac{LLR}{\text{Fill infiltration rate}}$$

Sand fill infiltration rate = 1.2 gpd/SF

Therefore, A = 8.6 ft.

Step 7. Determine Length of Absorption Area

$$B = \frac{\text{Design flow rate}}{LLR}$$

$$= \frac{5,000 \text{ gpd}}{10.3 \text{ gpd/ft.}}$$

$$B = 485 \text{ ft.}$$

Step 8. Mound Height

The fill depth and natural soil should provide at least 3 ft. of unsaturated soil below the distribution lines. Thus, a minimum of 21 in. of fill is necessary. Assuming a 1 in. pipe, the bed depth would be 9 inches. The cap and topsoil are 18 in. deep over the center of the absorption area and 12 in. deep over the edges. When allowances are made for the 15% slope, the mound height at the upslope edge of the absorption area is 3.5 ft., at the center is 4.6 ft., and at the downslope edge is 4.8 ft.

Step 9. Mound Width

Accounting for 3:1 side slopes

Upslope width

J = (Mound depth at upslope) (3) (slope correction factor)

= (3.5) (3) (0.66)

J = 6.9 ft.

Downslope width

I = (Mound depth at downslope) (3) (slope correction factor)

= (4.8) (3) (1.68)

I = 24.2 ft.

Since the absorption area is 8.6 ft. wide, the total mound width is 39.7 ft.

Thus, the toe of the mound will end 8.4 ft. upslope from the edge of the basal width in the C horizon (BW_C).

Because of its length, this mound should be constructed in several segments. At least 8.4 ft. should be allowed between the toe of one mound and the upslope edge of another to prevent overlapping of the basal areas. Additional space is to be included to allow maneuvering of construction equipment.

Step 10. In Contrast to a Mound System

Design of a mound system takes advantages of the horizontal movement of the wastewater as well as vertical movement. If a system were to be designed for this site on the basis of vertical transport only, it would need to be much longer. The loading rate would be limited by the vertical acceptance rate of the C horizon. Assuming that the width of the absorption area is the same as for the mound system, then:

$$LLR = (VAR_C)\ (A)$$

$$= (0.25\ gpd/SF)\ (8.6\ ft.)$$

$$LLR = 2.15\ gpd/ft.$$

Since the wastewater flow rate is 5,000 gpd, the total system length would be 2,326 ft.

Section Eleven

WORLDWIDE WEATHER CONDITIONS

11.1 TEMPERATURE DISTRIBUTION

The distribution of the temperature over the world and its variations through the year depend primarily on the amount of radiant energy received from the sun in different regions. This, in turn, depends mainly on latitude, but it is greatly modified by the locations of continents and oceans, prevailing winds, oceanic circulation, topography, and other factors.

Maps showing average temperature over the surface of the earth for January and for July are given in Figures 11.1 and 11.2. In the winter of the Northern Hemisphere (Figure 11.1), it will be noted that the poleward temperature gradient (the rate of fall in temperature) north of latitude 15 degrees is very steep over the interior of North America. This is shown by the fact that the lines indicating changes in temperature are very close together. The temperature gradient is also steep toward the cold pole over Asia—the area marked minus 40 degrees. In Western Europe, to the east of the Atlantic Ocean and the Gulf Stream (a region of prevailing westerly winds), the temperature gradient is much more gradual. This is indicated by the fact that the isotherms, or lines of equal temperature, are farther apart. In the winter of the Southern Hemisphere (Figure 11.2), the temperature change toward the South Pole is very gradual, and the isothermal departures from the east-west direction are not as evident because continental effects are largely absent.

In the summer of the two hemispheres—July in the north and January in the south—the temperature gradients are very much diminished as compared with those during the winter. This is especially marked over the middle and higher northern latitudes because of the greater warming of the extensive interiors of North America and Eurasia than of the smaller land areas in middle and higher southern latitudes.

11.2 DISTRIBUTION OF PRECIPITATION

Whether precipitation (Figure 11.3) occurs as rain or snow or in the rarer forms of hail or sleet depends largely on temperature. This may be influenced more by elevation than by latitude, as in the case of the perpetually snowcapped mountain peaks and glaciers near the equator in both South America and Africa.

The quantity of precipitation is governed by the amount of water vapor in the air and the nature of the process that leads to its condensation into liquid or solid form through cooling, usually by a moist air mass being forced to rise. Air may ascend to great elevations through local convection (as in thunderstorms or tropical rains); it may be forced up to cover topographical elevations by the prevailing wind (as on the southern or windward slopes of the Himalayas in the path of the southwestern monsoon of India), or it may ascend gradually in migratory low-pressure formations in which the frontal meeting of warm and cool air masses causes the moist air to rise and condense.

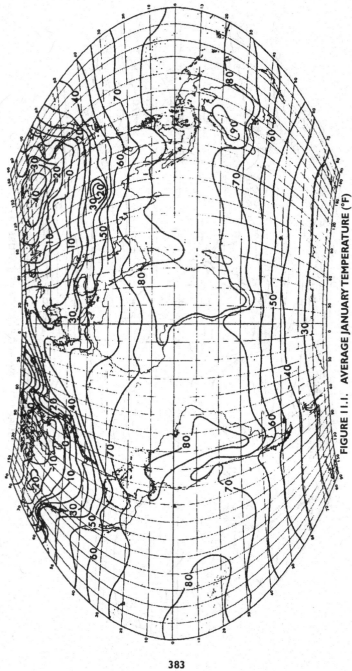

FIGURE 11.1. AVERAGE JANUARY TEMPERATURE (°F)

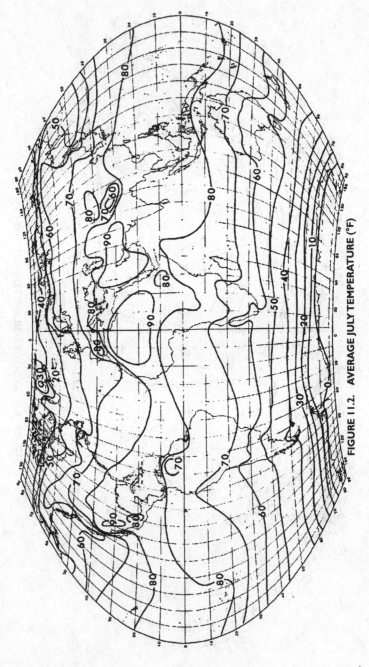

FIGURE 11.2. AVERAGE JULY TEMPERATURE (°F)

384

The areas of heaviest precipitation are generally located in tropical regions, where the greatest amount of water is present in the atmosphere and the greatest evaporation takes place, although rainfall can occur only where conditions favor condensation. Certain regions in high latitudes, such as southern Alaska and southern Chile, also show great amounts of precipitation where relatively warm, moist winds blowing off the ocean undergo a forced ascent over coastal mountain ranges.

In the dry polar regions the water-vapor content of the air is always very low because of the low temperature and very limited evaporation. The other major dry areas are the subtropical belts of high atmospheric pressure (in the vicinity of latitude 30 degrees on all continents, especially from the extreme western Sahara over a broad, somewhat broken belt to the Gobi Desert) and the arid strips on the lee sides of mountains whose windward slopes have heavy precipitation. These are caused by conditions that, even though the temperature may be high, are unfavorable to the condensation of whatever water vapor may be present in the atmosphere.

11.3   NORTH AMERICA

North America is nearly all within middle and northern latitudes, and therefore much of its interior displays the continental type of climate with notable seasonable temperature extremes. Along the coast of southern Alaska, western Canada, and the northwestern part of the United States, moderate midsummer temperatures are in marked contrast to those prevailing in the interior east of the mountains. (Note, for example, the great southward dip of the 60°F isotherm along the west coast in Fig. 11.2.) Similarly, the mild midwinter temperatures in the coastal areas stand out against the severe conditions to be found from the Great Lakes region northward and northwestward (Fig. 11.1). In the Caribbean region, temperature conditions are subtropical while in Mexico and Central America, climatic zones depend on elevation, ranging from subtropical to temperatures in the higher levels.

The prevailing westerly wind movement carries the continental type of climate eastward over the United States, so that the region of maritime climate along the Atlantic Ocean is very narrow. The northern areas are very cold; however, the midwinter low temperatures fall far short of the records set in the cold-pole area of northeastern Siberia, where, because the land area is more vast, it becomes much colder than the smaller land area of northern Canada.

From the Aleutian Peninsula to northern California, west of the crest of the mountains, there is a narrow strip where annual precipitation is over 40 in., even exceeding 100 in. locally on the coast of British Columbia. East of this belt there is an abrupt decrease in precipitation to less than 20 in. annually from the Gulf of California northward, and to less than 5 in. in parts of the southwestern United States.

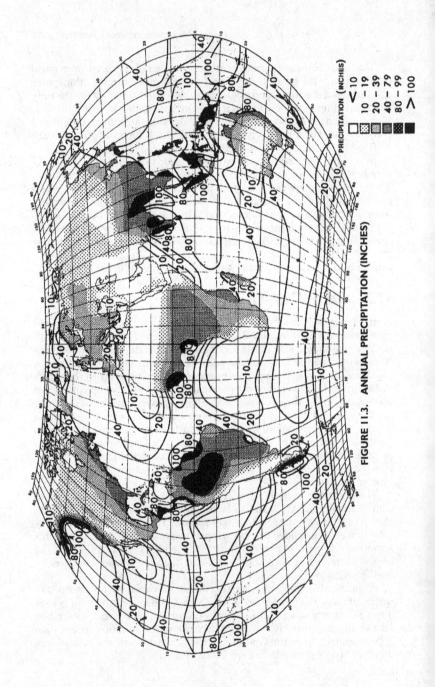

FIGURE 11.3. ANNUAL PRECIPITATION (INCHES)

PRECIPITATION (INCHES)

∨ 10
10 – 19
20 – 39
40 – 79
80 – 99
∧ 100

In the eastern part of the continent from the southern United States northeastward to Newfoundland, the average annual precipitation is more than 40 in. Rainfall in the Caribbean, southern Mexico, and Central America is generally abundant. It varies widely, however, even within short distances, especially from the windward to the leeward sides of the mountains.

11.4   SOUTH AMERICA

A large part of South America lies within the tropics and has a characteristically tropical climate. The remaining narrow southern portion is not subject to the extremes of heat and cold that are found where wide land areas give full sway to the continental type of climate with its hot summers and cold winters. Temperature anomalies unusual for a given latitude are to be found mainly at the high elevations of the Andes stretching from the Isthmus of Panama to Cape Horn.

The Antarctic Current and its cool Humboldt branch skirting the western shores northward to the equator, together with the prevailing onshore winds, exert a strong cooling influence over the coastal regions of all of the western countries of South America except Colombia. Along the eastern coast, the southerly moving Brazil current from the tropics has a warming effect except along Argentina's southern coast. In the northern countries of South America, the sharply contrasted dry and wet seasons are the result of the trade winds. In the dry season (corresponding to winter in the Northern Hemisphere), these winds sweep the entire region, while during the wet season (corresponding to summer in the Northern Hemisphere), calm and variable winds prevail. In the basin of the Amazon River, the rainfall is related to the equatorial belt of low pressure and to the trade winds, which give the maximum amounts of rainfall in the extreme west, where the winds ascend the Andean slopes. The desert areas on the west coast of South America, extending from the equator southward to the latitude of Santiago, are due primarily to the cold Humboldt or Peruvian Current and upwelling cold coastal water. The moist, cool ocean air is warmed in passing over the land, with a consequent decrease in relative humidity, so that the dew point is not reached and condensation of vapor does not occur until the incoming air has reached high elevations in the Andes, where temperatures are much lower than along the coast. In southern Chile, the summer season has moderate rainfall, and winters are excessively wet. The land does not get as much heating, and so the prevailing winds remain cooler and condensation of moisture from the ocean progresses from the shores up to the crest of the Andes. By the time the air passes these elevations, however, the moisture has been depleted so that the winds on the leeward slopes are dry, becoming more so as they are warmed on reaching lower levels. The mountains cast a great "rain shadow" (an area of little rain) over southern Argentina.

11.5 EUROPE

In Europe there is no extensive north-south mountain system such as is found in the Americas. The general east-west direction of the ranges in the south allows the conditions in the maritime west to change rather gradually toward Asia. Generally, rainfall is heaviest on the western coast, where locally it exceeds 60 in. annually, and diminishes toward the east to less than 20 in. in eastern Russia. Exceptions to this rule occur in the elevated regions of the Alps and the Caucasus. There is a well-defined rain shadow in Scandinavia, with over 60 in. of rain in western Norway and less than 20 in. in eastern Sweden. Over much of Europe precipitation is both abundant and evenly distributed throughout the year. In those areas where there is a seasonal distribution, namely around the Mediterranean, most of the precipitation comes in the winter, with very dry summers.

Isothermal lines generally parallel the lines of latitude except in winter, when the modifying influence of the Gulf Stream causes a pronounced east-west gradient. There are few marked dips in isotherms due to elevation or a continental type of climate such as are found in North America. In Scandinavia, however, the winter map shows an abrupt fall in temperature from the western coast of Norway to the eastern coast of Sweden. The coastal mountains eliminate the modification of the ocean, thus allowing a greater range of temperatures.

11.6 ASIA

The vast extent of Asia gives full opportunity for continental conditions to develop with a cold area of high barometric pressure in winter and a hot low-pressure area in summer. The former is northeast of the Himalayas and the latter stretches widely from west to east in the latitude of northwestern China. These distributions of pressure give southeastern Asia the well-known monsoon seasons.

In winter, the air circulation is outward over the land from the cold pole, and precipitation is very light over the entire continent. In summer, however, there is an inflow of air from the oceans, and the southeastern trade winds cross the equator and merge into the southwestern monsoon, which crosses India. This saturated air produces abundant rain over most of that country, with excessively heavy amounts where the air is forced to rise, even to moderate elevations. At Cherrapunji (4,455 ft.), on the southern side of the Khasi Hills in Assam, the average rainfall in a winter month is about 1 in., while in both June and July it is approximately 100 in. per month.

Farther north, the summer rainfall meets an impassable barrier in the Himalaya Mountains. A much weaker summer monsoon brings rainfall to Japan and eastern Asia but does not extend far into China because of lower elevations. Consequently, while the southeastern quadrant of Asia has heavy-to-excessive annual rainfall, the remainder of the continent is dry, with vast areas receiving less than 10 in. annually. North of the Himalayas the low plains are extremely cold in winter, although temperatures rise considerably in summer. At Verkhoyansk, which is just north of the Arctic Circle, the mean maximum temperature in January is about −54°F, and in July approximately 66°F. The extreme records are a maximum of 98°F and a minimum of −90°F.

In Southwestern Asia (west of India), the winter temperature control is still the interior high-pressure area, and temperatures are generally low as a result of the cool-air outflow. Summers are excessively hot and, because the area is beyond the path of the monsoon flow, precipitation is minimal year-round.

11.7 AFRICA

Africa, like South America, lies largely within the tropics and thus temperature distribution is determined mainly by altitude. Moreover, the cool Benguela Current sweeps the southwestern coast as do the warm tropical currents of the Indian Ocean on the southeastern coast, creating conditions closely paralleling those found around the South American continent. In the tropical climatic areas of Africa, conditions are characterized by prevailing low barometric pressure, with conventional rainfall and seasonal northward and southward movement of the heat equator, while in both the north and the south, the ruling influences are the belts of high barometric pressure.

Desert conditions typified by the Sahara extend from the Atlantic to the Red Sea and from the Mediterranean southward well beyond the Tropic of Cancer. The Atlas Mountains, in the northwest of this region, have the most precipitation, as they sit in the path of the moist trade winds. South of the Sahara, rainfall is much more abundant, especially from the west coast to the East African Rift Valley, with annual maximums of over 100 in. in the region around Sierra Leone. East of the Rift Valley precipitation is much less, varying from about 40 in. to less than 10 in. on the coast of Somalia. Moving south toward the Tropic of Capricorn, the rainfall totals are also much less, especially on the west coast.

The heavy rainfall over sections of Ethiopia from June to October, when more than 40 in. may fall, is one of the earth's outstanding examples of seasonal distribution of rainfall.

Moist equatorial climate is typified by conditions in Zaire, arid torrid climate by those of Egypt and the Sahara, and moderate plateau climate by those found in parts of Ethiopia, Kenya, and Tanzania.

11.8 AUSTRALIA

In the southern winter, high pressure sits over the interior of Australia, and all except the southernmost parts of the continent are dry. In summer, this pressure belt moves south of the continent with dry conditions persisting over the southern and western areas. Thus, much of the continent has less than 20 in. of precipitation each year, except in the extreme southwest and in a strip circling from the southeast to northwest. The average annual precipitation in most of the interior, especially to the south, is less than 10 in.

On the southern coast, the winter precipitation is the cyclonic type, the heavy summer rains of the north are of monsoon origin, and those of the east are in the large part orographic, owing to the mountains along the coast. On the ocean-facing eastern slopes of these mountains, the mean annual rainfall is over 40 in., and in many localities over 60 in. This is also true for areas affected by the monsoon rains in the north.

Because of the location of Australia, on both sides of the Tropic of Capricorn, temperatures far below freezing are to be found only in the south at high elevations. In the arid interior, extreme maximum temperatures rank with those of the hottest regions of the earth.

11.9 TEMPERATURE DATA FOR REPRESENTATIVE WORLDWIDE STATIONS

Country and Station	Lat °	'	Long °	'	Elevation (Ft.)	Record Length (Yr)	January Max (°F)	Min (°F)	April Max (°F)	Min (°F)	July Max (°F)	Min (°F)	October Max (°F)	Min (°F)	Extreme Max (°F)	Min (°F)
North America																
United States (Conterminous):																
Albuquerque, NM	35	3 N	106	37 W	5,311	30	46	24	69	42	91	66	71	45	105	-17
Asheville, NC	35	26 N	82	32 W	2,140	30	48	28	67	42	84	61	68	45	100	-16
Atlanta, GA	33	39 N	84	26 W	1,010	30	52	37	70	50	87	71	72	52	105	-8
Austin, TX	30	18 N	97	42 W	597	30	60	41	78	57	95	74	82	60	109	-2
Birmingham, Al	33	34 N	86	45 W	620	30	57	36	76	50	93	71	79	52	106	-10
Bismarck, ND	46	46 N	100	45 W	1,647	30	20	0	55	32	86	58	59	34	114	-45
Boise, ID	43	34 N	116	43 W	2,838	30	36	22	63	37	91	59	65	38	112	-28
Brownsville, TX	25	54 N	97	26 W	16	30	71	52	82	66	93	76	85	67	106	12
Buffalo, NY	42	56 N	78	44 W	705	30	31	18	53	34	80	59	60	41	99	-21
Cheyenne, WY	41	9 N	104	49 W	6,126	30	37	14	56	30	85	55	63	32	100	-38
Chicago, IL	41	47 N	87	45 W	607	30	33	19	57	41	84	67	63	47	105	-27

(continued)

Country and Station	Latitude		Longitude		Elevation (Ft.)	Record Length^a (Yr)	Temperature									
							Average Daily								Extreme	
							January		April		July		October			
							Maximum (°F)	Minimum (°F)	Maximum (°F)	Minimum (°F)	Maximum (°F)	Minimum (°F)	Maximum (°F)	Minimum (°F)	Maximum (°F)	Minimum (°F)
Des Moines, IA	41	32 N	93	39 W	938	30	29	11	59	38	87	65	66	43	110	-30
Dodge City, KS	37	46 N	99	58 W	2,582	30	42	20	66	41	93	68	71	46	109	-26
El Paso, TX	31	48 N	106	24 W	3,918	30	56	30	78	49	95	69	79	50	112	-8
Indianapolis, IN	39	44 N	86	17 W	792	30	37	21	61	40	86	64	67	44	107	-25
Jacksonville, FL	30	25 N	81	39 W	20	30	67	45	80	58	92	73	80	62	105	-7
Kansas City, MO	39	7 N	94	36 W	742	30	40	23	66	46	92	71	72	49	113	-23
Las Vegas, NV	36	5 N	115	10 W	2,162	30	54	32	78	51	104	76	80	53	117	-8
Los Angeles, CA	33	56 N	118	23 W	97	30	64	45	67	52	76	62	73	57	110	23
Louisville, KY	38	11 N	85	44 W	477	30	44	27	66	43	89	67	70	46	107	-20
Miami, FL	25	48 N	80	16 W	7	30	76	58	83	66	89	75	85	71	100	28
Minneapolis, MN	44	53 N	93	13 W	834	30	22	2	56	33	84	61	61	37	108	-34
Missoula, MT	46	55 N	114	5 W	3,190	30	28	10	57	31	85	49	58	30	105	-33
Nashville, TN	36	7 N	86	41 W	590	30	49	31	71	48	91	70	74	49	107	-17
New Orleans, LA	29	59 N	90	15 W	3	30	64	45	78	58	91	73	80	61	102	-7
New York, NY	40	47 N	73	58 W	132	30	40	27	60	43	85	68	66	50	106	-15
Oklahoma City, OK	35	24 N	97	36 W	1,285	30	46	28	71	49	93	72	74	52	113	-17

	Lat. (deg)	Lat. (min)	Long. (deg)	Long. (min)	Elev.											
Phoenix, AZ	33	26 N	112	01 W	1,117	30	64	35	84	50	105	75	87	55	122	16
Pittsburgh, PA	40	27 N	80	00 W	747	30	40	25	63	42	85	65	65	45	103	-20
Portland, ME	43	39 N	70	19 W	47	30	32	12	53	32	80	57	60	37	103	-39
Portland, OR	45	36 N	122	36 W	21	30	44	33	62	42	79	56	63	45	107	-3
Reno, NV	39	30 N	119	47 W	4,404	30	45	16	65	31	89	46	69	29	106	-19
Salt Lake City, UT	40	46 N	111	58 W	4,220	30	37	18	63	36	94	60	65	38	107	-30
San Francisco, CA	37	37 N	122	23 W	8	30	55	42	64	47	72	54	71	51	109	20
Sault Ste. Marie, M	46	28 N	84	22 W	721	30	23	8	46	30	76	54	55	38	98	-37
Seattle, WA	47	27 N	122	18 W	400	30	44	33	58	40	76	54	60	44	100	0
Sheridan, WY	44	46 N	106	58 W	3,964	30	34	9	56	31	87	56	62	33	106	-41
Spokane, WA	47	38 N	117	32 W	2,356	30	31	19	59	36	86	55	60	38	108	-30
Washington, DC	38	51 N	77	03 W	14	30	44	30	66	46	87	69	68	50	104	-18
Wilmington, NC	34	16 N	77	55 W	28	30	58	37	74	51	89	71	76	55	104	0
United States (Alaska):																
Anchorage	61	13 N	149	52 W	85	30	21	4	44	28	65	50	42	28	86	-38
Annette	55	2 N	131	34 W	110	30	38	30	50	37	63	51	51	42	90	-4
Barrow	71	18 N	156	47 W	31	30	-9	-23	7	-7	45	33	21	12	78	-56
Bethel	60	47 N	161	48 W	125	30	11	-4	34	18	62	48	38	25	90	-52
Cold Bay	55	12 N	162	43 W	96	30	33	23	38	28	54	45	45	36	78	-13
Fairbanks	64	49 N	147	52 W	436	30	-1	-21	42	17	72	48	35	17	99	-66
Juneau	58	22 N	134	35 W	12	30	30	20	45	31	63	48	47	37	90	-22
King Salmon	58	41 N	156	39 W	49	30	21	6	41	25	63	47	43	29	88	-48
Nome	64	30 N	165	26 W	13	30	12	-3	28	14	55	44	35	24	86	-54
St. Paul Island	57	9 N	170	13 W	22	30	30	21	33	24	49	42	41	33	66	-26

(continued)

11.9 TEMPERATURE DATA FOR REPRESENTATIVE WORLDWIDE STATIONS (continued)

Country and Station	Latitude		Longitude		Elevation (Ft.)	Record Length (Yr)	Temperature									
							January		April		July		October		Extreme	
	°	'	°	'			Max (°F)	Min (°F)	Max (°F)	Min (°F)	Max (°F)	Min (°F)	Max (°F)	Min (°F)	Max (°F)	Min (°F)
Shemya	52	43 N	174	6 E	122	30	34	29	38	33	49	44	42	38	63	16
Yakutat	59	31 N	139	40 W	28	30	34	20	45	29	61	48	49	35	86	-24
Canada:																
Aklavik, N.W.T.	68	14 N	135	0 W	30	22	-10	-26	19	-2	66	47	25	15	93	-62
Alert, N.W.T.	82	31 N	62	20 W	95	9	-19	-29	-8	-18	44	36	2	-7	67	-53
Calgary, Alta.	51	6 N	114	1 W	3,540	55	24	2	53	27	76	47	54	29	97	-49
Charlottetown, P.E.I.	46	17 N	63	8 W	181	65	26	10	43	30	73	58	54	41	98	-27
Chatham, N.B.	47	0 N	65	27 W	109	50	23	2	47	28	77	56	55	37	102	-43
Churchhill, Man.	58	45 N	94	4 W	94	30	-11	-27	24	4	64	43	34	20	96	-57
Edmonton, Alta.	53	34 N	113	31 W	2,219	71	16	-3	52	28	74	50	51	30	99	-57
Fort Nelson, B.C.	58	50 N	122	35 W	1,253	12	1	-15	47	25	74	51	43	25	98	-61
Fort Simpson, N.W.T.	61	45 N	121	14 W	554	42	-10	-27	38	14	74	50	36	21	97	-70
Frobisher Bay, N.W.T.	63	45 N	68	33 W	110	18	-9	-23	16	-1	53	39	29	18	76	-49
Gander, Nfld.	48	57 N	54	34 W	496	14	27	13	40	27	71	52	51	37	96	-17
Halifax, N.S.	44	39 N	63	34 W	86	75	32	15	47	31	74	55	57	41	99	-21
Kapuskasing. Ont.	49	25 N	82	28 W	743	19	10	-14	43	19	75	50	47	31	101	-53

Station																		
Knob Lake, Que.	54	48	N	66	49	W	1,712	30	−3	−21	30	12	64	46	37	25	88	−59
Montreal, Que.	45	30	N	73	34	W	187	67	21	6	50	33	78	61	54	40	97	−35
North Bay, Ont.	46	21	N	79	25	W	1,216	17	22	2	48	28	78	56	49	36	99	−46
Ottawa, Ont.	45	19	N	75	40	W	374	65	21	3	51	31	81	58	54	37	102	−38
Penticton, B.C.	49	28	N	119	36	W	1,129	32	32	21	61	35	84	53	59	38	105	−16
Port Arthur, Ont.	48	22	N	89	19	W	644	62	17	−4	44	26	74	52	50	34	104	−42
Prince Georgia, B.C.	53	53	N	122	41	W	2,218	27	23	3	54	27	75	44	52	30	102	−58
Prince Rupert, B.C.	54	17	N	130	23	W	170	26	39	30	50	37	62	49	53	42	90	−3
Quebec, Que.	48	48	N	71	23	W	239	72	18	2	44	29	76	57	51	37	97	−34
Regina, Sask.	50	26	N	104	40	W	1,884	55	10	−11	50	26	79	51	52	27	110	−56
Resolute, N.W.T.	74	43	N	94	59	W	220	13	−20	−33	−1	−16	45	35	11	0	61	−61
St. John, N.B.	45	17	N	66	4	W	119	61	28	11	43	32	69	54	54	41	93	−24
St. Johns, Nfld.	47	32	N	52	44	W	211	68	30	18	41	29	69	51	53	40	93	−21
Saskatoon, Sask.	52	8	N	106	38	W	1,690	38	9	−11	49	26	77	52	51	27	104	−55
The Pas, Man.	53	49	N	101	15	W	890	27	1	−18	45	21	75	54	45	26	100	−54
Toronto, Ont.	43	40	N	79	24	W	379	105	30	16	50	34	79	59	56	40	105	−26
Vancouver, B.C.	49	17	N	123	5	W	127	43	41	32	58	40	74	54	57	44	92	2
Whitehorse, Y.T.	60	43	N	135	4	W	2,303	10	13	−3	41	22	67	45	41	28	91	−62
Winnipeg, Man.	49	54	N	97	14	W	783	66	7	−13	48	27	79	55	51	31	108	−54
Yellow Knife, N.W.T.	62	28	N	114	27	W	674	13	−8	−23	29	9	69	52	36	26	90	−60
Greenland:																		
Angmagssalik	65	36	N	37	33	W	95	30	23	10	35	16	54	37	35	25	77	−26
Danmarkshaven	76	46	N	19	0	W	7	2	−1	−15	6	−13	47	34	13	2	63	−42
Eismitte	70	53	N	40	42	W	9,843	1	−33	−53	−14	−37	19	1	−23	−42	27	−85

(continued)

11.9 TEMPERATURE DATA FOR REPRESENTATIVE WORLDWIDE STATIONS (continued)

Country and Station	Latitude ° '		Longitude ° '	Elevation (Ft.)	Record Length (Yr)	January Maximum (°F)	January Minimum (°F)	April Maximum (°F)	April Minimum (°F)	July Maximum (°F)	July Minimum (°F)	October Maximum (°F)	October Minimum (°F)	Extreme Maximum (°F)	Extreme Minimum (°F)
Godthaab	64 10 N	51	43 W	66	40	19	10	31	20	52	38	35	26	76	−20
Ivigtut	61 12 N	48	10 W	98	48	24	12	38	24	57	42	40	29	86	−20
Jacobshavn	69 13 N	51	2 W	104	32	8	−7	24	6	51	40	31	20	71	−46
Nord	81 36 N	16	40 W	118	8	−15	−28	−5	−18	44	35	3	−6	61	−60
Scoresbysund	70 29 N	21	58 W	56	12	12	−3	22	6	49	36	25	15	63	−42
Thule	76 31 N	68	44 W	251	12	−4	−17	10	−7	46	38	19	8	63	−44
Upernivik	72 47 N	56	7 W	59	40	−1	−13	15	−1	48	35	26	21	69	−44
Mexico:															
Acapulco	16 50 N	99	56 W	10	8	85	70	87	71	89	75	88	74	97	60
Chihuahua	28 42 N	105	57 W	4,429	9	65	36	81	51	89	66	79	51	102	12
Guada Lajara	20 41 N	103	20 W	5,194	26	73	45	85	53	79	60	78	56	101	26
Guaymas	27 57 N	110	55 W	58	9	74	57	84	65	96	82	91	75	117	41
La Paz	24 7 N	110	17 W	85	9	75	54	86	58	96	73	90	68	108	31
Lerdo	25 30 N	103	32 W	3,740	10	72	45	86	57	90	68	82	58	105	23
Manzanillo	19 4 N	104	20 W	26	17	86	68	87	67	93	76	91	76	103	54
Mazatlan	23 11 N	106	25 W	256	10	71	61	76	65	86	77	85	76	93	52

(continued)

	Lat.		Long.		Elev.											
Merida	20	58 N	89	38 W	72	22	83	62	92	69	92	73	87	71	106	51
Mexico City	19	26 N	99	4 W	7,340	42	66	42	78	52	74	54	70	50	92	24
Monterrey	25	40 N	100	18 W	1,732	11	68	48	84	62	90	71	80	64	107	25
Salina Cruz	16	12 N	95	12 W	184	10	85	72	88	76	89	76	87	75	98	62
Tampico	22	16 N	97	51 W	78	12	75	59	83	69	89	75	85	71	104	34
Vera Cruz	19	12 N	96	8 W	52	10	77	66	83	72	87	74	85	73	98	53
Central America																
Belize:																
Beliza City	17	31 N	88	11 W	17	27	81	67	86	74	87	75	86	72	97	49
Costa Rica:																
San Jose	9	56 N	84	8 W	3,760	8	75	58	79	62	77	62	77	60	92	49
El Salvador:																
San Salvador	13	42 N	89	13 W	2,238	39	90	60	93	65	89	65	87	65	105	45
Guatemala:																
Guatemala City	14	37 N	90	31 W	4,855	6	73	53	82	58	78	60	76	60	90	41
Honduras:																
Tela	15	46 N	87	27 W	41	4	82	67	87	72	88	73	86	71	96	58
Panama:																
Balboa Heights	8	57 N	79	33 W	118	34	88	71	90	74	87	74	85	73	97	63
Cristobal	9	21 N	79	54 W	35	36	84	76	86	77	85	76	86	75	97	66
West Indies																
Bridgetown, Barbados	13	8 N	59	36 W	181	35	83	70	86	72	86	74	86	73	95	61
Basse-Terre, Guadaloupe	16	1 N	61	42 W	1,750	19	77	64	79	65	81	68	81	68	92	54
Fort-de-France, Martinique	14	37 N	61	5 W	13	22	83	69	86	71	86	74	87	73	96	56

11.9 TEMPERATURE DATA FOR REPRESENTATIVE WORLDWIDE STATIONS (continued)

Country and Station	Latitude			Longitude			Elevation (Ft.)	Record (Yr)	Temperature										
									Average Daily									Extreme	
									January		April		July		October				
									Max (°F)	Min (°F)	Max (°F)	Min (°F)	Max (°F)	Min (°F)	Max (°F)	Min (°F)		Max (°F)	Min (°F)
	°	'		°	'		(Ft.)	(Yr)	(°F)	(°F)	(°F)	(°F)	(°F)	(°F)	(°F)	(°F)		(°F)	(°F)
Hamilton, Bermuda	32	17 N		64	46 W		151	59	68	58	71	59	85	73	79	69		99	40
Havana, Cuba	23	8 N		82	21 W		80	25	79	65	84	69	89	75	85	73		104	43
Kingston, Jamaica	17	58 N		76	48 W		110	33	86	67	87	70	90	73	88	73		97	56
La Guerite, St. Christopher (St. Kitts)	17	20 N		62	45 W		157	19	80	71	83	73	86	76	85	75		91	61
Nassau, Bahamas	25	5 N		77	21 W		12	35	77	65	81	69	88	75	85	73		94	41
Port-au-Prince, Haiti	18	33 N		72	20 W		121	42	87	68	89	71	94	74	90	72		101	58
Port of Spain, Trinidad	10	40 N		61	31 W		67	49	87	69	90	69	88	71	89	71		101	52
St. Thomas, Virgin Is.	18	20 N		64	58 W		11	9	82	71	85	74	88	77	87	76		92	63
San Juan, Puerto Rico	18	26 N		66	00 W		13	30	81	67	84	69	87	74	87	73		94	60
Santo Domingo, Dominican Republic	18	29 N		69	54 W		57	26	84	66	85	69	88	72	87	72		98	59
South America																			
Argentina:																			
Bahia Blanca	38	43 S		62	16 W		95	33	88	62	71	51	57	39	71	48		109	18
Buenos Aires	34	35 S		58	29 W		89	23	85	63	72	53	57	42	69	50		104	22

Cipolletti	38	57 S	67	59 W	889	9	89	56	72	40	55	29	72	43	107	9
Corrientes	27	28 S	58	50 W	177	39	93	71	81	63	71	53	82	60	112	30
La Quiaca	22	6 S	65	36 W	11,345	23	70	41	69	32	60	16	71	32	95	0
Mendoza	32	53 S	68	49 W	2,625	23	90	60	73	47	59	35	76	50	109	15
Parana	31	44 S	60	31 W	210	12	91	67	77	58	62	45	75	54	113	21
Puerto Madryn	42	47 S	65	1 W	26	50	81	57	70	46	55	36	68	45	104	10
Santa Cruz	50	1 S	68	32 W	39	12	70	48	57	39	41	28	58	39	94	1
Santiago Del Estero	27	46 S	64	18 W	653	28	97	69	82	59	70	44	87	59	116	19
Ushuaia	54	50 S	68	20 W	26	16	57	41	48	33	39	25	52	35	85	−6
Bolivia:																
Conception	16	15 S	62	3 W	1,607	5	85	66	86	62	81	54	88	62	101	32
La Paz	16	30 S	68	8 W	12,001	31	63	43	65	40	62	33	66	40	80	26
Sucre	19	3 S	65	17 W	9,344	5	63	48	63	45	61	37	65	46	88	25
Brazil:																
Barra do Corda	5	35 S	45	28 W	266	9	89	71	89	71	92	64	94	72	103	45
Bela Vista	22	6 S	56	22 W	525	13	91	67	85	61	77	49	87	61	108	20
Belem	1	27 S	48	29 W	42	16	87	72	87	73	88	71	89	71	98	61
Brasilia	15	51 S	47	56 W	3,481	3	80	65	82	62	78	51	82	64	93	46
Conceicao do Araguai	8	15 S	49	12 W	53	5	88	70	91	68	95	63	93	68	102	55
Corumda	19	0 S	57	39 W	381	8	94	73	92	73	84	64	93	70	106	33
Florianopolis	27	35 S	48	33 W	96	17	83	72	74	64	68	57	73	63	102	32
Goias	15	58 S	50	4 W	1,706	11	86	63	91	63	89	56	94	63	104	41
Guarapuava	25	16 S	51	30 W	3,592	10	79	61	73	55	66	47	74	53	94	23
Manaus	3	8 S	60	1 W	144	11	88	75	87	75	89	75	92	76	101	63

(continued)

11.9 TEMPERATURE DATA FOR REPRESENTATIVE WORLDWIDE STATIONS (continued)

Country and Station	Latitude		Longitude		Elevation (Ft.)	Record Length (Yr.)	Temperature									
							Average Daily								Extreme	
							January		April		July		October			
							Maximum (°F)	Minimum (°F)	Maximum (°F)	Minimum (°F)	Maximum (°F)	Minimum (°F)	Maximum (°F)	Minimum (°F)	Maximum (°F)	Minimum (°F)
	°	'	°	'												
Parana	12	26 S	48	6 W	853	19	90	58	90	58	91	48	94	58	105	37
Porto Alegre	30	2 S	51	13 W	33	22	87	67	78	60	66	49	74	57	105	25
Quizeramobin	5	12 S	39	18 W	653	9	92	79	86	76	88	74	93	77	100	63
Recife	8	4 S	34	53 W	97	27	86	77	85	75	80	71	84	75	94	50
Rio de Janeiro	22	55 S	43	12 W	201	38	84	73	80	69	75	63	77	66	102	46
Salvador (Bahia)	13	0 S	38	30 W	154	25	86	74	84	74	79	69	83	71	100	50
Santarem	2	30 S	54	42 W	66	22	86	73	85	73	87	71	91	73	99	65
Sao Paulo	23	37 S	46	39 W	2,628	44	77	63	73	59	66	53	68	57	100	32
Sena Madureira	9	4 S	68	39 W	443	12	92	69	91	68	91	63	93	69	100	41
Uaupes	0	8 S	67	5 W	272	15	88	72	88	72	85	70	89	71	100	52
Uruguaiana	29	46 S	57	7 W	246	15	91	69	78	59	66	48	77	55	108	27
Chile:																
Ancud	41	47 S	73	52 W	184	30	62	51	57	47	50	42	55	45	82	30
Antofagasta	23	42 S	70	24 W	308	22	76	63	70	58	63	51	66	55	86	37
Arica	18	28 S	70	20 W	95	15	78	64	74	60	66	54	69	53	93	39

Cabo Raper	46	50 S	75	38 W	131	8	58	46	54	44	47	38	51	40	72	28
Los Evangelistas	52	23 S	75	7 W	190	16	50	44	48	41	43	36	45	39	66	19
Potrerillos	26	30 S	69	27 W	9,350	7	65	49	63	47	57	40	61	44	75	20
Puerto Aisen	42	24 S	72	42 W	33	8	63	50	55	43	45	37	55	42	93	18
Punta Arenas	53	10 S	70	54 W	26	15	58	45	50	39	40	31	51	38	86	11
Santiago	33	27 S	70	42 W	1,706	14	85	53	74	45	59	37	72	45	99	24
Valdivia	39	48 S	73	14 W	16	29	73	52	62	46	52	41	63	44	97	19
Valparaiso	33	1 S	71	38 W	135	30	72	56	67	52	60	47	65	50	94	32
Colombia:																
Andagoya	5	6 N	76	40 W	197	8	90	75	90	75	89	74	90	74	97	62
Bogota	4	42 N	74	8 W	8,355	10	67	48	67	51	64	50	66	50	75	30
Cartegena	10	28 N	75	30 W	39	6	84	73	87	76	88	78	87	77	98	61
Ipiales	0	50 N	77	42 W	9,680	9	61	50	60	49	57	42	62	49	77	32
Tumaco	1	49 N	78	45 W	7	10	82	75	84	76	82	75	82	75	90	64
Ecuador:																
Cuenca	2	53 S	78	39 W	8,301	7	69	50	69	50	65	47	70	49	81	29
Guayaquil	2	10 S	79	53 W	20	5	87	72	88	72	84	67	86	68	98	52
Quito	0	8 S	78	29 W	9,222	54	67	46	69	47	71	44	71	46	86	25
French Guiana:																
Cayenne	4	56 N	52	27 W	20	38	84	74	86	75	88	73	91	74	97	65
Guyana:																
Georgetown	6	50 N	58	12 W	6	54	84	74	85	76	85	75	87	76	93	68
Lethem	3	24 N	59	38 W	270	3	91	73	91	74	87	73	92	76	97	63

(continued)

11.9 TEMPERATURE DATA FOR REPRESENTATIVE WORLDWIDE STATIONS (continued)

Country and Station	Latitude °	'		Longitude °	'		Elevation (Ft.)	Record Length (Yr)	Temperature Average Daily								Extreme	
									January Max (°F)	January Min (°F)	April Max (°F)	April Min (°F)	July Max (°F)	July Min (°F)	October Max (°F)	October Min (°F)	Max (°F)	Min (°F)
Paraguay:																			
Asuncion	25	17	S	57	30	W	456	15	95	71	84	65	74	53	86	62	110	29
Bahia Negra	20	14	S	58	10	W	318	20	92	74	87	68	79	61	90	69	106	35
Peru:																			
Arequipa	16	21	S	71	34	W	8,460	13	67	49	67	48	67	47	68	47	82	25
Cajamarca	7	9	S	78	30	W	8,662	9	71	48	70	47	70	41	71	47	79	25
Cusco	13	33	S	71	59	W	10,866	13	68	45	71	40	70	31	72	43	86	16
Iquitos	3	45	S	73	13	W	384	5	90	71	87	71	88	68	90	70	100	54
Lima	12	5	S	77	3	W	394	15	82	66	80	63	67	57	71	58	93	49
Mollendo	17	0	S	72	7	W	80	10	79	66	76	63	67	57	70	59	90	50
Surinam:																			
Paramaribo	5	49	N	55	9	W	12	35	85	72	86	73	87	73	91	73	99	62
Uruguay:																			
Artigas	30	24	S	56	23	W	384	13	91	65	77	55	65	45	75	54	107	24
Montevideo	34	52	S	56	12	W	72	56	83	62	71	53	58	43	68	49	109	25

Venezuela:

Caracas	10	30 N	66	56 W	3,418	30	75	56	81	60	78	61	79	61	91	45
Cuidad Bolivar	8	7 N	63	32 W	197	10	90	72	93	75	90	75	93	75	100	64
Maracaibo	10	39 N	71	36 W	20	12	90	73	92	76	94	76	92	76	102	66
Merida	8	36 N	71	10 W	5,293	14	73	56	75	60	76	59	75	60	90	48
Santa Elena	4	36 N	61	7 W	2,976	10	82	61	82	63	81	61	84	61	95	48
Pacific Islands																
Easter Island (Isla de Pascua)	27	10 S	109	26 W	98	4	77	64	78	63	70	58	73	58	88	46
Mas a Tierra (Juan Fernandez)	33	37 S	78	52 W	20	25	72	60	68	57	60	50	61	51	86	39
Seymour Island (Galapagos Island)	0	28 S	90	18 W	36	3	86	72	87	75	81	69	81	67	93	58
Atlantic Islands																
Fernando de Noronha	3	50 S	32	25 W	148	32	84	75	82	75	81	73	82	75	93	63
Cumberland Bay, South Georgia	54	16 S	36	30 W	8	23	48	35	42	29	34	23	41	28	84	-3
Laurie Is., S. Orkneys	60	44 S	44	44 W	13	48	35	29	31	21	20	4	30	19	54	-40
Stanley, Falkland Is.	51	42 S	57	51 W	6	25	56	42	49	37	40	31	48	35	76	-12
Europe																
Albania:																
Durres	41	19 N	19	28 E	23	10	51	42	63	55	83	74	68	58	95	21
Andorra:																
Les Escaldes	42	30 N	1	31 E	3,543	5	43	29	59	39	78	55	61	42	91	0

(continued)

11.9 TEMPERATURE DATA FOR REPRESENTATIVE WORLDWIDE STATIONS (continued)

Country and Station	Latitude °	Latitude '	Longitude °	Longitude '	Elevation (Ft.)	Record (Yr)	January Max (°F)	January Min (°F)	April Max (°F)	April Min (°F)	July Max (°F)	July Min (°F)	October Max (°F)	October Min (°F)	Extreme Max (°F)	Extreme Min (°F)
Austria:																
Innsbruck	47	16 N	11	24 E	1,909	34	34	20	60	39	78	55	58	40	97	−16
Wien (Vienna)	48	15 N	16	22 E	664	50	34	26	57	41	75	59	55	44	98	−14
Bulgaria:																
Sofiya (Sofia)	42	42 N	23	20 E	1,805	30	34	22	62	41	82	57	63	42	99	−17
Varna	43	12 N	27	55 E	115	30	40	30	59	43	84	63	67	50	107	−12
Cyprus:																
Nicosia	35	9 N	33	17 E	716	40	58	42	74	50	97	69	81	58	116	23
Czechoslovakia:																
Praha (Prague)	50	5 N	14	25 E	662	40	34	25	55	40	74	58	54	44	98	−16
Prerov	49	27 N	17	27 E	702	20	34	25	57	38	77	55	56	40	100	−23
Denmark:																
Arhus	56	8 N	10	12 E	161	21	35	27	51	37	70	54	53	42	87	−12
Kobenhavn (Copenhagen)	55	41 N	12	33 E	43	30	36	29	50	37	72	55	53	42	91	−3
Estonia:																
Tallinn	59	26 N	24	48 E	146	15	27	18	42	31	70	55	47	38	89	−19

Finland:																	
Helsinki	60	10 N	24	57 E	30	20	27	17	43	31	71	57	45	37	89	−23
Kuusamo	65	57 N	29	12 E	843	20	17	2	35	18	68	50	36	27	90	−40
Vaasa	63	5 N	21	36 E	13	18	26	16	41	28	69	55	44	36	89	−29
France:																	
Ajaccio	41	52 N	8	35 E	243	46	56	40	66	48	85	64	72	55	103	23
Bordeaux	44	50 N	0	43 W	157	51	48	35	63	44	80	58	66	47	102	9
Brest	48	19 N	4	47 W	56	56	49	40	57	44	70	56	61	49	95	7
Cherbourg	49	39 N	1	38 W	30	47	47	40	54	43	67	57	59	50	91	14
Lille	50	35 N	3	5 E	141	40	42	33	58	40	75	55	59	45	96	0
Lyon	45	42 N	4	47 E	938	70	41	30	61	42	80	58	61	45	105	−13
Marseille	43	18 N	4	23 E	246	72	53	38	59	41	78	58	76	57	101	9
Paris	48	49 N	2	29 E	164	66	42	32	60	41	76	55	59	44	105	1
Strasbourg	48	35 N	7	46 E	465	20	40	31	59	41	78	57	58	43	101	−8
Toulouse	43	33 N	1	23 E	538	47	47	35	62	43	82	59	66	48	111	1
Germany:																	
Berlin	52	27 N	13	18 E	187	50	35	26	55	38	74	55	55	41	96	−15
Bremen	53	5 N	8	47 E	52	50	37	30	53	38	71	55	54	43	94	−7
Frankfurt A/M	50	7 N	8	40 E	338	50	37	29	58	41	75	56	56	43	100	−7
Hamburg	53	33 N	9	58 E	66	50	35	28	51	39	69	56	53	44	92	−4
Munchen (Munich)	48	9 N	11	34 E	1,739	50	33	23	54	37	72	54	53	40	92	−14
Munster	51	58 N	7	38 E	207	50	39	29	56	38	73	54	56	42	96	−17
Nurnberg	49	27 N	11	3 E	1,050	50	35	26	56	38	74	55	55	41	99	−18

(continued)

11.9 TEMPERATURE DATA FOR REPRESENTATIVE WORLDWIDE STATIONS (continued)

Country and Station	Lat °	'		Long °	'		Elevation (Ft.)	Record (Yr)	January Max (°F)	Min (°F)	April Max (°F)	Min (°F)	July Max (°F)	Min (°F)	October Max (°F)	Min (°F)	Extreme Max (°F)	Min (°F)
Gibraltar:																		
Windmill Hill	36	6	N	5	21	W	400	12	58	50	64	55	77	66	70	61	97	35
Greece:																		
Athinai (Athens)	37	58	N	23	43	E	351	72	54	42	67	52	90	72	74	60	109	20
Iraklion	35	20	N	25	8	E	98	21	60	48	70	54	85	72	77	62	114	32
Rodhos (Rhodes)	36	26	N	28	15	E	289	10	59	51	67	59	83	74	76	68	104	30
Thessaloniki (Salonika)	40	37	N	22	57	E	78	9	49	37	66	49	90	70	73	56	107	15
Hungary:																		
Budapest	47	31	N	19	2	E	394	50	35	26	62	44	82	61	61	45	103	-10
Debrecen	47	36	N	21	39	E	430	50	33	21	61	39	81	57	60	41	102	-22
Iceland:																		
Akureyri	65	41	N	18	5	W	16	23	34	26	40	30	57	47	43	34	83	-8
Reykjavik	64	9	N	21	56	W	92	25	36	28	43	33	58	48	44	36	74	4
Ireland:																		
Cork	51	54	N	8	29	W	56	27	48	38	55	41	68	53	58	44	85	15
Dublin	53	22	N	6	21	W	155	30	47	35	54	38	67	51	57	43	86	8

Location																
Shannon Airport	54	41 N	8	55 W	8	9	46	36	55	41	66	53	58	45	87	12
Italy:																
Ancona	43	37 N	13	32 E	52	30	46	36	62	50	83	68	67	55	102	18
Cagliari	39	15 N	9	3 E	3	30	56	43	66	50	86	67	72	58	102	25
Genova (Genoa)	44	24 N	8	55 E	318	10	50	41	65	53	82	70	73	58	100	18
Napoli (Naples)	40	51 N	14	15 E	82	30	54	40	65	52	84	70	71	60	101	24
Palermo	38	7 N	13	19 E	354	10	58	47	67	53	86	71	75	62	113	31
Roma (Rome)	41	48 N	12	36 E	377	10	54	39	68	46	88	64	73	53	104	20
Taranto	40	28 N	17	17 E	56	10	55	43	59	50	89	70	73	58	108	26
Venezia (Venice)	45	26 N	12	23 E	82	10	43	33	63	49	82	67	65	52	97	14
Latvia:																
Riga	56	57 N	24	6 E	67	30	29	20	48	35	72	56	49	39	93	-20
Lithuania:																
Kaunas	54	54 N	23	53 E	118	19	26	18	49	34	72	53	50	38	96	-23
Luxembourg:																
Luxembourg	49	37 N	6	3 E	1,096	7	36	29	58	40	74	55	56	43	99	-10
Malta:																
Valletta	35	54 N	14	31 E	233	90	59	51	66	56	84	72	76	66	105	34
Monaco:																
Monaco	43	44 N	7	25 E	180	60	54	46	61	53	77	70	67	60	93	27
Netherlands:																
Amsterdam	52	23 N	4	55 E	5	29	40	34	52	43	69	59	56	48	95	3
Norway:																
Bergen	60	24 N	5	19 E	141	49	43	27	55	34	72	51	57	38	89	3

(continued)

11.9 TEMPERATURE DATA FOR REPRESENTATIVE WORLDWIDE STATIONS (continued)

Country and Station	Latitude °	'	Longitude °	'	Elevation (Ft.)	Record length (Yr)	January Max (°F)	January Min (°F)	April Max (°F)	April Min (°F)	July Max (°F)	July Min (°F)	October Max (°F)	October Min (°F)	Extreme Max (°F)	Extreme Min (°F)
Kristiansand	58	10 N	7	59 E	175	11	32	25	50	35	71	53	53	39	90	−14
Oslo	56	56 N	10	44 E	308	44	30	20	50	34	73	56	49	37	93	−21
Tromso	69	39 N	18	57 E	335	47	30	22	37	27	59	48	40	33	83	−1
Trondheim	63	25 N	10	27 E	417	44	31	22	45	32	66	51	46	36	95	−22
Vardo	70	22 N	31	6 E	43	40	27	19	34	26	53	44	38	32	80	−11
Poland:																
Gdansk	54	24 N	18	40 E	36	36	33	25	49	37	70	56	53	42	94	−16
Krakow	50	4 N	19	57 E	723	35	32	22	55	38	76	57	56	41	97	−28
Warsaw	52	13 N	21	2 E	294	25	30	21	54	38	75	56	54	41	98	−22
Wroclaw	51	7 N	17	5 E	482	50	35	25	55	39	74	57	55	42	98	−26
Portugal:																
Braganca	41	49 N	6	47 W	2,395	11	46	31	59	39	80	54	62	42	103	10
Lagos	37	6 N	8	38 W	46	21	61	47	67	52	83	64	73	58	107	28
Lisbon	38	43 N	9	8 W	313	75	56	46	64	52	79	63	69	57	103	29
Romania:																
Bucuresti (Bucharest)	44	25 N	26	6 E	269	41	33	20	63	41	86	61	65	44	105	−18

	Lat °	Lat ′	Lon °	Lon	Elev.											
Cluj-Napoca	46	47 N	23	40 E	1,286	15	31	18	58	38	79	56	60	41	100	−26
Constanta	44	11 N	28	39 E	13	20	37	25	55	42	79	63	62	49	101	−13
Spain:																
Almeria	36	51 N	2	28 E	213	20	61	47	69	54	85	69	76	62	108	34
Barcelona	41	24 N	2	9 E	312	20	56	42	64	51	81	69	71	58	98	24
Burgos	42	20 N	3	42 W	2,825	29	42	30	57	38	77	53	61	43	99	0
Madrid	40	25 N	3	41 W	2,188	30	47	33	64	44	87	62	66	48	102	14
Sevilla	37	29 N	5	59 W	98	26	59	41	73	51	96	67	78	57	117	27
Valencia	39	28 N	0	23 W	79	26	58	41	67	51	83	68	73	57	107	20
Sweden:																
Abisko	68	21 N	18	49 E	1,273	11	20	6	33	19	61	45	35	24	82	−30
Goteberg	57	42 N	11	58 E	55	39	35	27	48	36	69	56	51	42	88	−13
Haparanda	65	50 N	24	9 E	30	20	22	10	38	23	71	53	39	30	89	−34
Karlstad	59	23 N	13	30 E	164	30	30	20	49	32	73	56	49	38	93	−21
Sarna	61	41 N	13	7 E	1,504	20	19	4	42	23	69	46	42	28	91	−51
Stockholm	59	21 N	18	4 E	146	30	31	23	45	32	70	55	48	39	97	−26
Visby	57	39 N	18	18 E	36	30	35	28	44	33	67	55	50	41	88	1
Switzerland:																
Berne	46	57 N	7	26 E	1,877	30	35	26	56	39	74	56	55	42	96	−9
Geneve (Geneva)	46	12 N	6	9 E	1,329	30	39	29	58	41	77	58	58	44	101	−1
Zurich	47	23 N	8	33 E	1,617	23	38	28	57	39	76	55	57	42	98	−12
United Kingdom:																
Belfast	54	35 N	5	56 W	57	7	42	34	53	38	65	52	55	44	82	14
Birmingham	52	29 N	1	56 W	535	30	42	35	53	40	69	54	55	45	92	11

(continued)

11.9 TEMPERATURE DATA FOR REPRESENTATIVE WORLDWIDE STATIONS (continued)

Country and Station	Latitude ° '	Longitude ° '	Elevation (Ft.)	Record Length (Yr)	Temperature — Average Daily								Extreme	
					January Max (°F)	January Min (°F)	April Max (°F)	April Min (°F)	July Max (°F)	July Min (°F)	October Max (°F)	October Min (°F)	Max (°F)	Min (°F)
Cardiff	51 28 N	3 10 W	203	30	45	36	55	41	69	54	57	45	91	2
Edinburgh	55 55 N	3 11 W	441	30	43	35	50	39	65	52	53	44	83	15
London	51 29 N	0 0	149	30	44	35	56	40	73	55	58	44	99	9
Liverpool	53 24 N	3 4 W	198	30	44	36	52	41	66	55	55	46	87	15
Perth	56 24 N	3 27 W	77	30	43	32	53	38	68	51	55	41	89	0
Plymouth	50 21 N	4 7 W	87	30	47	40	54	43	66	55	58	49	88	16
Wick	58 26 N	3 5 W	119	30	42	35	48	38	59	50	52	43	80	8
U.S.S.R. (Former):														
Arkhangelsk	64 33 N	40 32 E	22	23	9	2	36	23	64	51	36	30	91	−49
Astrakhan	46 21 N	48 2 E	45	10	23	14	57	40	85	69	56	40	99	−22
Dnepropetrovsk	48 27 N	35 4 E	259	18	25	16	53	39	80	62	56	40	101	−25
Kursk	51 45 N	36 12 E	773	15	19	11	47	35	74	58	48	36	91	−23
Lvov	49 50 N	24 1 E	978	9	31	22	53	38	77	59	55	43	97	−29
Minsk	53 54 N	27 33 E	738	12	22	13	47	33	70	54	47	36	92	−27
Moskva (Moscow)	55 46 N	37 40 E	505	15	21	9	47	31	76	55	46	34	96	−27
Odessa	46 29 N	30 44 E	214	20	28	22	52	41	79	65	57	47	99	−13

St. Petersburg (Leningrad)	59	56 N	30	16 E	16	26	23	12	45	31	71	57	45	37	91	-36
Saratov	51	32 N	46	3 E	197	14	15	7	50	35	82	64	48	36	102	-27
Sevastropol	44	37 N	33	31 E	75	20	39	30	55	42	79	65	63	50	97	-4
Stavropol	45	2 N	41	58 E	1,886	18	26	17	50	37	76	60	55	42	95	-22
Tbilisi	41	43 N	44	48 E	1,325	10	39	26	61	44	83	65	64	48	95	6
Ufa	54	43 N	55	56 E	571	20	6	-3	44	30	75	58	41	31	99	-42
Ust' Shchugor	64	16 N	57	34 E	279	15	4	-14	35	17	65	49	33	23	90	-67
Volograd (Stalingrad)	48	42 N	44	31 E	136	8	15	4	52	36	84	65	53	37	106	-30
Vyatka (Kirov)	58	36 N	49	41 E	594	20	6	-2	41	27	72	55	37	29	92	-43
Yugoslavia:																
Beograd (Belgrade)	44	84 N	20	28 E	453	16	37	27	64	45	84	61	65	47	107	-14
Skopje	41	59 N	21	28 E	787	10	40	26	67	42	88	60	65	43	105	-11
Split	43	31 N	16	26 E	420	14	51	29	65	50	87	68	69	55	100	17
Ocean Islands																
Barentsburg, Spitzbergen	78	2 N	14	15 E	23	19	10	-4	15	-3	46	38	25	17	60	-57
Bjornoya (Bear Island)	74	31 N	19	1 E	49	10	26	17	27	16	44	36	36	29	71	-25
Horta, Azores	38	32 N	28	38 W	200	30	62	54	64	55	76	65	71	62	88	38
Jan Mayen	71	1 N	8	28 W	131	5	31	21	31	22	46	38	39	29	60	-18
Lerwick, Shetland Is.	60	8 N	1	11 W	269	30	42	35	46	37	58	49	50	42	71	17
Matochkin Shar, Novaya Zemlya	73	16 N	56	24 E	61	9	8	-6	13	-1	47	36	30	21	68	-41
Ponta Delgada, Azores	37	45 N	25	40 W	118	30	62	54	64	55	76	64	71	61	85	37
Stornoway, Herbrides	58	11 N	6	21 W	34	30	44	37	49	39	61	51	53	44	78	11
Thorshavn, Faroes	62	2 N	6	45 W	82	50	42	33	45	36	56	47	58	40	70	8

(continued)

11.9 TEMPERATURE DATA FOR REPRESENTATIVE WORLDWIDE STATIONS (continued)

Country and Station	Latitude		Longitude		Elevation (Ft.)	Record Length (Yr)	Temperature									
							Average Daily								Extreme	
							January		April		July		October			
							Max (°F)	Min (°F)	Max (°F)	Min (°F)	Max (°F)	Min (°F)	Max (°F)	Min (°F)	Max Max (°F)	Min Min (°F)
	°	'	°	'												
Africa																
Algeria:																
Adrar	27	52 N	0	17 W	938	15	69	39	92	60	115	82	92	63	124	25
Alger (Algiers)	36	46 N	3	3 E	194	25	59	49	68	55	83	70	74	63	107	32
Annaba (Bone)	36	54 N	7	46 E	66	26	59	46	67	52	85	69	75	61	115	32
El Golea	30	35 N	2	53 E	1,247	15	63	37	84	56	107	79	87	60	120	23
Hassi Tanezrouft	28	6 N	6	42 E	1,224	15	67	38	90	59	110	78	92	63	124	19
Tamanrasset	22	42 N	5	31 E	4,593	15	67	39	86	56	95	71	85	59	102	20
Touggourt	33	7 N	6	4 E	226	26	62	38	83	55	107	77	84	59	122	26
Angola:																
Cangamba	13	41 S	19	52 E	4,331	6	84	62	89	58	82	46	87	59	109	20
Huambo	12	48 S	15	45 E	5,577	14	78	58	78	57	77	47	81	58	90	36
Luanda	8	49 S	13	13 E	194	27	83	74	85	75	74	65	79	71	98	58
Namibe	15	12 S	12	9 E	10	15	79	65	82	66	68	56	74	61	102	44
Benin:																
Cotonou	6	21 N	2	26 E	23	5	80	74	83	78	78	74	80	75	95	65

Station																
Botswana:																
Francistown	21	13 S	27	30 E	3,294	20	88	65	83	56	75	41	90	61	107	24
Mann	19	59 S	23	25 E	3,091	20	90	66	87	58	77	42	95	64	100	24
Tshabong	26	3 S	22	27 E	3,156	10	94	65	83	51	71	34	88	54	107	15
Burkina Faso:																
Bobo Dioulasso	11	10 N	4	15 W	1,411	11	92	58	99	71	87	69	90	70	115	46
Quagadougou	12	22 N	1	31 W	991	10	92	60	103	79	91	74	95	74	118	48
Cameroon:																
Ngaoundere	7	17 N	13	19 E	3,601	9	87	55	87	64	82	63	82	61	102	56
Yaounde	3	53 N	11	32 E	2,526	11	85	67	85	66	80	66	81	65	96	57
Central African Republic:																
Bangui	4	22 N	18	34 E	1,270	5	90	68	91	71	85	69	87	69	101	57
Ndele	8	24 N	20	39 E	1,939	3	99	67	98	73	86	69	90	68	109	58
Chad:																
Am Timan	11	2 N	20	17 E	1,430	3	98	56	105	68	89	70	96	67	113	43
Faya-Largeau	18	0 N	19	10 E	837	5	84	54	104	69	109	76	103	72	121	37
N'Djamena	12	7 N	15	2 E	968	5	93	57	107	74	92	72	97	70	114	47
Congo, Republic of:																
Brazzaville	4	15 S	15	15 E	1,043	15	88	69	91	71	82	63	89	70	98	54
Quesso	1	37 N	16	4 E	1,132	4	88	69	91	71	85	69	87	69	106	60
Pointe Noire	4	39 S	11	48 E	164	7	85	73	87	74	78	66	83	72	93	59
Cote D'Ivoire:																
Abidjan	5	19 N	4	1 W	65	13	88	73	90	75	83	73	85	74	96	59
Bouake	7	42 N	5	0 W	1,194	12	91	68	92	70	85	68	89	68	104	57

(continued)

11.9 TEMPERATURE DATA FOR REPRESENTATIVE WORLDWIDE STATIONS (continued)

Country and Station	Latitude		Longitude		Elevation (Ft.)	Record Length (Yr)	Temperature										
							January		April		July		October		Extreme	
	°	'	°	'			Max (°F)	Min (°F)	Max (°F)	Min (°F)	Max (°F)	Min (°F)	Max (°F)	Min (°F)	Max (°F)	Min (°F)
Djibouti:																
Djibouti	11	36 N	43	9 E	23	16	84	73	90	79	106	87	92	80	117	63
Egypt:																
Alexandria	31	12 N	29	53 E	105	45	65	51	74	59	85	73	83	68	111	37
Aswan	24	2 N	32	53 E	366	46	74	50	96	66	106	79	98	71	124	35
Cairo	29	52 N	31	20 E	381	42	65	47	83	57	96	70	86	65	117	34
Ethiopia:																
Addis Ababa	9	20 N	38	45 E	8,038	15	75	43	77	50	69	50	75	45	94	32
Asmera	15	17 N	38	55 E	7,628	9	74	44	78	51	71	53	72	53	88	31
Diredawa	9	2 N	41	45 E	3,937	8	81	58	91	69	90	68	89	67	100	49
Gambela	8	15 N	34	35 E	1,345	26	98	64	98	71	87	69	92	67	111	48
Gabon:																
Libreville	0	23 N	9	26 E	115	11	87	73	89	73	83	68	86	71	99	62
Mayoumba	3	2.5 S	10	38 E	200	8	84	73	86	73	78	68	82	72	91	60
Gambia:																
Banjul	13	21 N	16	40 W	90	9	88	59	91	65	86	74	89	72	106	45

		Lat.		Long.													
Ghana:																	
Accra	5	33 N	0	12 W	88	17	87	73	88	76	81	73	85	74	100	59
Kumasi	6	40 N	1	37 W	942	10	88	66	89	71	82	70	86	70	100	51
Guinea:																	
Conarkry	9	31 N	13	43 W	23	7	88	72	90	73	83	72	87	73	86	63
Kouroussa	10	39 N	9	53 W	1,217	9	93	60	99	73	87	69	90	69	109	39
Guinea Bissau:																	
Bolama	11	34 N	15	26 W	62	31	88	67	91	73	84	74	87	74	106	59
Kenya:																	
Mombasa	4	3 S	39	39 E	52	45	87	75	86	76	81	71	84	74	96	61
Nairobi	1	16 S	36	48 E	5,971	15	77	54	75	58	69	51	76	55	87	41
Liberia:																	
Monrovia	6	18 N	10	48 W	75	6	89	71	90	72	80	72	86	72	97	62
Libya:																	
Al Jawf	24	12 N	23	21 E	1,276	7	69	43	90	62	101	75	90	64	122	26
Banghasi (Bhenghazi)	32	6 N	20	4 E	82	46	63	50	74	58	84	71	80	66	109	37
Sabha	27	1 N	14	26 E	1,457	3	64	41	89	60	102	74	91	64	120	24
Tarabulus (Tripoli)	32	54 N	13	11 E	72	47	1	47	72	57	85	71	80	65	114	33
Madagascar:																	
Antananarivo	18	55 S	47	33 E	4,500	44	79	61	76	58	68	48	80	54	95	34
Diego Suarez	12	17 S	49	17 E	100	11	88	75	88	75	84	69	86	72	98	63
Toliari	23	20 S	43	41 E	20	27	92	72	89	64	81	58	86	65	108	43
Malawi:																	
Karonga	9	57 S	33	56 E	1,596	8	86	71	85	70	81	59	91	66	99	51

(continued)

11.9 TEMPERATURE DATA FOR REPRESENTATIVE WORLDWIDE STATIONS (continued)

Country and Station	Latitude		Longitude		Elevation (Ft.)	Record Length (Yr)	Temperature									
							January		April		July		October		Extreme	
	°	'	°	'			Max (°F)	Min (°F)	Max (°F)	Min (°F)	Max (°F)	Min (°F)	Max (°F)	Min (°F)	Max (°F)	Min (°F)
Zomba	15	23 S	35	19 E	3,141	27	80	65	78	62	72	53	85	64	95	41
Mali:																
Araouane	18	54 N	3	33 W	935	8	81	48	110	67	111	79	103	70	130	37
Bamako	12	39 N	7	58 W	1,116	11	91	61	103	76	89	71	93	71	117	47
Gao	16	16 N	0	3 W	902	15	83	58	105	77	97	80	100	78	116	44
Mauritania:																
Atar	20	31 N	13	4 W	761	7	84	54	97	67	106	81	98	72	117	39
Nema	16	36 N	7	16 W	883	9	86	62	105	79	99	78	101	79	120	47
Nouakchott	18	7 N	15	36 W	69	5	85	57	90	64	89	74	91	71	115	44
Morocco:																
Casablanca	33	35 N	7	39 W	164	48	63	45	69	52	79	65	76	58	110	31
Marrakech	31	36 N	8	1 W	1,509	35	65	40	79	52	101	67	83	57	120	27
Rabat	34	0 N	6	50 W	213	35	63	46	71	52	82	63	77	58	118	32
Sibi Ifni	29	27 N	10	11 W	148	14	66	52	71	59	75	64	75	62	124	40
Tangier	35	48 N	5	49 W	239	35	60	47	65	51	80	64	72	59	106	28

City	Lat °		Long °		Elev.												
Mozambique:																	
Beira	19	50 S	34	51 E	28	37	89	75	86	71	77	61	87	71	109	48
Chioco	15	36 S	32	21 E	899	8	96	65	93	63	86	55	101	68	117	32
Maputo	25	58 S	32	36 E	194	42	86	71	83	66	76	55	82	64	114	45
Namibia:																	
Keetmanshoop	26	35 S	18	8 E	3,295	17	95	65	85	57	70	42	87	55	108	26
Windhoek	22	34 S	17	6 E	5,669	30	85	63	77	55	68	43	84	59	97	25
Niger:																	
Agadez	16	59 N	7	59 E	1,706	8	86	50	105	70	104	75	101	68	115	40
Bilma	18	41 N	12	55 E	1,171	9	81	45	101	63	108	75	101	62	116	29
Niamey	13	31 N	2	6 E	709	10	93	58	108	77	94	74	101	74	114	47
Nigeria:																	
Enugu	6	27 N	7	29 E	763	11	90	72	91	74	83	71	87	71	99	55
Kaduna	10	35 N	6	26 E	2,113	18	89	59	95	72	83	68	89	66	105	46
Lagos	6	27 N	3	24 E	10	32	88	74	89	77	83	74	85	74	104	60
Maiduguri	11	51 N	13	5 E	1,162	15	90	54	104	72	90	73	96	68	112	43
Senegal:																	
Dakar	14	42 N	17	29 W	131	25	79	64	81	65	88	76	89	76	109	53
Kaolack	14	8 N	16	4 W	20	9	93	60	103	68	91	75	93	74	114	48
Sierra Leone:																	
Freetown/Lungi	8	37 N	13	12 W	92	8	87	73	88	76	82	73	85	72	98	62
Somalia:																	
Berbera	10	26 N	45	2 E	45	30	84	68	89	77	107	88	92	76	117	58
Mogdisho (Mogadishu)	2	2 N	45	21 E	39	13	86	73	90	78	83	73	86	76	97	59

(continued)

11.9 TEMPERATURE DATA FOR REPRESENTATIVE WORLDWIDE STATIONS (continued)

Country and Station	Latitude °	'	Longitude °	'	Elevation (Ft.)	Record length (Yr)	January Max (°F)	Min (°F)	April Max (°F)	Min (°F)	July Max (°F)	Min (°F)	October Max (°F)	Min (°F)	Extreme Max (°F)	Min (°F)
South Africa, Republic of:																
Capetown	33	54 S	18	32 E	56	19	78	60	72	53	63	45	70	52	103	28
Durban	29	50 S	31	2 E	16	15	81	69	78	64	72	52	75	62	107	39
Kimberly	28	48 S	24	46 E	3,927	19	91	64	77	52	65	36	83	54	103	20
Port Elizabeth	33	59 S	25	36 E	190	14	78	61	73	55	67	45	70	54	104	31
Port Nolloth	29	14 S	16	52 E	23	20	67	53	66	50	62	45	64	49	107	31
Pretoria	25	45 S	28	14 E	4,491	13	81	60	75	50	66	37	80	55	96	24
Walvis Bay	22	56 S	14	30 E	24	20	73	59	75	55	70	47	67	51	104	25
Sudan:																
El Fasher	13	38 N	25	21 E	2,395	17	88	50	102	64	96	70	99	64	113	33
Khartoum	15	37 N	32	33 E	1,279	46	90	59	105	72	101	77	104	75	118	41
Port Sudan	19	37 N	37	13 E	18	30	81	68	89	71	106	83	93	76	117	50
Wadi Halfa	21	55 N	31	20 E	410	39	75	46	98	62	106	74	98	67	127	28
Wau	7	42 N	28	3 E	1,443	38	96	64	99	72	89	69	93	69	115	50
Tanzania:																
Dar es Salaam	6	50 S	39	18 E	47	44	83	77	86	73	83	66	85	69	96	59

Iringa	7	47 S	35	42 E	5,330	14	76	59	75	59	72	52	80	57	90	42
Kigoma	4	53 S	29	38 E	2,903	26	80	67	81	67	83	63	84	69	100	53
Togo:																
Lome	6	10 N	1	15 E	72	5	85	72	86	74	80	71	83	72	94	58
Tunisia:																
Gabes	33	53 N	10	7 E	7	50	61	43	74	54	89	71	81	62	122	27
Tunis	36	47 N	10	12 E	217	50	58	43	70	51	90	68	77	59	118	30
Uganda:																
Kampala	0	20 N	32	36 E	4,304	15	83	65	79	64	77	62	81	63	97	53
Lira	2	15 N	32	54 E	3,560	14	91	61	86	64	81	61	86	61	100	50
Western Sahara:																
Ad Dakhla	23	42 N	15	52 W	35	12	71	56	74	60	78	65	80	65	107	48
Smara	26	46 N	11	31 W	1,509	6	73	47	88	58	99	66	88	61	121	37
Zaire:																
Kalemie	5	54 S	29	12 E	2,493	5	85	66	83	67	82	58	87	67	92	50
Kananga	5	54 S	22	25 E	2,198	3	85	68	86	68	85	63	85	68	94	57
Kinshasa	4	20 S	15	18 E	1,066	8	87	70	89	71	81	64	88	70	97	58
Kisangani	0	26 N	25	14 E	1,370	8	88	69	88	70	84	67	86	68	97	61
Zambia:																
Balovale	13	34 S	23	6 E	3,577	8	82	65	84	61	81	47	91	64	108	38
Kasama	10	12 S	31	11 E	4,544	10	79	61	79	60	76	50	87	62	95	39
Lukasa	15	25 S	28	19 E	4,191	10	78	63	79	59	73	49	88	64	100	39
Zimbabwe:																
Bulawayo	20	9 S	28	37 E	4,405	15	81	61	79	56	70	45	85	59	99	28
Harare	17	50 S	31	8 E	4,831	15	78	60	78	55	70	44	83	58	95	32

(continued)

11.9 TEMPERATURE DATA FOR REPRESENTATIVE WORLDWIDE STATIONS (continued)

Country and Station	Latitude			Longitude			Elevation (Ft.)	Record Length (Yr)	Temperature — Average Daily								Extreme	
									January		April		July		October			
	Lat.			Long.					Max (°F)	Min (°F)	Max (°F)	Min (°F)	Max (°F)	Min (°F)	Max (°F)	Min (°F)	Max (°F)	Min (°F)
	°	′		°	′													
Atlantic Islands																		
Funchal, Madeira Island	32	38 N		16	55 W		82	30	66	56	67	58	75	66	74	65	103	40
Georgetown, Ascension Is.	7	56 S		14	25 W		55	29	85	73	88	75	84	72	83	71	95	65
Hutts Gate, St. Helena	15	57 S		5	40 W		2,062	30	68	60	69	61	62	55	61	54	82	50
Las Palmas, Canary Is.	28	11 N		15	28 W		20	45	70	58	71	61	77	67	79	67	99	46
Porto Da Praia, Cape Verde	14	54 N		23	31 W		112	25	77	68	79	69	83	75	85	76	94	56
Santa Isabel, Bioko	3	46 N		8	46 E		—	2	87	67	89	70	84	69	86	70	102	61
Sao Tome, Sao Tome	0	20 N		6	43 E		16	10	86	73	86	73	82	69	84	71	91	56
Tristan da Cunha	37	3 S		12	19 W		75	5	66	59	64	57	57	50	59	51	75	38
Indian Ocean Islands																		
Agalega Island	10	26 S		56	40 E		10	3	86	77	87	77	83	75	84	75	91	69
Cocos (Keeling) Island	12	5 S		96	53 E		15	36	86	77	85	78	82	76	84	76	94	68
Heard Island	53	1 S		73	23 E		16	5	41	35	39	33	34	27	35	28	58	13
Hellburg, Reunion Is.	21	4 S		55	22 E		3,070	5	74	59	73	56	65	48	69	51	84	40
Port Victoria, Seychelles	4	37 S		55	27 E		15	60	83	76	86	77	81	75	83	75	92	67

	20	6 S	57	32 E	181	40	86	73	82	70	75	62	80	64	95	50
Mauritius																

Asia—Far East

China:

Canton	23	10 N	113	20 E	59	26	65	49	77	65	91	77	85	67	101	31
Chanaska	28	15 N	112	8 E	161	14	45	35	70	56	94	78	75	59	109	16
Chungking	29	30 N	106	33 E	855	27	51	42	73	59	93	76	71	61	111	28
Hankou	30	35 N	114	17 E	75	29	46	34	69	55	93	78	74	60	108	9
Harbin (Ha-erh-pin)	45	45 N	126	38 E	476	35	7	-14	54	31	84	65	54	31	102	-43
Kashi (Kaxgar)	39	24 N	76	7 E	4,296	27	33	12	71	48	92	68	71	43	106	-15
Kunming	25	2 N	102	43 E	6,211	32	61	37	76	51	77	62	70	53	91	22
Lanzhou	36	6 N	103	55 E	5,105	8	33	7	65	40	84	61	62	39	100	-3
Mukden (Shen-yang)	41	47 N	123	24 E	138	40	20	-2	60	36	87	69	62	39	103	-28
Shanghai	31	12 N	121	26 E	16	56	47	32	67	49	91	75	75	56	104	10
Tientsin	39	10 N	117	10 E	13	24	33	16	68	45	90	73	68	48	109	-3
Urumqi	43	45 N	87	40 E	2,972	6	13	-7	60	36	82	58	50	31	112	-30
Hong Kong	22	18 N	114	10 E	109	50	64	56	75	67	87	78	81	73	97	32

Japan:

Kushiro	43	2 N	144	12 E	315	41	30	8	44	31	66	55	58	40	87	-19
Miyako	39	38 N	141	59 E	98	30	43	23	58	37	77	62	66	46	99	1
Nagasaki	32	44 N	129	53 E	436	59	49	36	66	50	85	73	72	58	98	22
Osaka	34	47 N	135	26 E	49	60	47	32	65	47	87	73	72	55	102	19
Tokyo	35	41 N	139	46 E	19	60	47	29	63	46	83	70	69	55	101	17

(continued)

11.9 TEMPERATURE DATA FOR REPRESENTATIVE WORLDWIDE STATIONS (continued)

Country and Station	Latitude ° '	Longitude ° '	Elevation (Ft.)	Record length (Yr)	Temperature January Max (°F)	January Min (°F)	April Max (°F)	April Min (°F)	July Max (°F)	July Min (°F)	October Max (°F)	October Min (°F)	Extreme Max (°F)	Extreme Min (°F)
Mongolia:														
Ulan Bator	47 54 N	106 56 E	4,287	13	-2	-27	45	18	71	50	44	17	97	-48
North Korea:														
P'yongyang	39 1 N	125 49 E	94	43	27	8	61	38	84	69	65	43	100	-19
South Korea:														
Pusan	35 10 N	129 7 E	6	36	43	29	62	47	81	71	70	54	97	7
Seoul	37 31 N	126 55 E	34	22	32	15	62	41	84	70	67	45	99	-12
Taiwan:														
Tainan	22 57 N	120 12 E	53	13	72	55	82	67	89	77	86	70	95	39
Taipei	25 4 N	121 32 E	21	12	66	53	77	64	92	76	80	68	101	32
U.S.S.R. (Former):														
Alma-Ata	43 16 N	76 53 E	2,543	19	23	7	56	38	81	60	55	35	100	-30
Chita (Tchita)	52 2 N	113 30 E	2,218	10	-10	-27	42	19	75	51	38	18	99	-52
Dudinka	69 7 N	87 0 E	141	5	-23	-31	6	-10	59	47	19	11	84	-62
Ekaterinburg (Sverdlovsk)	56 49 N	60 38 E	897	21	6	-5	42	26	70	54	37	28	94	-45
Irkutsk	52 16 N	104 19 E	1,532	10	3	-15	42	20	70	50	41	21	98	-58

	Lat			Lon														
Kazalinsk	45	46 N	62	6 E	207	10	16	5	58	27	90	65	57	35	108	-27
Khabarovsk	48	28 N	135	3 E	165	7	-2	-13	41	28	75	63	48	34	91	-46
Kirensk	57	47 N	108	7 E	938	18	-14	-28	38	15	74	51	10	-4	95	-71
Krasnoyarsk	56	1 N	92	52 E	498	10	3	-10	34	23	67	55	34	26	103	-47
Markovo	64	45 N	170	50 E	85	15	-19	-29	5	-8	59	47	16	9	84	-72
Narym	58	50 N	81	39 E	197	13	7	-18	35	19	71	56	35	25	94	-61
Okhotsk	59	21 N	143	17 E	18	19	-6	-17	29	10	57	48	33	21	78	-50
Omsk	54	58 N	73	20 E	279	19	-1	-14	39	21	74	56	40	27	102	-56
Petropavlovsk	52	53 N	158	42 E	286	7	23	11	35	25	56	47	46	34	84	-29
Salehkard	66	31 N	66	35 E	60	18	-13	-21	18	4	61	49	26	20	85	-65
Semipalatinsk	50	24 N	80	13 E	709	10	8	-7	45	26	81	57	46	30	101	-47
Tashkent	41	20 N	69	18 E	1,569	19	37	21	65	47	92	64	65	41	106	-19
Verkhoyansk	67	34 N	133	51 E	328	24	-54	-63	19	-10	66	47	12	-3	98	-90
Vladivostok	43	7 N	131	55 E	94	14	13	0	46	34	71	60	55	41	92	-22
Yakutsk	62	1 N	129	43 E	535	19	-45	-53	27	6	73	54	23	11	97	-84
Brunei:																
Bandar Seri Begawan	4	55 N	114	55 E	10	5	85	76	87	77	87	76	86	77	99	70
Cambodia:																
Phnom Penh	11	33 N	104	51 E	39	37	88	71	95	76	90	76	87	76	105	55
Indonesia:																
Jakarta	6	11 S	106	50 E	26	80	84	74	87	75	87	73	87	74	98	66
Manokwari	0	53 S	134	3 E	10	5	86	73	86	74	87	74	87	74	93	68
Mapanget	1	32 N	124	55 E	264	21	85	73	86	73	87	73	89	72	97	65
Penfui	10	10 S	123	39 E	335	21	87	75	89	72	88	70	92	72	101	58

(continued)

11.9 TEMPERATURE DATA FOR REPRESENTATIVE WORLDWIDE STATIONS (continued)

Country and Station	Latitude		Longitude		Elevation (Ft.)	Record Length (Yr)	January		April		July		October		Extreme	
	°	'	°	'			Max (°F)	Min (°F)	Max (°F)	Min (°F)	Max (°F)	Min (°F)	Max (°F)	Min (°F)	Max (°F)	Min (°F)
Pontianak	0	0 N	109	20 E	13	20	87	74	89	75	89	74	89	75	96	68
Tabing	0	52 S	100	21 E	19	21	87	74	87	75	87	74	86	74	94	68
Tarakan	3	19 N	117	33 E	20	19	85	73	86	75	87	74	87	74	94	67
Laos:																
Vientiane	17	58 N	102	34 E	229	13	83	58	95	73	89	75	88	71	108	32
Malaysia:																
Kuala Lumpur	3	6 N	101	42 E	111	19	90	72	91	74	90	72	89	73	99	64
Kuching	1	29 N	110	20 E	85	5	85	72	90	73	90	72	89	73	98	64
Sandakan	5	54 N	118	3 E	38	45	85	74	89	76	89	75	88	75	99	70
Myanmar (Burma):																
Mandalay	21	59 N	96	6 E	252	20	82	55	101	77	93	78	89	73	111	44
Moulmein	16	26 N	97	39 E	150	43	89	65	95	77	83	74	88	75	103	52
Philippines:																
Davao	7	7 N.	125	38 E	88	15	87	72	91	73	88	73	89	73	97	65
Manila	14	31 N	121	0 E	49	61	86	69	93	73	88	75	88	74	101	58

Singapore:															
Singapore	1	18 N	103	50 E	33	39	86	73	88	75	88	75	87	74	97	66
Thailand:															
Bangkok	13	44 N	100	30 E	53	10	89	67	95	78	90	76	88	76	104	50
Vietnam:															
Hanoi	21	3 N	105	52 E	20	12	68	58	80	70	92	79	84	72	108	41
Ho Chi Minh City	10	49 N	106	39 E	33	31	89	70	95	76	88	75	88	74	104	57
Asia—Middle East															
Afghanistan:															
Kabul	34	30 N	69	13 E	5,955	9	36	18	66	43	92	61	73	42	104	-6
Kandhar	31	36 N	65	40 E	3,462	7	56	31	83	50	102	66	85	44	111	14
Bangladesh:															
Dacca	23	46 N	90	23 E	24	60	77	56	92	74	88	79	88	75	108	43
India:															
Ahmadabad	23	3 N	72	37 E	180	45	85	58	104	75	93	79	97	73	118	36
Bangalore	12	57 N	77	40 E	2,937	60	80	57	93	69	81	66	82	65	102	46
Bombay	19	6 N	72	51 E	27	60	88	62	93	74	88	75	93	73	110	46
Calcuta	22	32 N	88	20 E	21	60	80	55	97	76	90	79	89	74	111	44
Cherrapunji	25	15 N	91	44 E	4,309	35	60	46	71	59	72	65	72	61	87	33
Hyderabad	17	27 N	78	28 E	1,741	50	85	59	101	75	87	73	88	68	112	43
Jalpaiguri	26	32 N	88	43 E	272	50	74	50	90	68	89	77	87	70	104	36
Lucknow	26	45 N	80	52 E	400	60	74	47	101	71	92	80	91	67	119	34
Madras	13	4 N	80	15 E	51	60	85	67	95	78	96	79	90	75	113	57
Mormugao	15	22 N	73	49 E	157	10	86	70	88	79	83	75	86	75	98	59

(continued)

425

11.9 TEMPERATURE DATA FOR REPRESENTATIVE WORLDWIDE STATIONS (continued)

Country and Station	Latitude °	'	Longitude °	'	Elevation (Ft.)	Record Length (Yr)	January Max (°F)	Min (°F)	April Max (°F)	Min (°F)	July Max (°F)	Min (°F)	October Max (°F)	Min (°F)	Extreme Max (°F)	Min (°F)
New Delhi	28	35 N	77	12 E	695	10	71	43	97	68	95	80	93	64	115	31
Silchar	24	49 N	92	48 E	95	60	78	52	88	69	90	77	88	72	103	41
Srinagar	33	58 N	74	46 E	5,458	50	41	24	67	45	88	64	74	41	106	-4
Iran:																
Abadan	30	21 N	48	13 E	10	12	64	44	90	62	112	81	98	63	127	24
Esfahan (Isfahan)	32	37 N	51	41 E	5,238	45	47	25	72	46	98	67	78	47	108	-4
Kermanshah	34	19 N	47	7 E	4,331	15	45	23	68	38	99	56	79	38	108	-13
Rezaiyeh	37	32 N	45	5 E	4,364	5	32	17	67	45	91	64	67	47	99	-11
Tehran	35	41 N	51	19 E	3,937	24	45	27	71	49	99	72	76	53	109	-5
Iraq:																
Baghdad	33	20 N	44	24 E	111	15	60	39	85	57	110	76	92	61	121	18
Basra	30	34 N	47	47 E	8	10	64	45	85	63	104	81	94	64	123	24
Mosul	36	19 N	43	9 E	730	26	54	35	77	49	109	72	88	51	124	12
Israel:																
Haifa	32	48 N	35	2 E	23	16	65	49	77	58	88	75	85	68	112	27
Jerusalem	31	47 N	35	13 E	2,654	19	55	41	73	50	87	63	81	59	107	26

Tel Aviv	32	6 N	34	46 E	33	10	64	50	70	57	82	72	79	65	102	34
Jordan:																
Amman	31	58 N	35	59 E	2,547	25	54	39	73	49	89	65	81	57	109	21
Kuwait:																
Kuwait	29	21 N	48	0 E	16	14	61	49	83	68	103	86	91	73	119	33
Lebanon:																
Beirut	33	54 N	35	28 E	111	62	62	51	72	58	87	73	81	69	107	30
Nepal:																
Katmandu	27	42 N	85	22 E	4,423	27	65	36	84	53	84	69	80	56	99	27
Oman:																
Muscat	23	37 N	58	35 E	15	23	77	66	90	78	97	87	93	80	116	51
Pakistan:																
Karachi	24	48 N	66	59 E	13	43	77	55	90	73	91	81	91	72	118	39
Multan	30	11 N	71	25 E	400	60	68	42	95	68	102	86	94	64	122	29
Rawalpindi	33	35 N	73	3 E	1,676	60	62	38	86	59	98	77	89	57	118	25
Saudi Arabia:																
Dhahran	26	16 N	50	10 E	78	10	69	54	90	69	107	86	95	73	120	40
Jidda	21	28 N	39	10 E	20	5	84	66	91	70	99	79	95	73	117	49
Riyadh	24	39 N	46	42 E	1,938	3	70	46	89	64	107	78	94	61	113	19
Sri Lanka:																
Colombo	6	54 N	79	52 E	22	25	86	72	88	76	85	77	85	75	99	59
Syria:																
Deir Ez Zor	35	21 N	40	9 E	699	5	53	35	80	52	105	78	86	56	114	16
Dimashq (Damascus)	33	30 N	36	20 E	2,362	13	53	36	75	49	96	64	81	54	113	21

(continued)

11.9 TEMPERATURE DATA FOR REPRESENTATIVE WORLDWIDE STATIONS (continued)

Country and Station	Latitude		Longitude		Elevation	Record Length	Temperature									
							Average Daily								Extreme	
							January		April		July		October			
							Max	Min	Max	Min	Max	Min	Max	Min	Max	Min
	°	'	°	'	(Ft.)	(Yr)	(°F)	(°F)	(°F)	(°F)	(°F)	(°F)	(°F)	(°F)	(°F)	(°F)
Halab (Aleppo)	36	14 N	37	8 E	1,280	8	50	34	75	48	97	69	81	54	117	9
Turkey:																
Adana	36	59 N	35	18 E	82	21	57	39	74	51	93	71	84	58	109	19
Ankara	39	57 N	32	53 E	2,825	26	39	24	63	40	86	59	69	44	104	-13
Edirne	41	39 N	26	34 E	154	18	41	28	66	44	88	63	70	49	107	-8
Erzurum	39	54 N	41	16 E	6,402	16	24	8	50	32	78	53	59	37	93	-22
Istanbul	40	58 N	28	50 E	59	18	45	36	61	45	81	65	67	54	100	17
Izmir (Smyrna)	38	27 N	27	15 E	92	39	55	39	70	49	92	69	76	55	108	12
Samsun	41	17 N	36	19 E	131	24	50	38	59	45	79	65	69	56	103	20
United Arab Emirates:																
Sharjah	25	20 N	55	24 E	18	11	74	54	86	65	100	82	92	71	118	37
Yemen:																
Kamaran I.	15	20 N	42	37 E	20	26	82	74	89	79	98	85	93	82	105	66
Riyan	14	39 N	49	19 E	83	13	82	67	88	74	92	77	88	72	111	57
Indian Ocean Islands																
Port Blair, Andaman Island	11	40 N	92	43 E	261	60	84	72	89	75	84	75	84	74	97	62

428

Location																
Amindivi Island	11	7 N	72	44 E	13	29	86	74	92	80	86	77	86	77	99	65
Minicoy Island	8	18 N	73	0 E	9	20	85	73	87	80	85	76	85	76	98	63
Car Nicobar, Nicobar Island	9	9 N	92	49 E	47	13	86	77	90	77	86	77	85	75	95	66

Oceania

Australia:

Location																
Adelaide	34	57 S	138	32 E	20	86	86	61	73	55	59	45	73	51	118	32
Alice Springs	23	48 S	133	53 E	1,791	62	97	70	81	54	67	39	88	58	111	19
Bourke	30	5 S	145	58 E	361	63	99	70	82	55	65	40	85	56	125	25
Brisbane	27	25 S	153	5 E	17	53	85	69	79	61	68	49	80	60	110	35
Broome	17	57 S	122	13 E	56	41	92	79	93	72	82	58	91	72	113	40
Burketown	17	45 S	139	33 E	30	31	93	77	91	69	82	55	93	70	110	40
Canberra	35	18 S	149	11 E	1,886	23	82	55	67	44	52	33	68	43	109	14
Carnarvon	24	53 S	113	40 E	13	43	88	72	84	66	71	51	78	61	118	37
Cloncurry	20	40 S	140	30 E	622	32	99	77	90	67	77	51	95	68	127	37
Esperance	33	50 S	121	55 E	14	44	77	60	72	54	62	45	68	50	117	35
Hobart	42	53 S	147	20 E	177	70	71	53	63	48	52	40	63	46	105	31
Laverton	28	40 S	122	23 E	1,510	30	96	69	81	57	64	41	82	55	115	28
Melbourne	37	49 S	144	58 E	115	88	78	57	68	51	56	42	67	48	114	25
Mundiwindi	23	52 S	120	10 E	1,840	15	101	64	87	61	70	41	89	58	112	27
Perth	31	56 S	115	58 E	64	44	85	63	76	57	63	48	70	53	112	22
Port Darwin	15	25 S	130	52 E	104	58	90	77	92	76	87	67	93	77	105	31
Sydney	33	52 S	151	2 E	62	87	78	65	71	58	60	46	71	56	114	55
Thursday Island	10	35 S	142	13 E	200	31	87	77	86	77	82	73	86	76	98	64
Townsville	19	15 S	146	46 E	18	31	87	76	84	70	75	59	83	71	110	39

(continued)

11.9 TEMPERATURE DATA FOR REPRESENTATIVE WORLDWIDE STATIONS (continued)

Country and Station	Latitude			Longitude			Elevation (Ft.)	Record Length (Yr)	Temperature — Average Daily								Extreme	
									January		April		July		October			
	°	′		°	′				Max (°F)	Min (°F)	Max (°F)	Min (°F)	Max (°F)	Min (°F)	Max (°F)	Min (°F)	Max (°F)	Min (°F)
William Creek	28	55 S		136	21 E		247	39	96	69	80	55	65	41	84	56	119	25
Windorah	25	26 S		142	36 E		390	29	101	74	86	59	70	43	91	61	116	26
New Zealand:																		
Aukland	37	0 S		174	47 E		23	36	73	60	67	56	56	46	63	52	90	33
Christchruch	43	29 S		172	32 E		118	52	70	53	62	45	50	35	62	44	96	21
Dunebin	45	55 S		170	12 E		4	77	66	50	59	45	48	37	59	42	94	23
Wellington	41	17 S		174	46 E		415	66	69	56	63	51	53	42	60	48	88	29
Pacific Islands																		
Guam, Marinas Island	13	33 N		144	50 E		361	30	84	72	86	73	87	72	86	73	95	54
Honolulu, Hawaii	21	20 N		157	55 W		7	30	79	66	80	68	85	73	84	72	93	56
Iwo Jima, Bonin Island	24	47 N		141	19 E		353	15	71	64	77	69	86	78	84	76	95	46
Kanton, Phoenix Island	2	46 S		171	43 W		9	12	88	78	89	78	89	78	90	78	98	70
Madang, Papua New Guinea	5	12 S		145	47 E		19	12	87	75	88	74	88	74	88	75	98	62

Location																
Midway Island	28	13 N	177	23 W	29	21	69	62	71	64	81	74	79	72	92	46
Naha, Okinawa	26	12 N	127	39 E	96	30	67	56	76	64	89	77	81	69	96	41
Noumea, New Caledonia	22	16 S	166	27 E	246	24	86	72	83	70	76	62	80	65	99	52
Pago Pago, Samoa	14	19 S	170	43 W	29	2	87	75	87	76	83	74	85	75	98	67
Ponape, Senyavin Island	6	58 N	158	13 E	123	30	86	75	86	75	87	73	87	72	96	67
Port Moresby, Papua, New Guinea	9	29 S	147	9 E	126	20	89	76	87	75	83	73	86	75	98	64
Rabaul, Papua, New Guinea	4	13 S	152	11 E	28	19	90	73	90	73	89	73	92	73	100	65
Suva, Fiji	18	8 S	178	26 E	20	43	86	74	84	73	79	68	81	70	98	55
Tahiti, Society Island	17	33 S	149	36 W	7	23	89	72	89	72	86	68	87	70	93	61
Tulagi, Solomon Island	9	5 S	160	10 E	8	20	88	76	88	76	86	76	87	76	96	68
Wake Island	19	17 N	166	39 E	11	30	82	73	83	74	87	77	86	77	92	64
Yap Islands	9	31 N	138	8 E	62	30	85	76	87	77	88	75	88	75	97	69
Antarctica																
Byrd Station	80	1 S	119	32 W	5,095	6	10	-2	-11	-30	-25	-45	-15	-33	31	-82
Ellsworth	77	44 S	41	7 W	139	6	22	12	-10	-25	-21	-35	-2	-15	36	-70
McMurdo Station	77	53 S	166	48 E	8	10	30	21	-1	-13	-9	-24	2	-12	42	-59
South Pole Station	89	59 S	0	0 W	9,186	5	-16	-23	-66	-79	-67	-81	-55	-64	6	-107
Wilkes	66	16 S	110	31 E	31	7	34	28	17	9	8	-3	16	6	46	-35

[a] Record Length refers to average daily maximum and minimum temperature. A standard period of the 30 years from 1931 to 1960 had been used for locations in the United States and some other countries. The length of record of extreme maximum and minimum temperatures includes all available years of data for a given location and is usually for a longer period.

11.10 PRECIPITATION DATA FOR REPRESENTATIVE WORLDWIDE STATIONS

Country and Station	Record Length (yr)	Average Precipitation (in)												
		Jan.	Feb.	Mar.	Apr.	May	June	July	Aug.	Sept.	Oct.	Nov.	Dec.	Year
North America														
United States (Conterminous):														
Albuquerque, NM	30	0.4	0.4	0.5	0.5	0.8	0.6	1.2	1.3	1.0	0.8	0.4	0.5	8.4
Asheville, NC	30	4.2	4.0	4.8	4.0	3.7	3.5	5.9	4.9	3.6	3.1	2.8	3.6	48.1
Atlanta, GA	30	4.4	4.5	5.4	4.5	3.2	3.8	4.7	3.6	3.3	2.4	3.0	4.4	47.2
Austin, TX	30	2.4	2.6	2.1	3.6	3.7	3.2	2.2	1.9	3.4	2.8	2.1	2.5	32.5
Birmingham, AL	30	5.0	5.3	6.0	4.5	3.4	4.0	5.2	4.9	3.3	3.0	3.5	5.0	53.1
Bismarck, ND	30	0.4	0.4	0.8	1.2	2.0	3.4	2.2	1.7	1.2	0.9	0.6	0.4	15.2
Boise, ID	30	1.3	1.3	1.3	1.2	1.3	0.9	0.2	0.2	0.4	0.8	1.2	1.3	11.4
Brownsville, TX	30	1.4	1.5	1.0	1.6	2.4	3.0	1.7	2.8	5.0	3.5	1.3	1.7	26.9
Buffalo, NY	30	2.8	2.7	3.2	3.0	3.0	2.5	2.6	3.1	3.1	3.0	3.6	3.0	35.6
Cheyenne, WY	30	0.5	0.6	1.2	1.9	2.5	2.1	1.8	1.4	1.1	0.8	0.6	0.5	15.0
Chicago, IL	30	1.9	1.6	2.7	3.0	3.7	4.1	3.4	3.2	2.7	2.8	2.2	1.9	33.2
Des Moines, IA	30	1.3	1.1	2.1	2.5	4.1	4.7	3.1	3.7	2.9	2.1	1.8	1.1	30.5
Dodge City, KS	30	0.6	0.7	1.2	1.8	3.2	3.0	2.3	2.4	1.5	1.4	0.6	0.5	19.2
El Paso, TX	30	0.5	0.4	0.4	0.3	0.4	0.7	1.3	1.2	1.1	0.9	0.3	0.5	8.0
Indianapolis, IN	30	3.1	2.3	3.4	3.7	4.0	4.6	3.5	3.0	3.2	2.6	3.1	2.7	39.2

Jacksonville, FL	30	2.5	2.9	3.5	3.6	3.5	6.3	7.7	6.9	7.6	5.2	1.7	2.2	53.6
Kansas City, MO	30	1.4	1.2	2.5	3.6	4.4	4.6	3.2	3.8	3.3	2.9	1.8	1.5	34.2
Las Vegas, NV	30	0.5	0.4	0.4	0.2	0.1	*b	0.5	0.5	0.3	0.2	0.3	0.4	3.8
Los Angeles, CA	30	2.7	2.9	1.8	1.1	0.1	0.1	*b	*b	.2	0.4	1.1	2.4	12.8
Louisville, KY	30	4.1	3.3	4.6	3.8	3.9	4.0	3.4	3.0	2.6	2.3	3.2	3.2	41.4
Miami, FL	30	2.0	1.9	2.3	3.9	6.4	7.4	6.8	7.0	9.5	8.2	2.8	1.7	59.9
Minneapolis, MN	30	0.7	0.8	1.5	1.9	3.2	4.0	3.3	3.2	2.4	1.6	1.4	0.9	24.9
Missoula, MT	30	0.9	0.9	0.7	1.0	1.9	1.9	0.9	0.7	1.0	1.0	0.9	1.1	12.9
Nashville, TN	30	5.5	4.5	5.2	3.7	3.7	3.3	3.7	2.9	2.9	2.3	3.3	4.2	45.2
New Orleans, LA	30	3.8	4.0	5.3	4.6	4.4	4.4	6.7	5.3	5.0	2.8	3.3	4.1	53.7
New York, NY	30	3.3	2.8	4.0	3.4	3.7	3.3	3.7	4.4	3.9	3.1	3.4	3.3	42.3
Oklahoma City, OK	30	1.3	1.4	2.0	3.1	5.2	4.5	2.4	2.5	3.0	2.5	1.6	1.4	30.9
Phoenix, AZ	30	0.7	0.9	0.7	0.3	0.1	0.1	0.8	1.1	0.7	0.5	0.5	0.9	7.3
Pittsburgh, PA	30	2.8	2.3	3.5	3.4	3.8	4.0	3.6	3.5	2.7	2.5	2.3	2.5	36.9
Portland, ME	30	4.4	3.8	4.3	3.7	3.4	3.2	2.9	2.4	3.5	3.2	4.2	3.9	42.9
Portland, OR	30	5.4	4.2	3.8	2.1	2.0	1.7	0.4	0.7	1.6	3.6	5.3	6.4	37.2
Reno, NV	30	1.2	1.0	0.7	0.5	0.5	0.4	0.3	0.2	0.2	0.5	0.6	1.1	7.2
Salt Lake City, UT	30	1.4	1.2	1.6	1.8	1.4	1.0	0.6	0.9	0.5	1.2	1.3	1.2	14.1
San Francisco, CA	30	4.0	3.5	2.7	1.3	0.5	0.1	*b	*b	0.2	0.7	1.6	4.1	18.7
Sault Ste. Marie, M	30	2.1	1.5	1.8	2.2	2.8	3.3	2.5	2.9	3.8	2.8	3.3	2.3	31.3
Seattle, WA	30	5.7	4.2	3.8	2.4	1.7	1.6	0.8	1.0	2.1	4.0	5.4	6.3	39.0
Sheridan, WY	30	0.6	0.7	1.4	2.2	2.6	2.6	1.2	0.9	1.2	1.1	0.8	0.6	15.9
Spokane, WA	30	2.4	1.9	1.5	0.9	1.2	1.5	0.4	0.4	0.8	1.6	2.2	2.4	17.2

(continued)

11.10 PRECIPITATION DATA FOR REPRESENTATIVE WORLDWIDE STATIONS (continued)

Country and Station	Record Length yr	Jan.	Feb.	Mar.	Apr.	May	June	July	Aug.	Sept.	Oct.	Nov.	Dec.	Year
Washington, DC	30	3.0	2.5	3.2	3.2	4.1	3.2	4.2	4.9	3.8	3.1	2.8	2.8	40.8
Wilmington, NC	30	2.9	3.4	4.0	2.9	3.5	4.3	7.7	6.9	6.3	3.0	3.1	3.4	51.4
United States (Alaska):														
Anchorage	30	0.8	0.7	0.5	0.4	.05	1.0	1.9	2.6	2.5	1.9	1.0	0.9	14.7
Annette	30	11.4	8.5	9.6	9.1	7.1	5.7	6.0	7.5	9.9	16.9	14.7	12.1	118.5
Barrow	30	0.2	0.2	0.1	0.1	.1	0.4	0.8	0.9	0.6	0.5	0.2	0.2	4.3
Bethel	30	1.1	1.1	1.0	0.6	1.0	1.2	2.0	4.2	2.6	1.5	1.1	1.0	18.4
Cold Bay	30	2.3	3.2	1.8	1.5	2.3	2.0	1.8	4.3	4.3	4.6	3.8	2.6	34.5
Fairbanks	30	0.9	0.5	0.4	0.3	.7	1.4	1.8	2.2	1.1	0.9	0.6	0.5	11.3
Juneau	30	4.0	3.1	3.3	2.9	3.2	3.4	4.5	5.0	6.7	8.3	6.1	4.2	54.7
King Salmon	30	1.1	1.0	1.0	0.6	1.0	1.4	2.1	3.4	3.1	2.2	1.5	1.0	19.4
Nome	30	1.0	0.9	0.9	0.8	.7	0.9	2.3	3.8	2.7	1.7	1.2	1.0	17.9
St. Paul Island	30	1.8	1.2	1.1	1.0	1.3	1.2	2.3	3.3	3.1	3.2	2.5	1.8	23.8
Shemya	30	2.5	2.3	2.6	2.1	2.4	1.3	2.2	2.1	2.3	2.8	2.7	2.1	27.4
Yakutat	30	10.9	8.2	8.7	7.2	8.0	5.1	8.4	10.9	16.6	19.6	16.1	12.3	132.0
Canada:														
Aklavik, N.W.T.	22	0.5	0.5	0.4	0.5	.5	0.8	1.4	0.9	0.9	0.9	0.8	0.4	9.0
Alert, N.W.T.	10	0.2	0.3	0.3	0.3	.5	0.6	0.5	1.1	1.0	0.9	0.2	0.4	6.3

Calgary, Alta.	55	0.5	0.5	0.8	1.0	2.3	3.1	2.5	2.3	1.5	0.7	0.7	0.6	16.7
Charlottetown, P.E.I.	65	3.8	3.0	3.2	2.8	2.7	2.6	3.0	3.4	3.4	4.1	3.8	4.0	39.8
Chatham, N.B.	50	3.4	2.7	3.3	3.0	3.2	3.6	3.9	4.0	3.1	4.0	3.4	3.2	40.8
Churchhill, Man.	30	0.5	0.6	0.9	0.9	0.9	1.9	2.2	2.7	2.3	1.4	1.0	0.7	16.0
Edmonton, Alta.	71	0.9	0.7	0.7	1.0	1.9	3.2	3.3	2.4	1.3	0.8	0.9	0.9	18.0
Fort Nelson, B.C.	13	0.9	1.2	0.7	0.8	1.4	2.5	2.4	1.5	1.3	1.0	1.4	1.2	16.3
Fort Simpson, N.W.T.	42	0.7	0.7	0.5	0.7	1.4	1.5	2.4	1.5	1.3	1.1	0.9	0.8	13.1
Frobisher Bay, N.W.T.	10	0.7	0.9	0.8	0.8	0.7	0.9	1.5	2.0	1.8	1.1	1.1	1.0	13.3
Gander, Nfld.	14	2.6	3.3	2.8	2.6	2.6	2.8	3.6	3.6	3.7	4.1	4.2	3.7	39.6
Halifax, N.S.	71	5.4	4.4	4.9	4.5	4.1	4.0	3.6	4.4	4.1	5.4	5.3	5.4	55.7
Kapuskasing, Ont.	19	2.0	1.1	1.6	1.8	2.1	2.3	3.4	2.9	3.5	2.5	2.4	1.9	27.5
Knob Lake, Que.	10	1.9	1.9	1.4	1.6	1.7	3.3	3.3	4.4	3.4	2.9	2.4	1.5	29.7
Montreal, Que.	77	3.8	3.0	3.5	2.6	3.1	3.4	3.7	3.5	3.7	3.4	3.5	3.6	40.8
North Bay, Ont.	23	2.0	1.5	1.8	2.2	2.5	3.2	3.2	2.7	3.7	3.2	2.7	2.1	30.8
Ottawa, Ont.	65	2.9	2.2	2.8	2.7	2.5	3.5	3.4	2.6	3.2	2.9	3.0	2.6	34.3
Penticton, B.C.	32	1.0	0.7	0.7	0.7	1.1	1.2	0.8	0.8	1.0	0.8	0.9	1.1	10.8
Port Arthur, Ont.	59	0.9	0.8	1.0	1.5	2.1	2.8	3.6	2.8	3.4	2.5	1.5	0.9	23.8
Prince Georgia, B.C.	27	1.8	1.2	1.4	0.8	1.3	2.1	1.6	1.9	2.0	2.0	1.9	1.9	19.9
Prince Rupert, B.C.	26	9.8	7.6	8.4	6.7	5.3	4.1	4.8	5.1	7.7	12.2	12.3	11.3	95.3
Quebec, Que.	72	3.5	2.7	3.0	2.4	3.1	3.7	4.0	4.0	3.6	3.4	3.2	3.2	39.8
Regina, Sask.	49	0.5	0.3	0.7	0.7	1.8	3.3	2.4	1.8	1.3	0.9	0.6	0.4	14.7
Resolute, N.W.T.	7	0.1	0.1	0.2	0.2	0.5	0.8	0.9	1.1	0.8	0.5	0.2	0.1	5.5
St. John, N.B.	61	4.1	3.1	3.7	3.2	3.1	3.2	3.1	3.6	3.7	4.1	3.9	3.8	42.60

(continued)

11.10 PRECIPITATION DATA FOR REPRESENTATIVE WORLDWIDE STATIONS *(continued)*

Country and Station	Record Length (yr)	Jan.	Feb.	Mar.	Apr.	May	June	July	Aug.	Sept.	Oct.	Nov.	Dec.	Year
						Average Precipitation (in)								
St. Johns, Nfld.	58	5.3	5.1	4.6	3.8	3.9	3.1	3.1	4.0	3.7	4.8	5.7	6.0	53.1
Saskatoon, Sask.	38	0.9	0.5	0.7	0.7	1.4	2.6	2.4	1.9	1.5	0.9	0.5	0.6	14.6
The Pas, Man.	27	0.6	0.5	0.7	0.8	1.4	2.2	2.2	2.1	2.0	1.2	1.0	0.8	15.5
Toronto, Ont.	105	2.7	2.4	2.6	2.5	2.9	2.7	3.0	2.7	2.9	2.4	2.8	2.6	32.2
Vancouver, B.C.	41	8.6	5.8	5.0	3.3	2.8	2.5	1.2	1.7	3.6	5.8	8.3	8.8	57.4
Whitehorse, Y.T.	10	0.6	0.5	0.6	0.4	0.6	1.0	1.6	1.5	1.3	0.7	1.0	0.8	10.6
Winnipeg, Man.	66	0.9	0.9	1.2	1.4	2.3	3.1	3.1	2.5	2.3	1.5	1.1	0.9	21.2
Yellow Knife, N.W.T.	13	0.8	0.6	0.7	0.4	0.7	0.6	1.5	1.4	1.0	1.3	1.0	0.8	10.8
Greenland:														
Angmagssalik	38	2.9	2.4	2.6	2.1	2.0	1.8	1.5	2.1	3.3	4.7	3.0	2.7	31.1
Danmarkshaven	2	1.2	0.7	0.7	0.1	0.2	0.2	0.5	0.6	0.3	0.3	1.0	0.7	6.0
Eismitte	1	0.6	0.2	0.3	0.2	0.1	0.1	0.1	0.4	0.3	0.5	0.5	1.0	4.3
Godthaab	45	1.4	1.7	1.6	1.2	1.7	1.4	2.2	3.1	3.3	2.5	1.9	1.5	23.5
Ivigtut	50	3.3	2.6	3.4	2.5	3.5	3.2	3.1	3.7	5.9	5.7	4.6	3.1	44.6
Jacobshavn	52	0.4	0.4	0.5	0.5	0.6	0.8	1.2	1.4	1.3	0.9	0.7	0.5	9.2
Nord	8	0.8	0.8	0.5	0.3	0.1	0.3	1.0	1.4	1.2	0.6	1.4	0.5	8.9
Scoresbysund	12	1.8	1.4	0.9	1.4	0.4	0.8	1.5	0.7	1.7	1.4	1.1	1.9	15.0
Thule	12	0.4	0.3	0.2	0.2	0.3	0.2	0.7	0.6	0.6	0.7	0.5	0.2	4.9

Upernivik	50	0.4	0.5	0.7	0.6	0.6	0.5	0.9	1.1	1.1	1.1	1.1	0.6	9.2
Mexico:														
Acapulco	40	0.3	*b	0.0	*b	1.4	12.8	9.1	9.3	13.9	6.7	1.2	0.4	55.1
Chihuahua	22	0.2	0.4	0.3	0.2	0.2	1.7	3.6	3.7	3.3	0.9	0.5	0.4	15.4
Guada Lajara	33	0.4	0.2	0.2	0.2	1.1	8.8	9.4	8.5	7.2	2.2	0.8	0.7	39.7
Guaymas	41	0.5	0.2	0.2	0.1	*b	0.1	1.7	2.7	2.1	0.7	0.3	0.8	9.4
La Paz	12	0.2	0.1	0.0	0.0	0.0	0.2	0.4	1.2	1.4	0.6	0.5	1.1	5.7
Lerdo	14	0.4	0.1	0.2	0.3	0.8	1.5	1.5	1.3	2.0	0.8	0.8	0.5	10.2
Manzanillo	17	0.1	0.2	*b	0.0	0.1	4.7	5.7	6.4	14.5	5.1	0.9	1.8	39.5
Mazatlan	46	0.8	0.5	0.2	0.1	0.1	1.5	5.9	8.3	8.0	2.6	0.9	1.3	30.2
Merida	40	1.2	0.9	0.7	0.8	3.2	5.6	5.2	5.6	6.8	3.8	1.3	1.3	36.5
Mexico City	48	0.2	0.3	0.5	0.7	1.9	4.1	4.5	4.3	4.1	1.6	0.5	0.3	23.0
Monterrey	33	0.6	0.7	0.8	1.3	1.3	3.0	2.3	2.4	5.2	3.0	1.5	0.8	22.9
Salina Cruz	22	*b	0.4	0.6	0.5	3.3	11.6	4.5	5.5	7.1	4.0	0.9	0.1	38.5
Tampico	12	1.5	1.2	1.0	1.5	1.9	8.7	4.9	4.8	10.8	5.0	2.0	1.6	44.9
Vera Cruz	40	0.9	0.6	0.6	0.8	2.6	10.4	4.1	11.1	13.9	6.9	3.0	1.0	65.7
Central America														
Belize:														
Beliza City	33	5.4	2.4	1.5	2.2	4.3	7.7	6.4	6.7	9.6	12.0	8.9	7.3	74.4
Costa Rica:														
San Jose	34	0.6	0.2	0.8	1.8	9.0	9.5	8.3	9.5	12.0	11.8	5.7	1.6	70.8
El Salvador:														
San Salvador	39	0.3	0.2	0.4	1.7	7.7	12.9	11.5	11.7	12.1	9.5	1.6	0.4	70.0

(continued)

11.10 PRECIPITATION DATA FOR REPRESENTATIVE WORLDWIDE STATIONS (continued)

Country and Station	Length of record yr	Average Precipitation (in)												
		Jan.	Feb.	Mar.	Apr.	May	June	July	Aug.	Sept.	Oct.	Nov.	Dec.	Year
Guatemala:														
Guatemala City	29	0.3	0.1	0.5	1.2	6.0	10.8	8.0	7.8	9.1	6.8	0.9	0.3	51.8
Honduras:														
Tela	20	8.9	5.1	2.6	3.3	4.3	5.0	6.4	9.4	7.7	13.5	15.9	14.0	96.1
Panama:														
Balboa Heights	46	1.0	0.4	0.7	2.9	8.0	8.4	7.1	7.9	8.2	10.1	10.2	4.8	69.7
Cristobal	73	3.4	1.5	1.5	4.1	12.5	13.9	15.6	15.3	12.7	15.8	22.3	11.7	130.3
West Indies														
Bridgetown, Barbados	22	2.6	1.1	1.3	1.4	2.3	4.4	5.8	5.8	6.7	7.0	8.1	3.8	50.3
Basse-Terre, Guadaloupe	21	9.2	6.1	8.1	7.3	11.5	14.1	17.6	15.3	16.4	12.4	12.3	10.1	140.4
Fort-de-France, Martinique	31	4.7	4.3	2.9	3.9	4.7	7.4	9.4	10.3	9.3	9.7	7.9	5.9	80.4
Hamilton, Bermuda	62	4.4	4.7	4.8	4.1	4.6	4.4	4.5	5.4	5.2	5.8	5.0	4.7	57.6
Havana, Cuba	72	2.8	1.8	1.8	2.3	4.7	6.5	4.9	5.3	5.9	6.8	3.1	2.3	48.2
Kingston, Jamaica	59	0.9	0.6	0.9	1.2	4.0	3.5	1.5	3.6	3.9	7.1	2.9	1.4	31.5
La Guerite, St. Christopher (St. Kitts)	21	4.1	2.0	2.3	2.3	3.8	3.6	4.4	5.2	6.0	5.4	7.3	4.5	50.9
Nassau, Bahamas	57	1.4	1.5	1.4	2.5	4.6	6.4	5.8	5.3	6.9	6.5	2.8	1.3	46.4
Port-au-Prince, Haiti	70	1.3	2.3	3.4	6.3	9.1	4.0	2.9	5.7	6.9	6.7	3.4	1.3	53.3

Port of Spain, Trinidad	97	2.7	1.6	1.8	2.1	3.7	7.6	8.6	9.7	7.6	6.7	7.2	4.9	64.2
St. Thomas, Virgin Is.	9	2.5	1.9	1.7	2.2	4.6	3.2	3.2	4.1	6.9	5.6	3.9	3.9	43.7
San Juan, Puerto Rico	30	4.7	2.9	2.2	3.7	7.1	5.7	6.3	7.1	6.8	5.8	6.5	5.4	64.2
Santo Domingo, Dominican Republic	25	2.4	1.4	1.9	3.9	6.8	6.2	6.4	6.3	7.3	6.0	4.8	2.4	55.8
South America														
Argentina:														
Bahia Blanca	46	1.7	2.2	2.5	2.3	1.2	0.9	1.0	1.0	1.6	2.2	2.1	1.9	20.6
Buenos Aires	70	3.1	2.8	4.3	3.5	3.0	2.4	2.2	2.4	3.1	3.4	3.3	3.9	37.4
Cipolletti	24	0.4	0.4	0.7	0.4	0.6	0.6	0.5	0.3	0.6	0.9	0.5	0.5	6.4
Corrientes	40	4.7	4.5	5.3	5.6	3.3	1.9	1.7	1.5	2.8	4.7	5.2	5.2	46.4
La Quiaca	25	3.5	2.6	1.8	0.3	*	0.0	*	*	0.1	0.3	1.0	2.7	12.3
Mendoza	46	0.9	1.2	1.1	0.5	0.4	0.3	0.2	0.3	0.5	0.7	0.7	0.7	7.5
Parana	23	3.1	3.1	3.9	4.9	2.6	1.2	1.2	1.6	2.4	2.8	3.7	4.5	35.0
Puerto Madryn	50	0.4	0.6	0.7	0.5	0.9	0.6	0.6	0.4	0.6	0.7	0.4	0.6	7.0
Santa Cruz	20	0.6	0.3	0.3	0.6	0.4	0.5	0.4	0.5	0.3	0.3	0.4	0.7	5.3
Santiago Del Estero	20	3.4	3.0	3.0	1.3	0.6	0.3	0.2	0.2	0.5	1.4	2.5	4.1	20.4
Ushuaia	21	2.0	2.6	1.9	2.1	1.5	1.2	1.2	1.1	1.3	1.6	1.3	1.9	19.9
Bolivia:														
Conception	16	7.2	4.7	4.4	1.8	2.0	1.5	1.1	0.9	1.2	2.9	5.0	5.9	38.6
La Paz	50	4.5	4.2	2.6	1.3	0.5	0.3	0.4	0.5	1.1	1.6	1.9	3.7	22.6
Sucre	52	7.3	4.9	3.7	1.6	0.2	0.1	0.2	0.3	1.0	1.6	2.6	4.3	27.8
Brazil:														
Barra do Corda	9	6.7	8.7	8.0	6.1	2.3	1.0	0.7	0.7	1.0	2.5	3.9	5.7	47.2

(continued)

11.10 PRECIPITATION DATA FOR REPRESENTATIVE WORLDWIDE STATIONS (continued)

Country and Station	Length of Record (yr)	Jan.	Feb.	Mar.	Apr.	May	June	July	Aug.	Sept.	Oct.	Nov.	Dec.	Year
						Average Precipitation (in)								
Bela Vista	20	6.6	4.9	4.4	4.3	5.0	2.8	1.3	1.8	2.9	5.4	5.8	7.0	52.2
Belem	20	12.5	14.1	14.1	12.6	10.2	6.7	5.9	4.4	3.5	3.3	2.6	6.1	96.0
Brasilia	3	9.0	7.8	4.8	3.4	1.4	*	0.0	*	1.3	4.9	9.7	11.7	54.0
Conceicao do Araguai	5	14.9	12.1	10.8	4.1	1.9	0.4	*	0.5	1.5	6.6	4.9	8.6	66.2
Corumda	11	7.3	5.9	5.1	4.6	2.9	1.9	0.3	1.2	2.6	4.0	5.6	7.1	48.5
Florianopolis	25	7.6	5.6	6.3	4.1	3.6	3.5	2.2	3.7	4.3	5.1	3.5	4.3	53.1
Goias	11	12.5	9.9	10.2	4.6	0.4	0.3	0.0	0.3	2.3	5.3	9.4	9.5	64.8
Guarapuava	5	8.7	5.8	5.4	4.5	4.6	6.5	2.7	3.6	4.6	6.9	5.6	6.1	65.8
Manaus	25	9.8	9.1	10.3	8.7	6.7	3.3	2.3	1.5	1.8	4.2	5.6	8.0	71.3
Natal	18	1.9	4.8	7.0	9.2	7.1	8.7	7.7	3.8	1.4	0.8	0.7	1.1	54.2
Parana	19	11.3	9.3	9.4	4.0	0.5	*b	0.1	0.2	1.1	5.0	9.1	12.2	62.3
Porto Alegre	22	3.5	3.2	3.9	4.1	4.5	5.1	4.5	5.0	5.2	3.4	3.1	3.5	49.1
Quizeramobin	13	0.7	5.0	6.6	5.0	7.0	1.7	0.7	0.6	0.4	0.6	0.7	0.6	29.6
Recife	56	2.1	3.3	6.3	8.7	10.5	10.9	10.0	6.0	2.5	1.0	1.0	1.1	63.4
Rio de Janeiro	84	4.9	4.8	5.1	4.2	3.1	2.1	1.6	1.7	2.6	3.1	4.1	5.4	42.6
Salvador (Bahia)	20	2.6	5.3	6.1	11.2	10.8	9.4	7.2	4.8	3.3	4.0	4.5	5.6	74.8
Santarem	22	6.8	10.9	13.2	12.9	11.3	6.9	4.1	1.7	1.5	1.9	2.3	4.1	77.9
Sao Paulo	24	8.8	7.8	6.0	2.2	3.0	2.4	1.5	2.1	3.5	4.6	6.0	9.4	57.3

Sena Madureira	17	11.2	11.3	10.2	9.4	4.1	2.2	1.1	1.5	4.0	7.0	7.5	11.7	81.2
Uaupes	10	10.3	7.7	10.0	10.6	12.0	9.2	8.8	7.2	5.1	6.9	7.2	10.4	105.4
Uruguaiana	12	3.6	3.6	5.6	5.1	3.7	4.2	3.2	2.8	3.6	4.1	2.9	4.1	46.6
Chile:														
Ancud	46	3.1	3.7	5.3	7.4	9.9	11.0	10.3	9.4	6.5	4.2	4.7	4.6	80.1
Antofagasta	32	0.0	0.0	0.0	*b	*b	0.1	0.2	0.1	*b	0.1	*b	0.0	0.5
Arica	25	*b	0.0	0.0	0.0	0.0	0.0	0.0	*b	0.0	7.0	0.0	*b	*b
Cabo Raper	10	7.8	5.8	7.1	7.7	7.5	7.9	9.5	7.5	5.6	7.0	6.7	7.0	87.1
Los Evangelistas	27	11.7	10.0	11.3	11.4	9.6	9.4	9.4	8.6	9.2	8.8	9.9	10.1	119.4
Potrerillos	7	*b	*b	0.3	*b	0.7	*b	0.5	0.3	0.2	0.2	0.0	*b	2.2
Puerto Aisen	11	7.8	7.8	8.3	7.5	14.7	10.4	11.1	11.1	6.5	7.8	7.0	7.9	107.9
Punta Arenas	15	1.5	0.9	1.3	1.4	1.3	1.6	1.1	1.2	0.9	1.1	0.7	1.4	14.4
Santiago	58	0.1	0.1	0.2	0.5	2.5	3.3	3.0	2.2	1.2	0.6	0.3	0.2	14.2
Valdivia	60	2.6	2.9	5.2	9.2	14.2	17.7	15.5	12.9	8.2	5.0	4.9	4.1	102.4
Valparaiso	41	0.1	*b	0.3	0.6	4.1	5.9	3.9	2.9	1.3	0.4	0.2	0.2	19.9
Colombia:														
Andagoya	15	25.0	21.4	19.5	26.1	25.5	25.8	23.3	25.3	24.6	22.7	22.4	19.5	281.1
Bogota	49	2.3	2.6	4.0	5.8	4.5	2.4	2.0	2.2	2.4	6.3	4.7	2.6	41.8
Cartegena	10	0.4	0.0	0.4	0.9	3.4	3.4	3.0	1.6	0.5	10.8	8.9	4.5	36.8
Ipiales	13	3.1	2.3	3.5	3.5	2.8	1.9	1.3	1.1	1.4	3.1	3.3	2.6	29.9
Tumaco	10	16.9	11.7	9.6	14.6	17.4	12.0	7.7	7.3	7.3	5.9	4.9	7.0	122.3
Ecuador:														
Cuenca	10	2.0	1.8	3.2	4.3	4.3	1.7	0.9	1.1	1.6	3.1	1.8	2.5	28.3
Guayaquil	10	8.3	11.4	11.5	8.1	2.1	0.4	0.2	*b	*b	*b	0.1	1.1	43.3

(continued)

Country and Station	Record Length[a] yr	Jan.	Feb.	Mar.	Apr.	May	June	July	Aug.	Sept.	Oct.	Nov.	Dec.	Year
						Average Precipitation (in)								
Quito	33	3.9	4.4	5.6	6.9	5.4	1.7	0.8	1.2	2.7	4.4	3.8	3.1	43.9
French Guiana:														
Cayenne	51	14.4	12.3	15.8	18.9	21.7	15.5	6.9	2.8	1.2	1.3	4.6	10.7	126.1
Guyana:														
Georgetown	35	8.0	4.5	6.9	5.5	11.4	11.9	10.0	6.9	3.2	3.0	6.1	11.3	88.7
Lethem	9	1.2	1.4	1.3	5.7	11.5	11.9	14.8	9.4	3.4	2.3	4.3	1.3	68.5
Paraguay:														
Asuncion	30	5.5	5.1	4.3	5.2	4.6	2.7	2.2	1.5	3.1	5.5	5.9	6.2	51.8
Bahia Negra	20	5.4	5.3	4.9	2.9	2.3	1.6	1.5	0.6	2.3	4.2	5.3	4.3	40.6
Peru:														
Arequipa	37	1.3	1.8	0.7	0.2	*b	*b	*b	*b	0.0	*b	*b	0.4	4.4
Cajamarca	9	3.6	5.2	4.6	3.4	1.7	9.5	9.2	9.3	2.3	2.3	1.9	3.2	28.2
Cusco	12	6.4	5.9	4.3	2.0	0.6	0.2	0.2	0.4	1.0	2.6	3.0	5.4	32.0
Iquitos	5	9.1	10.4	9.4	13.6	10.7	5.7	6.4	5.2	10.5	7.3	9.1	10.3	107.7
Lima	15	0.1	*b	*b	*b	0.2	0.2	0.3	0.3	0.3	0.1	0.1	*b	1.6
Mollendo	10	*b	0.1	*b	*b	0.1	0.1	*b	0.2	0.2	0.1	0.1	*b	0.9
Surinam:														
Paramaribo	75	8.4	6.5	7.9	9.0	12.2	11.9	9.1	6.2	3.1	3.0	4.9	8.8	91.0

		Jan	Feb	Mar	Apr	May	Jun	Jul	Aug	Sep	Oct	Nov	Dec	Year
Uruguay:														
Artigas	50	4.3	3.9	4.7	5.1	4.1	4.1	2.8	3.0	4.0	4.7	3.8	4.1	48.6
Montevideo	56	2.9	2.6	3.9	3.9	3.3	3.2	2.9	3.1	3.0	2.6	2.9	3.1	37.4
Venezuela:														
Caracas	46	0.9	0.4	0.6	1.3	3.1	4.0	4.3	4.3	4.2	4.3	3.7	1.8	32.9
Cuidad Bolivar	10	1.4	0.8	0.7	1.0	3.8	5.5	6.3	7.1	3.6	4.0	2.8	1.3	38.3
Maracaibo	36	0.1	*b	0.3	0.8	2.7	2.2	1.8	2.2	2.8	5.9	3.3	0.6	22.7
Merida	14	2.5	1.5	3.6	6.7	9.8	7.3	4.7	5.7	6.7	9.5	8.2	3.4	69.7
Santa Elena	10	3.2	3.2	3.2	5.7	9.6	9.5	9.1	7.6	5.3	4.9	4.9	4.5	70.7
Pacific Islands														
Easter Island (Isla de Pascua)	10	4.8	3.7	4.6	4.2	4.6	4.3	3.5	3.0	2.7	3.7	4.6	4.9	48.6
Mas a Tierra (Juan Fernandez)	29	0.8	1.2	1.6	3.4	5.9	6.4	5.8	4.4	2.9	1.9	1.6	1.0	36.9
Seymour Island (Galapagos Island)	3	0.8	1.4	1.1	0.7	*b	*b	*b	*b	*b	*b	*b	*b	4.0
Atlantic Islands														
Fernando de Noronha	32	1.7	4.7	7.4	10.5	10.5	7.3	5.4	1.9	0.7	0.3	0.4	0.5	51.3
Cumberland Bay, South Georgia	24	3.3	4.3	5.3	5.4	5.2	4.9	5.5	5.3	3.5	2.6	3.4	3.0	51.7
Laurie Is., S. Orkneys	46	1.4	1.5	1.9	1.6	1.2	1.0	1.3	1.3	1.1	1.1	1.3	1.0	15.7
Stanley, Falkland Is.	41	2.8	2.3	2.5	2.6	2.6	2.1	2.0	2.0	1.5	1.6	2.0	2.8	26.8
Europe														
Albania:														
Durres	10	3.0	3.3	3.9	2.2	1.6	1.9	0.5	1.9	1.7	7.1	8.5	7.3	42.9

(continued)

443

11.10 PRECIPITATION DATA FOR REPRESENTATIVE WORLDWIDE STATIONS (continued)

Country and Station	Length of Record (yr)	Jan.	Feb.	Mar.	Apr.	May	June	July	Aug.	Sept.	Oct.	Nov.	Dec.	Year
						Average Precipitation (in)								
Andorra:														
Les Escaldes	9	1.5	1.7	2.9	2.4	4.7	3.1	2.2	3.4	3.1	3.5	3.3	2.5	34.3
Austria:														
Innsbruck	35	2.1	1.8	1.5	2.2	2.9	4.1	5.1	4.5	3.1	2.4	2.2	1.9	33.8
Wien (Vienna)	100	1.5	1.4	1.8	2.0	2.8	2.7	3.0	2.7	2.0	2.0	1.9	1.8	25.6
Bulgaria:														
Sofiya (Sofia)	27	1.3	1.1	1.7	2.3	3.3	3.2	2.4	2.0	2.3	2.1	1.9	1.4	25.0
Varna	20	1.5	0.9	1.2	1.2	1.8	2.6	1.9	1.2	1.5	1.9	1.9	2.0	19.6
Cyprus:														
Nicosia	64	2.9	2.0	1.3	0.8	1.1	0.4	*b	*b	0.2	0.9	1.7	3.0	14.6
Czechoslovakia:														
Praha (Prague)	70	0.9	0.8	1.1	1.5	2.4	2.8	2.6	2.2	1.7	1.2	1.2	0.9	19.3
Prerov	21	1.3	1.1	1.1	2.0	2.4	2.9	3.5	3.2	2.0	2.4	1.5	1.4	24.8
Denmark:														
Arhus	21	2.3	1.5	1.4	1.8	1.2	2.2	2.5	3.3	3.2	2.6	2.5	2.1	26.6
Kobenhavn (Copenhagen)	30	1.6	1.3	1.2	1.7	1.7	2.1	2.2	3.2	1.9	2.1	2.2	2.1	23.3
Estonia:														
Tallinn	63	1.1	1.0	0.9	1.1	1.7	1.9	2.1	2.7	2.3	2.1	1.9	1.5	20.2

		Jan	Feb	Mar	Apr	May	Jun	Jul	Aug	Sep	Oct	Nov	Dec	Year
Finland:														
Helsinki	50	2.2	1.7	1.7	1.7	1.9	2.0	2.3	3.3	2.8	2.9	2.7	2.4	27.6
Kuusamo	20	1.1	1.1	1.1	1.1	1.4	2.3	2.8	3.0	2.1	2.1	1.6	1.1	20.8
Vaasa	19	1.1	0.8	0.8	1.0	1.4	1.8	2.4	2.5	2.7	2.3	1.7	1.1	19.6
France:														
Ajaccio	86	3.0	2.3	2.6	2.2	1.6	0.9	2.8	0.7	1.7	3.8	4.4	3.1	29.1
Bordeaux	47	2.7	2.8	2.9	2.6	2.5	2.3	2.0	1.9	2.2	3.0	3.9	3.9	32.7
Brest	56	3.5	3.0	2.5	2.5	1.9	2.0	2.0	2.2	2.3	3.6	4.2	4.4	34.1
Cherbourg	47	3.3	2.9	2.7	2.0	1.9	1.8	1.9	3.0	2.9	4.6	5.1	5.2	37.3
Lille	40	2.5	1.9	2.5	2.0	2.4	2.2	2.8	2.3	2.6	3.0	3.0	3.2	30.3
Lyon	70	1.4	1.4	1.8	2.1	2.8	2.9	2.8	2.9	3.1	3.1	2.6	1.9	28.8
Marseille	102	1.9	1.5	1.8	2.0	1.9	1.0	0.6	0.9	2.6	3.7	3.1	2.2	23.2
Paris	118	1.5	1.3	1.5	1.7	2.0	2.1	2.1	2.0	2.0	2.2	2.0	1.9	22.3
Strasbourg	20	1.6	1.4	1.7	2.6	2.6	3.1	3.4	3.4	3.1	2.7	2.0	1.9	29.5
Toulouse	47	1.9	1.7	2.3	2.7	2.9	2.4	1.5	2.1	2.3	2.2	2.4	2.3	26.7
Germany:														
Berlin	40	1.9	1.3	1.5	1.7	1.9	2.3	3.1	2.2	1.9	1.7	1.7	1.9	23.1
Bremen	80	1.9	1.6	1.8	1.5	2.1	2.6	3.2	2.8	2.1	2.2	2.0	2.2	26.0
Frankfurt A/M	80	1.7	1.3	1.6	1.5	2.0	2.5	2.8	2.6	1.9	2.2	2.0	2.0	24.1
Hamburg	80	2.1	1.9	2.0	1.8	2.1	2.7	3.4	3.2	2.5	2.6	2.1	2.5	28.9
Munchen (Munich)	80	1.7	1.4	1.9	2.7	3.7	4.6	4.7	4.2	3.2	2.2	1.9	1.9	34.1
Munster	40	2.6	1.9	2.2	2.0	2.2	2.7	3.3	3.1	2.5	2.7	2.4	2.9	30.5
Nurnberg	80	1.5	1.2	1.3	1.7	2.2	2.5	3.1	3.1	2.1	2.1	1.9	1.7	24.4

(continued)

445

11.10 PRECIPITATION DATA FOR REPRESENTATIVE WORLDWIDE STATIONS (continued)

Country and Station	Length of record (yr)	Jan.	Feb.	Mar.	Apr.	May	June	July	Aug.	Sept.	Oct.	Nov.	Dec.	Year
								Average Precipitation (in)						
Gibraltar:														
Windmill Hill	12	4.6	3.4	3.7	2.5	1.4	0.2	*b	0.1	0.8	3.5	4.1	5.4	29.7
Greece:														
Athinai (Athens)	80	2.2	1.6	1.4	0.8	0.8	0.6	0.2	0.4	0.6	1.7	2.8	2.8	15.8
Iraklion	22	3.7	3.0	1.6	0.9	0.7	0.1	*b	0.1	0.7	1.7	2.7	4.0	19.2
Rodhos (Rhodes)	6	5.7	3.9	2.6	1.7	0.5	0.3	0.0	*b	0.4	1.7	5.2	6.7	28.5
Thessaloniki (Salonika)	26	1.5	1.5	1.6	1.9	2.0	1.2	1.0	0.7	1.2	2.4	2.1	1.9	19.0
Hungary:														
Budapest	50	1.5	1.5	1.7	2.0	2.7	2.6	2.0	1.9	1.8	2.1	2.4	2.0	24.2
Debrecen	80	1.2	1.1	1.4	1.8	2.4	2.8	2.5	2.3	1.8	2.2	2.0	1.6	23.1
Iceland:														
Akureyri	26	1.7	1.5	1.7	1.3	0.6	0.9	1.3	1.6	1.9	2.3	1.9	1.9	18.6
Reykjavik	30	4.0	3.1	3.0	2.1	1.6	1.7	2.0	2.6	3.1	3.4	3.6	3.7	33.9
Ireland:														
Cork	35	4.9	3.6	3.3	2.6	2.9	2.0	2.9	3.1	2.9	3.9	4.5	4.7	41.3
Dublin	35	2.7	2.2	2.0	1.9	2.3	2.0	2.8	3.0	2.8	2.7	2.7	2.6	29.7
Shannon Airport	12	3.8	3.0	2.0	2.2	2.4	2.1	3.1	3.0	3.0	3.4	4.2	4.3	36.5

Italy:

Ancona	30	2.6	1.7	1.6	2.3	2.1	1.9	1.5	1.5	3.5	3.7	2.5	3.0	28.0
Cagliari	25	2.2	1.5	1.5	1.2	1.5	0.5	0.1	0.4	1.0	3.0	1.8	2.3	17.0
Genova (Genoa)	10	3.9	4.0	3.3	3.4	4.6	1.4	1.6	2.3	4.7	6.1	7.2	4.1	46.6
Napoli (Naples)	30	3.7	3.2	3.0	2.6	1.8	1.8	0.6	0.7	2.8	5.1	4.5	5.4	35.2
Palermo	30	3.8	3.4	2.4	1.9	1.1	0.6	0.2	0.6	2.0	3.7	4.1	4.5	28.3
Rome	30	3.3	2.9	2.0	2.0	1.9	0.7	0.4	0.7	2.8	4.3	4.4	4.1	29.5
Taranto	10	1.6	0.9	1.3	0.8	1.0	0.6	0.4	0.7	1.0	2.2	1.8	1.9	14.2
Venezia (Venice)	30	2.0	2.1	2.4	2.8	3.2	3.3	2.6	2.6	2.6	3.7	3.5	2.6	33.4

Latvia:

Riga	57	1.3	1.0	1.1	1.2	1.7	2.4	3.0	3.0	2.1	2.0	1.9	1.5	22.2

Lithuania:

Kaunas	19	1.6	1.3	1.3	1.8	2.0	3.2	3.3	3.5	1.9	1.9	1.6	1.6	25.0

Luxembourg:

Luxembourg	100	2.3	2.0	1.9	2.1	2.4	2.5	2.8	2.6	2.4	2.7	2.7	2.8	29.2

Malta:

Valletta	90	3.3	2.3	1.5	0.8	0.4	0.1	*b	0.2	1.3	2.7	3.6	3.9	20.3

Monaco:

Monaco	60	2.4	2.3	3.1	2.2	2.1	1.4	0.7	1.1	2.3	4.7	4.3	3.5	30.1

Netherlands:

Amsterdam	29	2.0	1.4	1.3	1.6	1.8	1.8	2.6	2.7	2.8	2.8	2.6	2.2	25.6

Norway:

Bergen	75	7.9	6.0	5.4	4.4	3.9	4.2	5.2	7.3	9.2	9.2	8.0	8.1	78.8
Kristiansand	56	5.0	3.6	3.6	2.7	2.5	2.8	3.5	5.3	4.7	6.2	5.7	6.4	52.0

(continued)

11.10 PRECIPITATION DATA FOR REPRESENTATIVE WORLDWIDE STATIONS (continued)

Country and Station	Length of record (yr)[a]	Average Precipitation (in)												
		Jan.	Feb.	Mar.	Apr.	May	June	July	Aug.	Sept.	Oct.	Nov.	Dec.	Year
Oslo	56	1.7	1.3	1.4	1.6	1.8	2.4	2.9	3.8	2.5	2.9	2.3	2.3	26.9
Tromso	75	4.1	3.8	3.3	2.4	2.1	2.1	2.3	2.9	4.7	4.5	4.0	3.9	40.1
Trondheim	65	3.1	2.7	2.6	2.0	1.7	1.9	2.4	3.0	3.4	3.7	2.8	2.8	32.1
Vardo	56	2.5	2.5	2.3	1.5	1.3	1.3	1.5	1.7	1.9	2.5	2.1	2.4	23.5
Poland:														
Gdansk	35	1.2	1.0	1.3	1.5	1.8	2.3	2.8	2.6	2.1	1.8	1.8	1.5	21.7
Krakow	35	1.1	1.3	1.4	1.8	2.8	4.0	4.5	3.8	2.7	2.2	1.7	1.3	28.6
Warsaw	113	1.2	1.1	1.3	1.5	1.9	2.6	3.0	3.0	1.9	1.7	1.4	1.4	22.0
Wroclaw	40	1.5	1.1	1.5	1.7	2.4	2.4	3.4	2.7	1.8	1.7	1.5	1.5	23.2
Portugal:														
Braganca	11	11.9	6.9	7.7	3.7	3.0	1.6	0.5	0.6	1.5	3.0	6.3	7.1	53.8
Lagos	17	3.2	2.6	2.8	1.4	0.8	0.2	*b	*b	0.4	1.5	2.6	2.8	18.3
Lisbon	75	3.3	3.2	3.1	2.4	1.7	0.7	0.2	0.2	1.4	3.1	4.2	3.6	27.0
Romania:														
Bucuresti (Bucharest)	41	1.5	1.1	1.7	1.6	2.5	3.8	2.3	1.8	1.5	1.6	1.9	1.5	22.8
Cluj-Napoca	16	1.3	1.2	1.0	2.1	3.3	3.3	2.6	3.3	2.0	1.7	1.0	1.2	24.0
Constanta	39	1.2	1.2	1.1	1.1	1.3	1.7	1.3	1.1	1.1	1.4	1.2	1.4	15.1

Spain:														
Almeria	20	0.9	1.0	0.7	0.9	0.7	0.2	* [b]	0.1	0.6	0.9	1.5	1.1	8.6
Barcelona	30	1.2	2.1	1.9	1.8	1.8	1.3	1.2	1.7	2.6	3.4	2.7	1.8	23.5
Burgos	29	1.5	1.5	2.1	1.9	2.4	1.7	0.8	0.7	1.4	2.0	2.2	2.0	20.2
Madrid	30	1.1	1.7	1.7	1.7	1.5	1.2	0.4	0.3	1.2	1.9	2.2	1.6	16.5
Sevilla	26	2.2	2.9	3.3	2.3	1.3	0.9	0.1	0.1	1.1	2.6	3.7	2.8	23.3
Valencia	29	0.9	1.5	0.9	1.2	1.1	1.3	0.4	0.5	2.2	1.6	2.5	1.3	15.4
Sweden:														
Abisko	11	0.7	0.6	0.5	0.5	0.7	1.8	1.6	1.8	1.2	1.0	0.6	0.6	11.7
Goteberg	61	2.5	2.0	2.0	1.7	1.9	2.2	2.8	3.7	3.1	3.1	2.7	2.8	30.5
Haparanda	20	2.2	1.6	1.2	1.5	1.4	1.7	2.1	2.8	2.6	2.8	2.5	2.0	24.4
Karlstad	30	1.9	1.2	1.2	1.4	1.9	1.9	2.6	3.1	2.9	2.4	2.4	1.9	24.8
Sarna	20	1.6	0.8	0.9	1.2	1.6	2.8	3.6	3.3	2.6	2.3	1.8	1.8	24.3
Stockholm	30	1.5	1.1	1.1	1.5	1.6	1.9	2.8	3.1	2.1	2.1	1.9	1.9	22.4
Visby	30	1.7	1.1	1.2	1.4	1.1	1.4	2.0	2.7	1.7	1.9	2.1	2.0	20.3
Switzerland:														
Berne	77	1.9	2.0	2.6	3.0	3.7	4.4	4.4	4.3	3.5	3.5	2.7	2.5	38.5
Geneve (Geneva)	125	1.9	1.8	2.2	2.5	3.0	3.1	2.9	3.6	3.6	3.8	3.1	2.4	33.9
Zurich	23	2.3	1.9	2.9	3.4	4.0	4.9	5.0	4.6	3.3	3.2	2.5	2.9	40.9
United Kingdom:														
Belfast	30	4.2	2.8	2.3	2.4	2.3	2.5	3.5	3.5	3.4	3.8	3.6	3.9	38.2
Birmingham	30	2.9	2.1	1.7	2.2	2.5	1.8	2.8	2.7	2.3	2.9	3.2	2.6	29.7
Cardiff	30	4.6	3.0	2.3	2.5	3.0	2.2	3.4	3.9	3.6	4.5	4.6	4.3	41.9

(continued)

11.10 PRECIPITATION DATA FOR REPRESENTATIVE WORLDWIDE STATIONS (continued)

Country and Station	Length of record (yr)	Jan.	Feb.	Mar.	Apr.	May	June	July	Aug.	Sept.	Oct.	Nov.	Dec.	Year
							Average Precipitation (in)							
Edinburgh	30	2.5	1.6	1.6	1.6	2.2	1.9	3.1	3.1	2.6	2.9	2.4	2.1	27.6
London	30	2.0	1.5	1.4	1.8	1.8	1.6	2.0	2.2	1.8	2.3	2.5	2.0	22.9
Liverpool	30	2.7	1.9	1.5	1.6	2.2	2.0	2.8	3.1	2.6	3.0	3.0	2.5	28.9
Perth	30	3.1	2.2	1.9	1.7	2.3	2.0	3.1	2.9	2.8	3.3	2.7	2.7	30.7
Plymouth	30	4.3	3.0	2.6	2.3	2.5	2.0	2.6	2.9	2.8	3.8	4.6	4.4	37.8
Wick	30	2.9	2.1	1.8	2.1	1.8	2.0	2.6	2.6	2.9	3.2	3.1	2.9	30.0
U.S.S.R. (Former):														
Arkhangelsk	25	1.2	1.1	1.1	0.7	1.3	1.9	2.6	2.7	2.2	1.9	1.6	1.3	19.8
Astrakhan	25	0.5	0.5	0.4	0.6	0.6	0.7	0.5	0.4	0.6	0.4	0.6	0.6	6.4
Dnepropetrovsk	17	1.4	1.1	1.2	1.4	1.8	3.0	1.9	1.6	1.0	1.8	1.6	1.6	19.4
Kursk	20	1.5	1.3	1.2	1.5	2.2	2.5	3.2	2.3	1.6	1.8	1.5	1.7	22.3
Lvov	35	1.3	1.5	1.8	2.0	2.8	3.7	4.1	3.1	2.4	2.1	0.8	1.6	28.2
Minsk	20	1.4	1.5	1.3	1.5	2.0	2.8	3.0	3.1	1.6	1.5	1.5	1.7	22.9
Moskva (Moscow)	11	1.5	1.4	1.1	1.9	2.2	2.9	3.0	2.9	1.9	2.7	1.7	1.6	24.8
Odessa	15	1.0	0.7	0.7	1.1	1.1	1.9	1.6	1.4	1.1	1.4	1.1	1.1	14.3
St. Petersburg (Leningrad)	95	1.0	0.9	0.9	1.0	1.6	2.0	2.5	2.8	2.1	1.8	1.4	1.2	19.2
Saratov	15	1.0	1.0	0.8	1.0	1.3	1.8	1.2	1.3	1.1	1.4	1.4	1.2	14.5
Sevastopol	30	1.1	1.1	1.1	0.9	0.6	1.1	0.8	0.6	1.1	1.5	1.2	1.1	12.2

Stavropol	41	1.4	1.1	1.5	2.4	3.0	4.1	3.0	2.0	2.5	2.3	1.8	1.8	26.9
Tbilisi	10	0.7	0.8	1.3	1.6	3.6	3.1	2.2	1.7	1.9	1.3	2.0	1.2	21.4
Ufa	23	1.6	1.3	1.2	0.9	1.6	2.4	2.6	2.2	1.8	2.3	2.2	2.3	22.5
Ust' Shchugor	15	1.1	0.8	0.8	0.7	1.4	2.2	3.0	3.2	2.4	2.2	1.5	1.3	20.6
Vologad (Stalingrad)	12	0.9	1.0	0.6	0.6	1.0	1.9	0.9	0.8	0.7	1.0	1.5	1.3	12.2
Vyatka (Kirov)	29	1.2	1.0	0.9	0.9	1.9	2.5	2.1	2.9	2.3	2.0	1.6	1.3	20.6
Yugoslavia:														
Beograd (Belgrade)	16	1.6	1.3	1.6	2.2	2.6	2.8	1.9	2.5	1.7	2.7	1.8	1.9	24.6
Skopje	10	1.5	1.2	1.3	1.5	1.9	1.9	1.3	1.1	1.1	2.6	2.3	1.8	19.5
Split	51	3.1	2.5	3.2	3.0	2.5	2.1	1.2	1.6	2.9	4.4	4.2	4.4	35.1
Ocean Islands														
Barentsburg, Spitzbergen	15	1.4	1.3	1.1	0.9	0.5	0.4	0.6	0.9	1.0	1.2	0.9	1.5	11.7
Bjornoya (Bear Island)	25	1.6	1.3	1.3	0.9	0.8	0.7	0.8	1.2	1.8	1.7	1.4	1.6	15.1
Horta, Azores	30	4.5	4.1	4.2	3.0	2.9	2.0	1.5	1.9	3.2	4.4	4.1	4.5	40.3
Jan Mayen	29	2.1	1.7	1.6	1.4	0.9	0.9	1.4	1.8	2.5	2.5	2.2	2.2	21.2
Lerwick, Shetland Is.	30	4.5	3.4	2.9	2.7	2.2	2.2	2.7	2.9	3.7	4.3	4.5	4.5	40.5
Matochkin Shar, Novaya Zemlya	9	0.6	0.6	0.6	0.4	0.3	0.4	0.4	1.5	1.5	0.6	0.6	0.4	8.9
Ponta Delgada, Azores	30	4.0	3.5	3.5	2.5	2.3	1.4	1.0	1.2	2.9	3.6	3.7	3.0	32.6
Stornoway, Herbrides	15	6.4	3.2	3.2	3.1	2.5	2.4	3.0	4.3	4.7	6.2	4.6	5.5	49.1
Thorshavn, Faroes	50	6.6	5.2	4.8	3.6	3.4	2.5	3.1	3.5	4.7	5.9	6.3	6.6	56.2
Africa														
Algeria:														
Adrar	15	*b	*b	0.1	*b	*b	*b	*	*b	*b	0.2	0.2	*b	0.6
Alger (Algiers)	25	4.4	3.3	2.9	1.6	1.8	0.6	*	0.2	1.6	3.1	5.1	5.4	30.0

(continued)

11.10 PRECIPITATION DATA FOR REPRESENTATIVE WORLDWIDE STATIONS (continued)

Country and Station	Length of record (yr)	Average Precipitation (in)												
		Jan.	Feb.	Mar.	Apr.	May	June	July	Aug.	Sept.	Oct.	Nov.	Dec.	Year
Annaba (Bone)	26	5.6	4.1	2.9	2.2	1.5	0.6	0.1	0.3	1.2	3.0	4.3	5.2	31.0
El Golea	15	0.1	0.3	0.5	*b	*b	*b	*b	*b	*b	0.3	0.4	0.3	1.9
Hassi Tanezrouft	15	0.3	0.1	0.1	0.2	*b	*b	0.0	*b	*b	*b	0.2	0.2	1.1
Tamanrasset	15	0.2	*b	*b	0.2	0.4	0.1	0.1	0.4	0.1	*b	*b	*b	1.5
Touggourt	26	0.2	0.4	0.5	0.2	0.2	0.2	*b	*b	0.1	0.3	0.5	0.3	2.9
Angola:														
Cangamba	7	8.9	7.4	6.8	1.8	0.1	0.0	0.0	0.2	0.2	1.6	5.1	8.5	40.6
Huambo	14	8.7	7.8	9.8	5.7	0.4	0.0	*b	*b	0.6	5.5	9.6	8.9	57.0
Luanda	59	1.0	1.4	3.0	4.6	0.5	*b	*b	*b	0.1	0.2	1.1	0.8	12.7
Namibe	21	0.3	0.4	0.7	0.5	*b	*b	*b	*b	*b	*b	0.1	0.1	2.1
Benin:														
Cotonou	10	1.3	1.3	4.6	4.9	10.0	14.4	3.5	1.5	2.6	5.3	2.3	0.5	52.4
Botswana:														
Francistown	28	4.2	3.1	2.8	0.7	0.2	0.1	*b	*b	*b	0.9	2.3	3.4	17.7
Mann	20	4.3	3.8	3.5	1.1	0.2	*b	0.0	0.0	*b	0.5	1.9	2.8	18.2
Tshabong	14	2.0	1.9	1.9	1.3	0.4	0.4	0.1	*b	0.2	0.7	1.1	1.5	11.5
Burkina Faso:														
Bobo Dioulasso	10	0.1	0.2	1.1	2.1	4.6	4.8	9.8	12.0	8.5	2.5	0.7	0.0	46.4

Quagadougou	15	*b	0.1	0.5	0.6	3.3	4.8	8.0	10.9	5.7	1.3	*b	0.0	35.2
Cameroon:														
Ngaoundere	10	*b	*b	1.1	5.5	7.0	8.4	10.6	9.6	9.2	5.3	0.5	*b	57.2
Yaounde	11	0.9	2.6	5.8	6.7	7.7	6.0	2.9	3.1	8.4	11.6	4.6	0.9	61.2
Central African Republic:														
Bangui	5	1.0	1.7	5.0	5.3	7.4	4.5	8.9	8.1	5.9	7.9	4.9	0.2	60.8
Ndele	3	0.2	1.3	0.6	1.7	8.4	6.1	8.3	10.1	10.7	7.8	0.6	0.0	55.8
Chad:														
Am Timan	3	0.0	0.0	0.1	1.2	4.3	5.0	7.3	12.3	5.8	1.2	0.0	0.0	37.2
Faya-Largeau	5	0.0	0.0	0.0	0.0	*b	0.0	*b	0.7	*b	0.0	0.0	0.0	0.7
N'Djamena	5	0.0	0.0	0.0	0.1	1.2	2.6	6.7	12.6	4.7	1.4	0.0	0.0	29.3
Congo, Republic of:														
Brazzaville	18	6.3	4.9	7.4	7.0	4.3	0.6	*b	*b	2.2	5.4	11.5	8.4	58.0
Quesso	4	2.4	3.6	6.4	3.2	5.8	4.6	2.9	3.7	7.9	10.0	5.7	2.4	58.6
Pointe Noire	7	5.4	6.7	6.4	8.0	3.9	0.0	0.0	0.0	0.4	4.1	6.6	6.6	48.1
Cote D'Ivoire:														
Abidjan	10	1.6	2.1	3.9	4.9	14.2	19.5	8.4	2.1	2.8	6.6	7.9	3.1	77.1
Bouake	10	0.4	1.5	4.1	5.8	5.3	6.0	3.1	4.6	8.2	5.2	1.5	1.0	46.7
Djibouti:														
Djibouti	46	0.4	0.5	1.0	0.5	0.2	*b	0.1	0.3	0.3	0.4	0.9	0.5	5.1
Egypt:														
Alexandria	61	1.9	0.9	0.4	0.1	*b	*b	*b	*b	*b	0.2	1.3	2.2	7.0

(continued)

453

11.10 PRECIPITATION DATA FOR REPRESENTATIVE WORLDWIDE STATIONS (continued)

Country and Station	Length of record (yr)	Average Precipitation (in)												
		Jan.	Feb.	Mar.	Apr.	May	June	July	Aug.	Sept.	Oct.	Nov.	Dec.	Year
Aswan	11	*b	*b	*b	*b	*b	*b	0.0	0.0	0.0	*b	*b	*b	*b
Cairo	42	0.2	0.2	0.2	0.1	0.1	*b	0.0	0.0	*b		0.1	0.2	1.1
Ethiopia:														
Addis Ababa	37	0.5	1.5	2.6	3.4	3.4	5.4	11.0	11.8	7.5	0.8	0.6	0.2	48.7
Asmera	17	*b	*b	0.4	1.5	1.5	1.3	6.7	5.0	1.3	0.3	0.4	*b	18.4
Diredawa	8	0.8	0.8	3.3	3.0	2.8	1.5	4.3	3.8	2.2	0.5	0.3	0.8	24.1
Gambela	30	0.2	0.4	1.4	3.2	5.9	6.7	8.5	9.5	7.3	3.5	1.8	0.4	48.8
Gabon:														
Libreville	21	9.8	9.3	13.2	13.4	9.6	0.5	0.1	0.7	4.1	13.6	14.7	9.8	98.8
Mayoumba	8	6.5	9.3	6.2	10.2	2.3	0.1	0.0	0.2	2.6	9.3	10.7	4.6	62.0
Gambia:														
Banjul	9	0.1	0.1	*b	*b	0.4	2.3	11.1	19.7	12.2	4.3	0.7	0.1	51.0
Ghana:														
Accra	65	0.6	1.3	2.2	3.2	5.6	7.0	1.8	0.6	1.4	2.5	1.4	0.9	28.5
Kumasi	10	0.8	2.3	5.7	5.1	7.5	7.9	4.3	3.1	6.8	7.1	3.7	0.8	55.2
Guinea:														
Conarkry	10	0.1	0.1	0.4	0.9	6.2	22.0	51.1	41.5	26.9	14.6	4.8	0.4	169.0
Kouroussa	10	0.4	0.3	0.9	2.8	5.3	9.7	11.7	13.6	13.4	6.6	1.3	0.7	66.4

Guinea Bissau:														
Bolama	37	*b	*b	*b	0.8	7.8	23.1	27.6	16.9	8.0	1.6	0.1		85.9
Kenya:														
Mombasa	54	1.0	0.7	2.5	7.7	12.6	4.7	3.5	2.5	2.5	3.4	3.8	2.4	47.3
Nairobi	17	1.5	2.5	4.9	8.3	6.2	1.8	0.6	0.9	1.2	2.1	4.3	3.4	37.7
Liberia:														
Monrovia	4	0.2	0.1	4.4	11.7	13.4	36.1	24.2	18.6	29.9	25.2	8.2	2.9	174.9
Libya:														
Al Jawf	7	*b	0.0	0.0	0.0	*b	0.0	0.0	0.0	0.0	0.0	0.0	*b	*b
Banghazi (Bhenghazi)	46	2.6	1.6	0.8	0.2	0.1	*b	*b	*b	0.1	0.7	1.8	2.6	10.5
Sabha	10	*b	*b	*b	*b	0.1	0.0	0.0	0.0	0.0	*b	*b	*b	0.3
Tarabulus (Tripoli)	56	3.2	1.8	1.1	0.4	0.2	0.1	*b	*b	0.4	1.6	2.6	3.7	15.1
Madagascar:														
Antananarivo	62	11.8	11.0	7.0	2.1	0.7	0.3	0.3	0.4	0.7	2.4	5.3	11.3	53.4
Diego Suarez	31	10.6	9.5	7.6	2.2	0.3	0.2	0.2	0.3	0.3	0.7	1.1	5.8	38.7
Toliari	15	3.1	3.2	1.4	0.3	0.7	0.4	0.1	0.2	0.3	0.7	1.4	1.7	13.5
Malawi:														
Karonga	8	7.1	7.0	10.8	6.2	1.7	0.1	*b	*b	0.0	0.3	0.3	4.7	38.3
Zomba	29	12.1	9.9	10.1	2.7	0.7	0.4	0.3	0.3	0.2	1.0	4.3	10.9	52.9
Mali:														
Araouane	10	*b	*b	0.0	0.0	0.0	0.2	0.2	0.5	0.6	0.1	0.1	*b	1.7
Bamako	10	*b	*b	0.1	0.6	2.9	5.4	11.0	13.7	8.1	1.7	0.6	*b	44.1
Gao	19	*b	0.0	*b	0.1	0.4	1.0	2.9	5.4	1.5	0.2	*b	0.0	11.5

(continued)

11.10 PRECIPITATION DATA FOR REPRESENTATIVE WORLDWIDE STATIONS (continued)

Country and Station	Length of record (yr)	Average Precipitation (in)												
		Jan.	Feb.	Mar.	Apr.	May	June	July	Aug.	Sept.	Oct.	Nov.	Dec.	Year
Mauritania:														
Atar	10	*b	0.0	*b	*b	*b	0.1	0.3	1.2	1.1	0.1	*b	*b	2.8
Nema	10	0.1	*b	*b	*b	0.7	1.1	2.3	4.7	2.1	0.7	*b	0.1	11.6
Nouakchott	10	*b	0.1	*b	*b	*b	0.1	0.5	4.1	0.9	0.4	0.1	*b	6.2
Morocco:														
Casablanca	40	2.1	1.9	2.2	1.4	0.9	0.2	0.0	*b	0.3	1.5	2.6	2.8	15.9
Marrakech	31	1.0	1.1	1.3	1.2	0.6	0.3	0.1	0.1	0.4	0.9	1.2	1.2	9.4
Rabat	29	2.6	2.5	2.6	1.7	1.1	0.3	*b	*b	0.4	1.9	3.3	3.4	19.8
Sibi Ifni	14	1.0	0.6	0.5	0.6	0.1	0.1	*b	*b	0.4	0.1	0.9	1.8	6.1
Tangier	35	4.5	4.2	4.8	3.5	1.7	0.6	*b	*b	0.9	3.9	5.8	5.4	35.3
Mozambique:														
Beira	39	10.9	8.4	10.1	4.2	2.2	1.3	1.2	1.1	0.8	5.2	5.3	9.2	59.9
Chioco	8	7.8	5.7	4.4	0.6	*b	*b	*b	*b	*b	1.1	2.6	5.2	27.4
Maputo	42	5.1	4.9	4.9	2.1	1.1	0.8	0.5	0.5	1.1	1.9	3.2	3.8	29.9
Namibia:														
Keetmanshoop	45	0.8	1.1	1.4	0.6	0.2	*b	*b	*b	0.1	0.2	0.3	0.4	5.2
Windhoek	60	3.0	2.9	3.1	1.6	0.3	*b	*b	*b	0.1	0.4	0.9	1.9	14.3

Niger:													
Agadez	10	0.0	0.0	*b	*b	0.2	0.3	1.9	3.7	0.7	0.0	0.0	0.0	6.8
Bilma	10	0.0	0.0	*b	*b	*b	0.0	0.1	0.5	0.3	0.0	0.0	0.0	0.9
Niamey	10	*b	*b	0.2	0.3	1.3	3.2	5.2	7.4	3.7	0.5	*b	0.0	21.6
Nigeria:														
Enugu	33	0.7	1.1	2.6	5.9	10.4	11.4	7.6	6.7	12.8	9.8	2.1	0.5	71.5
Kaduna	34	*b	0.1	0.5	2.5	5.9	7.1	8.5	11.9	10.6	2.9	0.1	*b	50.1
Lagos	47	1.1	1.8	4.0	5.9	10.6	18.1	11.0	2.5	5.5	8.1	2.7	1.0	72.3
Maiduguri	40	*b	*b	*b	0.3	1.6	2.7	7.1	8.7	4.2	0.7	*b	0.0	25.3
Senegal:														
Dakar	26	*b	*b	*b	*b	*b	0.7	3.5	10.0	5.2	1.5	0.1	0.3	21.3
Kaolack	10	*b	0.0	*b	*b	0.3	2.6	6.9	10.7	7.0	2.7	0.1	*b	30.3
Sierra Leone:														
Freetown/Lungi	8	0.4	0.2	1.2	3.1	9.5	14.3	29.2	36.5	22.3	14.2	5.5	1.2	137.6
Somalia:														
Berbera	30	0.3	0.1	0.2	0.5	0.3	*b	*b	0.1	*b	0.1	0.2	0.2	2.0
Moqdisho (Mogadishu)	21	*b	*b	*b	2.3	2.3	3.8	2.5	1.9	1.0	0.9	1.6	0.5	16.9
South Africa, Republic of:														
Capetown	18	0.6	0.3	0.7	1.9	3.1	3.3	3.5	2.6	1.7	1.2	0.7	0.4	20.0
Durban	78	4.3	4.8	5.1	3.0	2.0	1.3	1.1	1.5	2.8	4.3	4.8	4.7	39.7
Kimberly	57	2.4	2.5	3.1	1.5	0.7	0.2	0.2	0.3	0.6	1.0	1.6	2.0	16.1
Port Elizabeth	84	1.2	1.3	1.9	1.8	2.4	1.8	1.9	2.0	2.3	2.2	2.2	1.7	22.7
Port Nolloth	64	0.1	0.1	0.2	0.2	0.3	0.3	0.3	0.3	0.2	0.1	0.1	0.1	2.3

(continued)

11.10 PRECIPITATION DATA FOR REPRESENTATIVE WORLDWIDE STATIONS (continued)

Country and Station	Record Length (yr)	Average Precipitation (in)												
		Jan.	Feb.	Mar.	Apr.	May	June	July	Aug.	Sept.	Oct.	Nov.	Dec.	Year
Pretoria	12	5.0	4.3	4.5	1.7	0.9	0.3	0.3	0.2	0.8	2.2	5.2	5.2	30.9
Walvis Bay	20	*b	0.2	0.3	0.1	0.1	*b	*b	0.1	*b	*b	*b	*b	0.9
Sudan:														
El Fasher	17	*b	0.0	*b	*b	0.3	0.7	4.5	5.3	1.2	0.2	0.0	0.0	12.2
Khartoum	46	*b	*b	*b	*b	0.1	0.3	2.1	2.8	0.7	0.2	*b	0.0	6.2
Port Sudan	40	0.2	0.1	*b	*b	*b	*b	0.3	0.1	*b	0.4	1.7	0.9	3.7
Wadi Halfa	39	*b	*b	*b	*b	*b	0.0	*b	*b	*b	*b	*b	0.0	*b
Ewau	38	*b	0.2	0.9	2.6	5.3	6.5	7.5	8.2	6.6	4.9	0.6	*b	43.3
Tanzania:														
Dar es Salaam	49	2.6	2.6	5.1	11.4	7.4	1.3	1.2	1.0	1.2	1.6	2.9	3.6	41.9
Iringa	24	6.8	5.1	7.1	3.5	0.5	*b	*b	*b	0.1	0.2	1.5	4.5	29.3
Kigoma	18	4.8	5.0	5.9	5.1	1.7	0.2	0.1	0.2	0.7	1.9	5.6	5.3	36.5
Togo:														
Lome	15	0.6	0.9	1.9	4.6	5.7	8.8	2.8	0.4	1.4	2.4	1.1	0.4	31.0
Tunisia:														
Gabes	50	0.9	0.7	0.8	0.4	0.3	*b	*b	0.1	0.5	1.2	1.2	0.6	6.7
Tunis	50	2.5	2.0	1.6	1.4	0.7	0.3	0.1	0.3	1.3	2.0	1.9	2.4	16.5

Uganda:													
Kampala	15	1.8	2.4	5.1	6.9	5.8	2.9	1.8	3.4	3.6	-3.8	4.8	3.9	46.2
Lira	14	0.7	1.0	3.5	6.9	7.9	4.9	6.4	10.0	8.3	6.1	3.2	1.8	60.7
Western Sahara:													
Ad Dakhla	14	*b	*b	*b	*b	0.1	0.0	*b	0.2	1.4	0.1	0.2	1.0	3.0
Smara	6	0.1	*b	0.0	*b	*b	0.0	0.0	*b	1.0	*b	0.4	0.0	1.5
Zaire:													
Kalemie	20	4.2	4.7	6.3	8.4	3.3	0.3	0.1	0.3	0.8	2.8	7.9	6.3	45.4
Kananga	14	5.4	5.6	7.7	7.6	3.3	0.8	0.5	2.3	4.6	6.5	9.1	8.9	62.3
Kinshasa	12	5.3	5.7	7.7	7.7	6.2	0.3	0.1	0.1	1.2	4.7	8.7	5.6	53.3
Kisangani	14	2.1	3.3	7.0	6.2	5.4	4.5	5.2	6.5	7.2	8.6	7.8	3.3	67.1
Zambia:													
Balovale	9	8.5	6.9	5.8	1.2	*b	0.0	0.0	*b	0.3	2.3	4.4	8.9	38.3
Kasama	10	10.7	9.9	10.9	2.8	0.5	*b	*b	*b	*b	0.8	6.4	9.5	51.5
Lukasa	10	9.1	7.5	5.6	0.7	0.1	*b	*b	0.0	*b	0.4	3.6	5.9	32.9
Zimbabwe:													
Bulawayo	50	5.6	4.3	3.3	0.7	0.4	0.1	*b	*b	0.2	0.8	3.2	4.8	23.4
Harare	50	7.7	7.0	4.6	1.1	0.5	0.1	*b	0.1	0.2	1.1	3.8	6.4	32.6
Atlantic Islands													
Funchal, Madeira Island	30	2.5	2.9	3.1	1.3	0.7	0.2	*b	*b	1.0	3.0	3.5	3.3	21.5
Georgetown, Ascension Is.	45	0.2	0.4	0.7	1.1	0.5	0.5	0.5	0.4	0.3	0.3	0.2	0.1	5.2
Hutts Gate, St. Helena	30	2.1	3.1	4.2	3.1	2.8	3.2	4.3	2.6	2.2	1.7	1.2	1.6	32.1
Las Palmas, Canary Is.	48	1.4	0.9	0.9	0.5	0.2	*b	*b	*b	0.2	1.1	2.1	1.6	8.6
Porto Da Praria, Cape Verde	25	0.1	*b	*b	*b	0.0	*b	0.2	3.8	4.5	1.2	0.3	0.1	10.2

(continued)

Country and Station	Record length[a] yr	Jan.	Feb.	Mar.	Apr.	May	June	July	Aug.	Sept.	Oct.	Nov.	Dec.	Year
Santa Isabel, Bioko	16	1.3	2.5	4.2	7.2	9.4	11.1	7.4	6.6	9.6	10.4	3.5	1.7	74.9
Sao Tome, Sao Tome	10	3.2	4.2	5.9	5.0	5.3	1.1	*b	*b	0.9	4.3	4.6	3.5	38.0
Tristan da Cunha	5	3.5	3.5	6.4	4.7	7.1	5.9	6.1	6.9	7.9	5.8	4.3	4.0	66.1
Indian Ocean Islands														
Agalega Island	2	5.9	10.1	4.9	6.9	13.2	8.9	8.7	3.2	1.8	4.2	7.0	10.0	84.7
Cocos (Keeling) Island	38	5.4	7.7	8.5	10.4	7.9	9.0	8.7	4.8	3.7	3.3	4.2	4.6	78.2
Heard Island	5	5.8	5.8	5.7	6.1	5.8	3.9	3.6	2.2	2.5	3.7	4.0	5.1	54.3
Hellburg, Reunion Is.	11	22.4	8.0	16.4	7.2	5.3	4.4	3.1	3.0	2.0	2.3	3.5	12.9	90.5
Port Victoria, Seychelles	64	15.2	10.5	9.2	7.2	6.7	4.0	3.3	2.7	5.1	6.1	9.1	13.4	92.5
Royal Alfred Observ., Mauritius	43	8.5	7.8	8.7	5.0	3.8	2.6	2.3	2.5	1.4	1.6	1.8	4.6	50.6
Asia—Far East														
China:														
Canton	36	0.9	1.9	4.2	6.8	10.6	10.6	8.1	8.5	6.5	3.4	1.2	0.9	63.6
Chanaska	26	1.9	3.7	5.3	5.7	8.2	8.7	4.4	4.3	2.7	3.0	2.7	1.5	52.1
Chungking	60	0.7	0.8	1.5	3.8	5.7	7.1	5.6	4.7	5.8	4.3	1.9	0.8	42.9
Hankou	55	1.8	1.9	3.6	5.8	7.0	9.0	7.0	4.1	3.0	3.1	1.9	1.2	49.4
Harbin (Ha-erh-pin)	38	0.2	0.2	0.4	0.9	1.7	3.7	6.6	4.7	2.3	1.2	0.5	0.2	22.6
Kashi (Kaxgar)	18	0.6	0.1	0.5	0.2	0.3	0.2	0.4	0.3	0.1	0.1	0.2	0.3	3.2

Kunmig	31	0.4	0.5	0.7	0.8	4.3	6.3	8.8	8.6	5.0	3.0	1.7	0.4	40.5
Lanzhou	4	0.2	0.2	0.2	0.5	0.8	0.7	3.3	5.1	2.2	0.6	0.0	0.3	14.1
Mukden (Shen-yang)	42	0.2	0.2	0.7	1.2	2.6	3.8	7.0	6.3	2.9	1.7	0.9	0.4	28.2
Shanghai	81	1.9	2.4	3.3	3.6	3.8	7.0	5.8	5.5	5.2	2.9	2.1	1.5	45.0
Tientsin	25	0.2	0.1	0.4	0.5	1.1	2.4	7.6	6.0	1.7	0.6	0.4	0.2	21.0
Urumqi	6	0.6	0.3	0.5	1.5	1.1	1.5	0.7	1.0	0.6	1.7	1.6	0.4	11.5
Hong Kong	50	1.3	1.8	2.9	5.4	11.5	15.5	15.0	14.2	10.1	4.5	1.7	1.2	85.1
Japan:														
Kushiro	41	1.8	1.4	2.8	3.6	3.8	4.1	4.4	4.9	6.6	4.0	3.1	2.0	42.9
Miyako	30	2.9	3.0	3.2	3.5	4.5	5.0	5.0	7.2	9.5	6.8	3.0	2.6	56.2
Nagasaki	59	2.8	3.3	4.9	7.3	6.7	12.3	10.1	6.9	9.8	4.5	3.7	3.2	75.5
Osaka	60	1.7	2.3	3.8	5.2	4.9	7.4	5.9	4.4	7.0	5.1	3.0	1.9	52.6
Tokyo	60	1.9	2.9	4.2	5.3	5.8	6.5	5.6	6.0	9.2	8.2	3.8	2.2	61.6
Mongolia:														
Ulan Bator	15	*b	*b	0.1	0.2	0.3	1.0	2.9	1.9	0.8	0.2	0.2	0.1	7.7
North Korea:														
P'yongyang	43	0.6	0.4	1.0	1.8	2.6	3.0	9.3	9.0	4.4	1.8	1.6	0.8	36.4
South Korea:														
Pusan	36	1.7	1.4	2.7	5.5	5.2	7.9	11.6	5.1	6.8	2.9	1.6	1.2	53.6
Seoul	22	1.2	0.8	1.5	3.0	3.2	5.1	14.8	10.5	4.7	1.6	1.8	1.0	49.2
Taiwan:														
Tainan	13	0.7	0.7	1.1	3.2	6.3	15.6	16.0	15.8	8.4	1.2	0.9	0.6	70.5
Taipei	12	3.8	5.3	4.3	5.3	6.9	8.8	8.8	8.7	8.2	5.5	4.2	2.9	72.7

(continued)

11.10 PRECIPITATION DATA FOR REPRESENTATIVE WORLDWIDE STATIONS (continued)

Country and Station	Length of record yr	Average Precipitation (in)												
		Jan.	Feb.	Mar.	Apr.	May	June	July	Aug.	Sept.	Oct.	Nov.	Dec.	Year
U.S.S.R. (Former):														
Alma-Ata	27	1.3	0.9	2.2	4.0	3.7	2.6	1.4	1.2	1.0	2.0	1.9	1.3	23.5
Chita (Tchita)	24	0.1	0.1	0.1	0.4	1.1	1.8	3.3	3.3	1.2	0.5	0.2	0.2	12.3
Dudinka	5	0.3	0.4	0.2	0.3	0.6	1.9	1.5	1.5	1.8	0.9	0.4	0.3	10.7
Ekaterinburg (Sverdlovsk)	29	0.5	0.4	0.5	0.7	1.9	2.7	2.6	2.7	1.6	1.2	1.1	0.8	16.7
Irkutsk	38	0.5	0.4	0.3	0.6	1.3	2.2	3.1	2.8	1.7	0.7	0.6	0.6	14.9
Kazalinsk	19	0.4	0.4	0.5	0.5	0.6	0.2	0.2	0.3	0.3	0.4	0.5	0.6	4.9
Khabarovsk	8	0.3	0.2	0.3	0.7	2.0	3.5	4.1	3.3	3.0	0.7	0.6	0.5	19.2
Kirensk	19	0.8	0.5	0.5	0.5	1.0	1.8	2.1	2.1	1.7	1.0	1.0	1.0	14.0
Krasnoyarsk	8	0.1	0.2	0.1	0.2	1.0	1.4	1.2	2.1	1.7	0.9	0.5	0.4	9.8
Markovo	16	0.2	0.2	0.3	0.1	0.3	0.8	1.0	1.9	1.1	0.4	0.4	0.3	7.0
Narym	14	0.8	0.5	0.8	0.5	1.3	2.6	2.4	2.7	1.7	1.4	1.1	0.9	16.8
Okhotsk	25	0.1	0.1	0.2	0.4	0.9	1.6	2.2	2.6	2.4	1.0	0.2	0.1	11.8
Omsk	22	0.6	0.3	0.3	0.5	1.2	2.0	2.0	2.0	1.1	1.0	0.7	0.8	12.5
Petropavlovsk	35	3.0	2.2	3.4	2.5	2.2	2.0	3.1	3.2	3.8	3.9	3.6	3.0	35.9
Salehkard	37	0.3	0.3	0.3	0.3	0.7	1.3	1.9	2.0	1.5	0.7	1.1	0.4	10.2
Semipalatinsk	10	0.9	0.5	0.5	0.6	1.2	1.5	1.1	1.3	0.7	1.2	1.5	1.0	11.6
Tashkent	19	2.1	1.1	2.6	2.3	1.4	0.5	0.2	0.1	0.1	1.2	0.3	1.6	14.7

Verkhoyansk	44	0.2	0.2	0.1	0.2	0.3	0.9	1.1	1.0	0.5	0.3	0.3	0.2	5.3
Vladivostok	53	0.3	0.4	0.7	1.2	2.1	2.9	3.3	4.7	4.3	1.9	1.2	0.6	23.6
Yakutsk	22	0.3	0.2	0.1	0.3	0.4	1.1	1.6	1.3	1.1	0.5	0.4	0.3	7.4
Brunei:														
Bandar Seri Begawan	12	14.6	7.6	7.8	9.8	10.9	9.5	9.0	7.3	11.8	14.5	15.2	13.0	131.0
Cambodia:														
Phnom Penh	49	0.3	0.4	1.4	3.1	5.7	5.8	6.0	6.1	8.9	9.9	5.5	1.7	54.8
Indonesia:														
Jakarte	78	11.8	11.8	8.3	5.8	4.5	3.8	2.5	1.7	2.6	4.4	5.6	8.0	70.8
Manokwari	40	12.0	9.4	13.2	11.1	7.8	7.2	5.4	5.6	4.9	4.7	6.5	10.3	98.1
Mapanget	63	18.6	13.8	12.2	8.0	6.4	6.5	4.8	4.0	3.3	4.9	8.9	14.7	106.1
Penfui	63	15.2	13.7	9.2	2.6	1.2	0.4	0.2	0.0	0.0	0.7	3.3	9.1	55.7
Pontianak	63	10.8	8.2	9.5	10.9	11.1	8.7	6.5	8.0	9.0	14.4	15.3	12.7	125.1
Tabing	63	13.9	10.1	12.2	14.5	12.8	11.7	10.5	13.7	16.2	20.1	20.5	19.2	175.4
Tarakan	31	10.9	10.2	14.0	13.9	13.5	12.6	10.3	12.4	11.6	14.3	15.2	13.4	152.3
Laos:														
Vientiane	27	0.2	0.6	1.5	3.9	10.5	11.9	10.5	11.5	11.9	4.3	0.6	0.1	67.5
Malaysia:														
Kuala Lumpur	19	6.2	7.9	10.2	11.5	8.8	5.1	3.9	6.4	8.6	9.8	10.2	7.5	96.1
Kuching	19	24.0	20.1	12.9	11.0	10.3	7.1	7.7	9.2	8.6	10.5	14.1	8.2	153.7
Sandakan	46	19.0	10.9	8.6	4.5	6.2	7.4	6.7	7.9	9.3	10.2	14.5	8.5	123.7
Myanmar (Burma):														
Mandalay	20	0.1	0.1	0.2	1.2	5.8	6.3	2.7	4.1	5.4	4.3	2.0	0.4	32.6

(continued)

11.10 PRECIPITATION DATA FOR REPRESENTATIVE WORLDWIDE STATIONS (continued)

Country and Station	Record length yr	Average Precipitation (in)												
		Jan.	Feb.	Mar.	Apr.	May	June	July	Aug.	Sept.	Oct.	Nov.	Dec.	Year
Moulmein	60	0.2	0.2	0.5	3.0	19.9	37.1	47.5	44.2	27.1	8.5	1.7	0.9	190.2
Philippines:														
Davao	34	4.8	4.5	5.2	5.8	9.2	9.1	6.5	6.5	6.7	7.9	5.3	6.1	77.6
Manila	75	0.9	0.5	0.7	1.3	5.1	10.0	17.0	16.6	14.0	7.6	5.7	2.6	82.0
Singapore:														
Singapore	64	9.9	6.8	7.6	7.4	6.8	6.8	6.7	7.7	7.0	8.2	10.0	10.1	95.0
Thailand:														
Bangkok	10	0.2	1.1	1.1	2.3	5.2	6.0	6.9	9.2	14.0	9.9	1.8	0.1	57.8
Vietnam:														
Hanoi	12	0.8	1.2	2.5	3.6	4.1	11.2	11.9	15.2	10.0	3.5	2.6	2.8	69.4
Ho Chi Minh City	33	0.6	0.1	0.5	1.7	8.7	13.0	12.4	10.6	13.2	10.6	4.5	2.2	78.1
Asia—Middle East														
Afghanistan:														
Kabul	45	1.3	1.5	3.6	3.3	0.9	0.2	0.1	0.1	*b	0.4	0.6	0.6	12.6
Kandhar	7	3.1	1.7	0.8	0.3	0.2	*b	0.1	*b	0.0	*b	*b	0.8	7.0
Bangladesh:														
Dacca	61	0.3	1.2	2.4	5.4	9.6	12.4	13.0	13.3	9.8	5.3	1.0	0.2	73.9

India:

Ahmadabad	45	*b	0.1	0.1	*b	0.4	3.7	12.2	8.1	4.2	0.4	0.1	*b	29.3
Bangalore	60	0.2	0.3	0.4	1.6	4.2	2.9	3.9	5.0	6.7	5.9	2.7	0.4	34.2
Bombay	60	0.1	0.1	0.1	*b	0.7	19.1	24.3	13.4	10.4	2.5	0.5	0.1	71.2
Calcutta	60	0.4	1.2	1.4	1.7	5.5	11.7	12.8	12.9	9.9	4.5	0.8	0.2	63.0
Cherrapunji	35	0.7	2.1	7.3	26.2	50.4	106.1	96.3	70.1	43.3	19.4	2.7	0.5	425.1
Hyderabad	45	0.3	0.4	0.5	1.2	1.1	4.4	6.0	5.3	6.5	2.5	1.1	0.3	29.6
Jalpaiguri	55	0.3	0.7	1.3	3.7	11.8	25.9	32.2	25.3	21.2	5.6	0.5	0.2	128.7
Lucknow	60	0.8	0.7	0.3	0.3	0.8	4.5	12.0	11.5	7.4	1.3	0.2	0.3	40.1
Madras	60	1.4	0.4	0.3	0.6	1.0	1.9	3.6	4.6	4.7	12.0	14.0	5.5	50.0
Mormugao	30	*b	*b	0.6	0.7	2.6	29.6	31.2	15.9	9.5	3.8	1.3	0.2	94.8
New Delhi	75	0.9	0.7	0.3	0.3	0.5	2.9	7.1	6.8	4.6	0.4	0.1	0.4	25.2
Silchar	53	0.8	2.1	7.9	14.3	15.6	21.7	19.7	19.7	14.4	6.5	1.4	0.4	124.5
Srinagar	50	2.9	2.8	3.6	3.7	2.4	1.4	2.3	2.4	1.5	1.2	0.4	1.3	25.9

Iran:

Abadan	10	1.5	1.7	0.6	0.8	0.1	0.0	0.0	0.0	0.0	0.1	1.0	1.8	7.6
Esfahan (Isfahan)	45	0.7	0.6	0.8	0.6	0.3	*b	0.1	*b	*b	0.1	0.4	0.7	4.4
Kermanshah	15	2.6	2.3	2.8	2.2	1.6	*b	*b	*b	*b	0.4	2.0	2.4	16.4
Rezaiyeh	3	1.9	2.3	2.0	1.7	1.2	0.5	*b	0.1	0.2	1.5	0.8	1.6	13.8
Tehran	33	1.8	1.5	1.8	1.4	0.5	0.1	0.1	0.1	0.1	0.3	0.8	1.2	9.7

Iraq:

Baghdad	15	0.9	1.0	1.1	0.5	0.1	*b	*b	*b	*b	0.1	0.8	1.0	5.5
Basra	10	1.4	1.1	1.2	1.2	0.2	0.0	*b	*b	*b	*b	1.4	0.8	7.3

(continued)

11.10 PRECIPITATION DATA FOR REPRESENTATIVE WORLDWIDE STATIONS (continued)

Country and Station	Length of record[a] (yr)	Average Precipitation (in)												
		Jan.	Feb.	Mar.	Apr.	May	June	July	Aug.	Sept.	Oct.	Nov.	Dec.	Year
Mosul	29	2.8	3.1	2.1	1.9	0.7	*b	*b	*b	*b	0.2	1.9	2.4	15.2
Israel:														
Haifa	30	6.9	4.3	1.6	1.0	0.2	*b	*b	*b	0.1	1.0	3.7	7.3	26.2
Jerusalem	50	5.1	4.7	2.9	0.9	0.1	*b	0.0	0.0	*b	0.3	2.2	3.5	19.7
Tel Aviv	10	4.9	2.7	2.0	0.7	0.1	0.0	0.0	0.0	0.1	0.4	4.1	6.1	21.1
Jordan:														
Amman	25	2.7	2.9	1.2	0.6	0.2	0.0	0.0	0.0	*b	0.2	1.3	1.8	10.9
Kuwait:														
Kuwait	10	0.9	0.9	1.1	0.2	*b	0.0	0.0	0.0	0.0	0.1	0.6	1.1	5.1
Lebanon:														
Beirut	71	7.5	6.2	3.7	2.2	0.7	0.1	*b	*b	0.2	2.0	5.2	7.3	35.1
Nepal:														
Katmandu	9	0.6	1.6	0.9	2.3	4.8	9.7	14.7	13.6	6.1	1.5	0.3	0.1	56.2
Oman:														
Muscat	38	1.1	0.7	0.4	0.4	*b	0.1	*b	*b	0.0	0.1	0.4	0.7	3.9
Pakistan:														
Karachi	59	0.5	0.4	0.3	0.1	0.1	0.7	3.2	1.6	0.5	0.1	0.1	0.2	7.8
Multan	60	0.4	0.4	0.4	0.3	0.3	0.6	2.0	1.8	0.5	0.1	0.1	0.2	7.1

Rawalpindi	60	2.5	2.5	2.7	1.9	1.3	2.3	8.1	9.2	3.9	0.6	0.3	1.2	36.5
Saudi Arabia:														
Dhahran	10	1.1	0.6	0.4	0.2	0.1	0.0	0.0	0.0	0.0	0.0	0.2	0.9	3.5
Jidda	5	0.2	*b	*b	*b	*b	0.0	*b	*b	*b	*b	1.0	1.2	2.5
Riyadh	3	0.1	0.8	0.9	1.0	0.4	*b	0.0	*b	0.0	0.0	*b	*b	3.2
Sri Lanka:														
Colombo	40	3.5	2.7	5.8	9.1	14.6	8.8	5.3	4.3	6.3	13.7	12.4	5.8	92.3
Syria:														
Deir Ez Zor	8	1.6	0.8	0.3	0.8	0.1	*b	0.0	0.0	0.0	0.2	1.5	0.9	6.2
Dimashq (Damascus)	7	1.7	1.7	0.3	0.5	0.1	*b	*b	0.0	0.7	0.4	1.6	1.6	8.6
Halab (Aleppo)	10	3.5	2.5	1.5	1.1	0.3	0.1	0.0	*b	*b	1.0	2.2	3.3	15.5
Turkey:														
Adana	31	4.3	4.0	2.5	1.6	2.0	0.7	0.2	0.2	0.7	1.9	2.4	3.8	24.3
Ankara	24	1.3	1.2	1.3	1.3	1.9	1.0	0.5	0.4	0.7	0.9	1.2	1.9	13.6
Edirne	18	2.2	1.9	1.7	1.9	1.7	2.1	1.5	1.1	1.1	2.1	2.9	3.0	23.2
Erzurum	16	1.4	1.6	2.0	2.5	3.1	2.1	1.3	0.9	1.1	2.3	1.8	1.1	21.2
Istanbul	18	3.7	2.3	2.6	1.9	1.4	1.3	1.7	1.5	2.3	3.8	4.1	4.9	31.5
Izmir (Smyrna)	58	4.4	3.3	3.0	1.7	1.3	0.6	0.2	0.2	0.8	2.1	3.3	4.8	25.5
Samsun	27	2.9	2.6	2.7	2.3	1.8	1.5	1.5	1.3	2.4	3.2	3.5	2.4	29.1
United Arab Emirates:														
Sharjah	12	0.9	0.9	0.4	0.2	0.0	0.0	0.0	0.0	0.0	0.0	0.4	1.4	4.2
Yemen:														
Kamaran I.	21	0.2	0.2	0.1	0.1	0.1	*b	0.5	0.7	0.1	0.1	0.4	0.9	3.4

(continued)

11.10 PRECIPITATION DATA FOR REPRESENTATIVE WORLDWIDE STATIONS (continued)

Country and Station	Record Length yr	Jan.	Feb.	Mar.	Apr.	May	June	July	Aug.	Sept.	Oct.	Nov.	Dec.	Year
Riyan	13	0.3	0.1	0.6	0.2	*b	0.1	0.1	0.1	*b	*b	0.7	0.3	2.5
Indian Ocean Islands														
Port Blair, Andaman Island	60	1.8	1.1	1.1	2.4	15.1	21.7	15.4	16.3	17.4	12.5	10.5	7.9	123.2
Amindivi Island	30	0.7	*b	*b	1.5	3.7	14.3	12.0	7.7	6.3	5.8	2.6	1.3	56.0
Minicoy Island	50	1.8	0.7	0.9	2.3	7.0	11.6	8.9	7.8	6.3	7.3	5.5	3.4	63.5
Car Nicobar, Nicobar Island	30	3.9	1.2	2.1	3.5	12.5	12.4	9.3	10.2	12.9	11.6	11.4	7.8	98.8
Oceania														
Australia:														
Adelaide	104	0.8	0.7	1.0	1.8	2.7	3.0	2.6	2.6	2.1	1.7	1.1	1.0	21.1
Alice Spings	30	1.7	1.3	1.1	0.4	0.6	0.5	0.3	0.3	0.3	0.7	1.2	1.5	9.9
Bourke	72	1.4	1.5	1.1	1.1	1.0	1.1	0.9	0.8	0.8	0.9	1.2	1.4	13.2
Brisbane	91	6.4	6.3	5.7	3.7	2.8	2.6	2.2	1.9	1.9	2.5	3.7	5.0	44.7
Broome	50	6.3	5.8	3.9	1.2	0.6	0.9	0.2	0.1	*b	*b	0.6	3.3	22.9
Burketown	53	8.2	6.3	5.2	1.0	0.2	0.3	*b	*b	*b	0.4	1.5	4.4	27.5
Canberra	25	1.9	1.7	2.2	1.6	1.8	2.1	1.8	2.2	1.6	2.2	1.9	2.0	23.0
Carnarvon	57	0.4	0.7	0.7	0.6	1.5	2.4	1.6	0.7	0.2	0.1	*b	0.2	9.1
Cloncurry	59	4.4	4.2	2.4	0.7	0.5	0.6	0.3	0.1	0.3	0.5	1.3	2.7	18.0
Esperance	60	0.7	0.7	1.2	1.8	3.3	4.1	4.0	3.8	2.7	2.2	1.0	0.9	26.4

Hobart	100	1.9	1.5	1.8	1.9	1.8	2.2	2.1	1.9	2.1	2.3	2.4	2.1	24.0
Laverton	30	0.8	0.8	1.6	0.8	0.9	0.7	0.6	0.5	0.2	0.3	0.8	0.8	8.8
Melbourne	88	1.9	1.8	2.2	2.3	2.1	2.1	1.9	1.9	2.3	2.6	2.3	2.3	25.7
Mundiwindi	15	1.0	1.9	2.0	0.8	0.6	0.9	0.1	0.3	0.3	0.5	0.5	1.2	10.1
Perth	63	0.3	0.4	0.8	1.7	5.1	7.1	6.7	5.7	3.4	2.2	0.8	0.5	34.7
Port Darwin	70	15.2	12.3	10.0	3.8	0.6	0.1	*b	0.1	0.5	2.0	4.7	9.4	58.7
Sydney	87	3.5	4.0	5.0	5.3	5.0	4.6	4.6	3.0	2.9	2.8	2.9	2.9	46.5
Thursday Island	49	18.2	15.8	13.9	8.0	1.6	0.5	0.4	0.2	0.1	0.3	1.5	7.0	67.5
Townsville	67	10.9	11.2	7.2	3.3	1.3	1.4	0.6	0.5	0.7	1.3	1.9	5.4	45.7
William Creek	30	0.5	0.6	0.3	0.3	0.3	0.5	0.2	0.3	0.3	0.5	0.5	0.7	5.0
Windorah	50	1.4	1.6	1.6	0.9	0.8	0.8	0.5	0.4	0.5	0.6	0.9	1.4	11.4
New Zealand:														
Aukland	92	3.1	3.7	3.2	3.8	5.0	5.4	5.7	4.6	4.0	4.0	3.5	3.1	49.1
Christchruch	64	2.2	1.7	1.9	1.9	2.6	2.6	2.7	1.9	1.8	1.7	1.9	2.2	25.1
Dunebin	77	3.4	2.8	3.0	2.8	3.2	3.2	3.1	3.0	2.7	3.0	3.2	3.5	36.9
Wellington	79	3.2	3.2	3.2	3.8	4.6	4.6	5.4	4.6	3.8	4.0	3.5	3.5	47.4
Pacific Islands														
Guam, Marinas Island	30	4.6	3.5	2.6	3.0	4.2	5.9	9.0	12.8	13.4	13.1	10.3	6.1	88.5
Honolulu, Hawaii	30	3.8	3.3	2.9	1.3	1.0	0.3	0.4	0.9	1.0	1.8	2.2	3.0	21.9
Iwo Jima, Bonin Island	17	3.2	2.5	2.1	3.7	4.9	4.0	6.4	6.5	4.6	5.9	4.8	4.3	52.8
Kanton, Phoenix Island	30	2.6	2.2	2.5	3.6	4.3	2.6	2.6	2.5	1.2	1.1	1.6	2.6	29.4
Madang, Papua, New Guinea	20	12.1	11.9	14.9	16.9	15.1	10.8	7.6	4.8	5.3	10.0	13.3	14.5	137.2
Midway Island	20	4.6	3.7	3.1	2.5	1.9	1.3	2.9	3.9	3.7	3.7	3.6	4.2	40.7

(continued)

11.10 PRECIPITATION DATA FOR REPRESENTATIVE WORLDWIDE STATIONS (continued)

Country and Station	Length of Record (yr)	Average Precipitation (in)												
		Jan.	Feb.	Mar.	Apr.	May	June	July	Aug.	Sept.	Oct.	Nov.	Dec.	Year
Naha, Okinawa	30	5.3	5.4	6.1	6.1	8.9	10.0	7.1	10.0	7.1	6.6	5.9	4.3	82.8
Noumea, New Caledonia	52	3.7	5.1	5.7	5.2	4.4	3.7	3.6	2.6	2.5	2.0	2.4	2.6	43.5
Pago Pago, Samoa	41	24.5	20.5	19.2	16.5	15.4	12.3	10.0	8.2	13.1	14.9	19.2	19.8	193.6
Ponape, Senyavin Island	30	11.1	9.7	14.6	20.0	20.3	16.7	16.2	16.3	15.8	16.0	16.9	18.3	191.9
Port Moresby, Papua, New Guinea	38	7.0	7.6	6.7	4.2	2.5	1.3	1.1	0.7	1.0	1.4	1.9	4.4	39.8
Rabaul, Papua, New Guinea	24	14.8	10.4	10.2	10.0	5.2	3.3	5.4	3.7	3.5	5.1	7.1	10.1	88.8
Suva, Fiji	43	11.4	10.7	14.5	12.2	10.1	6.7	4.9	8.3	7.7	8.3	9.8	12.5	117.1
Tahiti, Society Island	27	13.2	11.5	6.5	6.8	4.9	3.2	2.6	1.9	2.3	3.4	6.5	11.9	74.7
Tulagi, Solomon Island	37	14.3	15.8	15.0	10.0	8.1	6.8	7.6	8.7	8.0	8.7	10.0	10.4	123.4
Wake Island	30	1.1	1.4	1.5	1.9	2.0	1.9	4.6	7.1	5.2	5.3	3.1	1.8	36.9
Yap Islands	30	7.9	4.6	5.4	6.4	9.5	10.7	13.8	14.7	14.0	13.2	11.2	10.2	121.6

Antarctica[c]

Byrd Station	6	0.4	0.4	0.2	0.3	0.4	0.5	0.7	0.7	0.3	0.7	0.0	0.3	4.9
Ellsworth	6	0.3	0.2	0.3	0.6	0.2	0.2	0.2	0.2	0.3	0.4	0.5	0.2	3.6
McMurdo Station	10	0.5	0.7	0.4	0.4	0.4	0.3	0.2	0.3	0.4	0.2	0.2	0.3	4.3
South Pole Station	5	*	0.1	0.0	0.0	0.0	0.0	0.0	0.0	0.0	*	0.0	*	0.1
Wilkes	7	0.5	0.4	1.7	1.1	1.4	1.2	1.3	0.8	1.5	1.2	0.8	0.3	12.2

[a] Record Length refers to average daily maximum and minimum precipitation. A standard period of the 30 years from 1931 to 1960 had been used for locations in the United States and some other countries.

[b] * = Less than 0.05" but more than 0".

[c] Except for Antarctica, amounts of solid precipitation such as snow or hail have been converted to their water equivalent. Because of the frequent occurrence of blowing snow, it has not been possible to determine the precise amount of precipitation actually falling in Antarctica. The values shown are the average amounts of solid snow accumulating in a given period as determined by snow markers. The liquid content of the accumulation is undetermined.

11.11 WORLDWIDE EXTREMES OF TEMPERATURE AND PRECIPITATION

Temperature Extremes Recorded by Continental Area

Key Number	Area	Highest/Lowest Temperatures (°F)	Place	Elevation (ft.)	Date
		Highest Temperatures			
1	Africa	136	Azizia, Libya	380	Sep. 13, 1992
2	North America	134	Death Valley, California	-178	July 10, 1913
3	Asia	129	Tirat Tsvi, Israel	-722	June 21, 1942
4	Australia	128	Cloncurry, Queensland	622	Jan. 16, 1889
5	Europe	122	Seville, Spain	26	Aug. 4, 1881
6	South America	120	Rivadavia, Argentina	676	Dec. 11, 1905
7	Oceania	108	Tuguegarao, Philippines	72	Apr. 29, 1912
8	Antartica	58	Esperanza, Palmer Peninsula	26	Oct. 20, 1956
		Lowest Temperatures			
9	Antartica	-127	Vostok	11,220	Aug. 24, 1960
10	Asia	-90	Oymykon, U.S.S.R.	2,265	Feb. 6, 1933
11	Greenland	-87	Northice	7,690	Jan. 9, 1954
12	North America	-81	Snag, Yukon, Canada	1,925	Feb. 3, 1947
13	Europe	-67	Ust' Shchugor, U.S.S.R.	279	January[a]
14	South America	-27	Sarmiento, Argentina	879	June 1, 1907
15	Africa	-11	Ifrane, Morocco	5,364	Feb. 11, 1935

472

| 16 | Australia | -8 | Charllote Pass, N.S.W. | —[b] | July 22, 1947[c] |
| 17 | Oceania | -14 | Haleakala Summit, Maui | 9,750 | Jan. 2, 1961 |

[a] Exact date unknown; lowest in 15 yr period.
[b] Elevation unknown.
[c] Earlier date.

Extremes of Average Annual Precipitation Recorded by Continental Area

Key No.	Area	Greatest/Least Amounts of Precipitation (in.)	Place	Elevation (ft.)	Years of Record
		Greatest Amounts			
18	Oceania	460	Mt. Waileale, Kauai, Hawaii	5,075	32
19	Asia	450	Cherrapunji, India	4,309	74
20	Africa	404.6	Debundscha, Cameroon	30	32
21	South America	353.9	Quibdo, Colombia	240	10–16
22	North America	262.1	Henderson Lake, B.C., Canada	12	14
23	Europe	182.8	Crkica, Yugoslavia	3,337	22
24	Australia	179.3	Tully, Queensland	—[a]	31
		Least Amounts			
25	South America	0.03	Arica, Chile	95	59
26	Africa	<0.1	Wadi Halfa, Sudan	410	39

(continued)

11.11 WORLDWIDE EXTREMES OF TEMPERATURE AND PRECIPITATION *(continued)*

Key No.	Area	Greatest/Least Amounts of Precipitation (in.)	Place	Elevation (ft.)	Years of Record
27	Antartica	0.8[b]	South Pole Station	9,186	10
28	North America	1.2	Batagues, Mexico	16	14
29	Asia	1.8	Aden, Yemen	22	50
30	Australia	4.05	Mulka, South Australia	—[a]	34
31	Europe	6.4	Astrakhan, Former U.S.S.R.	45	25
32	Oceania	8.93	Puako, Hawaii	5	13

[a] Elevation unknown.

[b] The value given is the average amount of solid snow accumulating in one year as indicated by snow markers. The liquid content of the snow is undetermined.

474

Appendix

MISCELLANEOUS CONVERSION FACTORS

Feet head (ft. hd.) × 0.433	= Pounds per square inch (PSI)
Pounds per square inch × 2.31	= Feet head
Meters × 3.28	= Feet head
Inches of mercury × 1.133	= Feet head
U.S. gallons per minute × 1.1337	= Cubic feet per minute
Cubic feet per minute × 7.48	= U.S. gallons per minute
British Imperial gallons × 1.201	= U.S. gallons
Acre inches per hour × 453	= Gallons per minute
Acre foot per day × 226	= Gallons per minute
1,000,000 gallons per day	= 694 Gallons per minute
U.S. gallons × 0.833	= British Imperial gallon
U.S. gallons × 8.336	= Pounds
Acre feet × 325,850	= U.S. gallons
U.S. gallons × 231	= Cubic inches
Horsepower (H.P.) × 746	= Watts
Horsepower × 0.746	= Kilowatts
Calories × 3.968	= British thermal units (B.T.U.)
Foot-pounds per second × 0.7373	= Watts
Kilowatts × 1.34	= Horsepower
Square feet × 144	= Square inches
Square yards × 9	= Square feet
Acres × 4,840	= Square yards
Acres × 43,560	= Square feet
Square mile (section) × 640	= Acres
Mile × 5,280	= Feet
Cubic yard × 27	= Cubic feet
Circumference of circle × 0.3183	= Diameter of circle
Diameter of circle × 3.1416	= Circumference of circle
Diameter of circle squared × 0.7854	= Area of circle
Radius of circle squared × 3.1416	= Area of circle
Cubic feet per second × 448.8	= Gallons per minute

$$\text{Cubic feet per second} = \frac{\text{Gallons per minute}}{449}$$

$$0.048 \times$$

$$\text{Velocity in feet per second} = \frac{\text{Gallons per minute}}{\text{Diameter of pipe squared}}$$

$$\text{or } 144 \, \frac{Q \left(\text{Flow in gallons per minute} \right)}{A_1 \left(\text{Pipe } ID_2 \right)}$$

Conversion

- All gallons per minute are shown in U.S. gallons.
- To convert to British Imperial gallons per minute, multiply by 0.833.
- To convert pounds per square inch to atmospheres, divide by 14.7.
- To convert pounds per square inch to kilograms per square centimeter, divide by 14.22.
- To convert feet to meters, divide by 3.28.

INDEX